IUTAM SYMPOSIUM ON RECENT DEVELOPMENTS IN NON-LINEAR OSCILLATIONS OF MECHANICAL SYSTEMS

SOLID MECHANICS AND ITS APPLICATIONS
Volume 77

Aims and Scope of the Series

The fundamental questions arising in mechanics are: *Why?, How?,* and *How much?* The aim of this series is to provide lucid accounts written bij authoritative researchers giving vision and insight in answering these questions on the subject of mechanics as it relates to solids.

The scope of the series covers the entire spectrum of solid mechanics. Thus it includes the foundation of mechanics; variational formulations; computational mechanics; statics, kinematics and dynamics of rigid and elastic bodies: vibrations of solids and structures; dynamical systems and chaos; the theories of elasticity, plasticity and viscoelasticity; composite materials; rods, beams, shells and membranes; structural control and stability; soils, rocks and geomechanics; fracture; tribology; experimental mechanics; biomechanics and machine design.

The median level of presentation is the first year graduate student. Some texts are monographs defining the current state of the field; others are accessible to final year undergraduates; but essentially the emphasis is on readability and clarity.

For a list of related mechanics titles, see final pages.

IUTAM Symposium on Recent Developments in Non-linear Oscillations of Mechanical Systems

Proceedings of the IUTAM Symposium held in Hanoi, Vietnam, March 2–5, 1999

Edited by

Nguyen Van Dao
Vietnam National University, Hanoi

and

E.J. Kreuzer
Technical University,
Hamburg-Harburg, Germany

KLUWER ACADEMIC PUBLISHERS
DORDRECHT / BOSTON / LONDON

A C.I.P. Catalogue record for this book is available from the Library of Congress.

ISBN 0-7923-6470-8

Published by Kluwer Academic Publishers,
P.O. Box 17, 3300 AA Dordrecht, The Netherlands.

Sold and distributed in North, Central and South America
by Kluwer Academic Publishers,
101 Philip Drive, Norwell, MA 02061, U.S.A.

In all other countries, sold and distributed
by Kluwer Academic Publishers,
P.O. Box 322, 3300 AH Dordrecht, The Netherlands.

Printed on acid-free paper

Printed in the Netherlands.

CONTENTS

Preface ix

Awrejcewicz J. and Tomczak K.
Stability Improvement of the Impact Dynamical Systems-Analytical and Numerical Methods. 1

Benedettini F.
Theoretical and Experimental Evidence of Symmetric Response Instability in the Finite, Planar Dynamics of a Circular Arch. 11

Bevilacqua L., Battista R.B. and Ebecken N.F.F.
Dynamical Analysis of an Offshore Platform with Vibration Absorbers. 23

Bishop S. R.
Complex Dynamics of a "Simple" Mechanical System: The Parametrically Excited Pendulum. 35

Chernousko F. L.
Control of Oscillations in Systems with Many Degrees of Freedom. 45

Czolczynski K., Kapitaniak T. and Brindley J.
Elimination of Instabilities in Rotors Supported in Gas Bearings. 55

Maisser P., Freudenberg H. and Pham Anh Tuan
Integrated Electromechanical Multi-Coordinate Drive with an Elastic Deformable Slide. 65

Grebogi C., Baptista M.S., Hunt B.R., Ott E. and Yorke J.A.
Control of Shipboard Cranes. 75

Guo-Kang Er and Vai Pan Iu
A Consistent and Effective Method for Nonlinear Random Oscillations of MDOF Systems. 85

Hu H.Y. and Wu Z.Q.
Stability and Hopf Bifurcation of a Four-wheel-steering Vehicle Involving Driver's Delay. 95

Kreuzer E. and Ellermann K.
Moored Crane Vessels in Regular Waves. 105

Lê Luong Tài
On the Stability of Systems of Differential Equations Unsolved for Derivatives. 115

Maisser P. and Jungnickel U.
Stability of Controlled Motion of a Gymnast in High-speed Mid Air Maneuvers. 121

Mitropolsky Y. A.
Asymptotic Methods for investigating Non-linear Wave Processes. 131

Nguyen Van Dao
Some Problems on Non-linear Oscillations. 141

Nguyen Van Dinh and Tran Kim Chi
Stationary and Transient Processes in Oscillating Systems with Two Degrees of Freedom Subjected to Parametric and Forced Excitations. 153

Nguyen Hai and Ostiguy G. L.
On the Application of Two Perturbation Methods to Non-linear Systems. 165

Nguyen Xuan Hung and Nguyen Xuan Hoang
New Method of Calculating the Vibration of Offshore Structure Subjected to Wave Action. 175

Nguyen Van Khang, Hoang Ha, Vu Van Khiem and Do Xuan Tho
On the Transverse Vibrations of Beam-Bridges Under the Action of Some Moving Bodies. 187

Nguyen Tien Khiem, Dao Nhu Mai, Nguyen Viet Khoa and Le Van Anh
Pile Soil System Identification by a Modal Testing Approach. 197

Nguyen Nhat Le, Dinh Van Phong and Do Sanh
On Numerical Methods for Constrained Mechanical Systems. 207

Nguyen Dong Anh and Ninh Quang Hai
A Technique for Solving Nonlinear Systems Subject to Random Excitation. 217

Popp K. and Bruns J. U.
On the Identification of Nonlinear Systems. 227

Rega G., Lacarbonara W. and Nayfeh A.H.
Reduction Methods for Non-linear Vibrations of Spatially Continuous Systems with Initial Curvature. 235

Samoilenko A.M.
Asymptotic Method for Investigation of *m*-frequency Oscillation Systems. 247

Schiehlen W. and Fritz A.
Nonlinear Oscillations of Vehicles in Convoy. 257

Schulz M.
Instability Phenomena of Foldable Structures with Unilateral Contact. 269

Spyrou K. J.
Similarities in the Yaw and Roll Dynamics of Ships in Extreme Astern Seas. 279

Albrecht B., Steindl A., and Troger H.
Non-linear three-dimensional Oscillations of Fluid Conveying Viscoelastic Tubes with an Additional Mass. 291

van de Vrande B. L. and van Campen D. H.
Nonlinear Dynamics of a Rigid Rotor on Compliant Journal Bearings. 301

Vestroni F. and Bernardini D.
Nonlinear Dynamic Behaviour of Shape Memory Alloy Oscillators. 311

Wauer J.
Nonlinear Waves in a Fluid-Filled Planar Duct with a flexible Wall. 321

Addresses of participants 333

List of Sponsors 341

PREFACE

This volume contains selected papers presented at the Symposium on "**Recent Developments in Non-linear Oscillations of Mechanical Systems**", held in Hanoi, Vietnam, from 2 - 5 March 1999. This Symposium was initiated and sponsored by the International Union of Theoretical and Applied Mechanics (IUTAM) and organised in conjunction with Vietnam National University, Hanoi. The purpose of the Symposium was to bring together scientists active in different fields of oscillations with the aim to review the recent progress in theory of oscillations and engineering applications and to outline the prospects in its further achievements to then co-ordinate and direct research in this field to further co-operation between scientists and various scientific institutions.

An International Scientific Committee was appointed by the Bureau of IUTAM with the following members:

Nguyen Van Dao (Vietnam, Co-Chairman)
E.J. Kreuzer (Germany, Co-Chairman)
D.H. van Campen (The Netherlands)
F.L. Chernousko (Russia)
A.H. Nayfeh (U.S.A)
Nguyen Xuan Hung (Vietnam)
W.O. Schiehlen (Germany)
J.M.T. Thompson (U.K)
Y. Ueda (Japan).

This Committee selected the participants to be invited and the papers to be presented at the Symposium. As a result of this procedure, 52 active scientists from 16 countries responded to the invitation, and 42 papers were presented in lecture and poster discussion sessions.

The scientific presentations were devoted to the following topics :

1. Non-linear Oscillations of beams, plates, vehicles and other dynamic systems;
2. Analysis and Control of Non-linear Systems;
3. Non-linear waves;
4. Dynamics of Offshore structures;
5. System Identification;
6. Mathematical and Numerical methods for investigating non-linear systems.

The papers of this volume are arranged in alphabetical order with respect to the family name of the first author.

The presentations and discussions, including the round table discussion, during the Symposium will certainly stimulate further theoretical and applied investigations in non-linear oscillations. The publication of the proceedings will promote this development.

The success of the Symposium would not have been possible without the excellent work of the Local Organising Committee. Members of that Committee were Do Sanh

(Chairman), Phan Nguyen Di, Nguyen Van Khang, Tran Quang Khoi, Nguyen Cao Menh, Dinh Van Phong, Nguyen Thi Trung and Vu Ngoc Tu. The secretarial work for both the Conference and this volume has been admirably undertaken by Mrs. Nguyen Thi Hong Hanh.

We are sincerely grateful to Prof. Werner Schiehlen, President of IUTAM for his encouragement to the success of this IUTAM Symposium and to Vietnam National University, Hanoi, the host of this significant scientific event.

The editors wish to thank the participants of this IUTAM Symposium, especially the authors of the papers, and all organisers for their enthusiastic and valuable contributions to the Symposium.

Sincere thanks are also due to Prof. Graham Gladwell, and Kluwer Academic Publishers for their help and co-operation.

We gratefully acknowledge financial support from IUTAM and the Vietnam Council for Natural Science.

Nguyen Van Dao **E.J. Kreuzer**

STABILITY IMPROVEMENT OF THE IMPACT DYNAMICAL SYSTEMS - ANALYTICAL AND NUMERICAL METHODS

J. AWREJCEWICZ and K. TOMCZAK
Division of Automatics and Biomechanics
Technical University of Łódź
1/15 Stefanowskiego St., 90-924 Łódź, POLAND

1. Introduction

Vibro-impact vibration problems with one-degree-of-freedom systems have a long history in mechanics. The problems like stationary subharmonic motions and their stability, the influence of damping and friction on vibro-impact dynamics, elastic and plastic type impacts, time histories and phase portraits of the vibro-impact systems have been considered [1-7].

In this work we propose an analytical approach to determine suitable delay loop coefficients to realise the required vibro-impact periodic dynamics for a non-resonance case. The obtained analytical formulas allow for a proper choice of the delay loop coefficients in order to achieve the required vibro-impact periodic motion quicker then in the case without a loop. When the vibro-impact periodic motion is achieved the delay loop is automatically switched off.

2. The analysed system

A feedback control with the delay loop is used in order to improve a stability of a vibro-impact periodic motion. It possesses the following properties:

- simple construction (a feedback loop with delay elements);

N. Van Dao and E.J. Kreuzer (eds.),
IUTAM Symposium on Recent Developments in Non-linear Oscillations of Mechanical Systems, 1–10.

- simple mathematical description;
- the only one information required about an object being analysed is a period of oscillations (besides, an object can be treated as a "black box").

The analysed system with the kinematic excitation is presented in Figure 1. The system dynamics (including a delay loop) is governed by the equation:

$$\ddot{x} + c\dot{x} + \alpha^2 x = P_0 \cos\omega t + S[x(t) - x(t-T)] + Q[\dot{x}(t) - \dot{x}(t-T)] \qquad \text{for} \quad x < s, \quad (1)$$

$$x_+ = x_-, \quad \dot{x}_+ = -R\dot{x}_- \qquad \text{for} \quad x = s,$$

where: $c = c_1/m$, $\alpha^2 = (k_1 + k_2)/m$, $P_0 = k_2 y_0/m$, $S = k_2 p/m$, $Q = k_2 q/m$ and $T = 2\pi k/\omega$ is the period of a stabilised periodic orbit, R denotes the restitution coefficient, and s is the constraint. The natural number k defines the number of excitation periods occurring during one impact. The indeces "+" and "–" define positions and velocities of the body just after and before an impact, correspondingly.

(a)

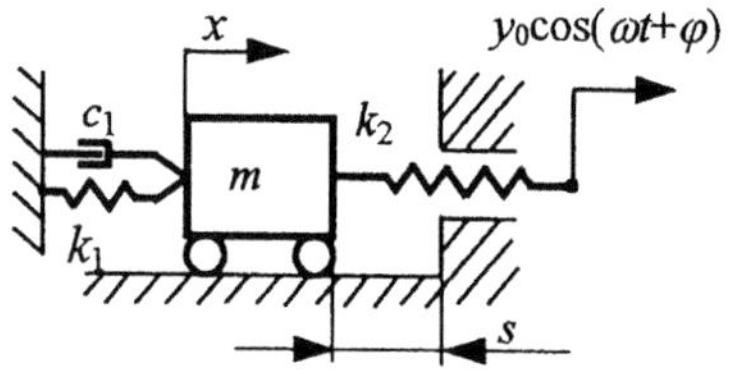

(b)

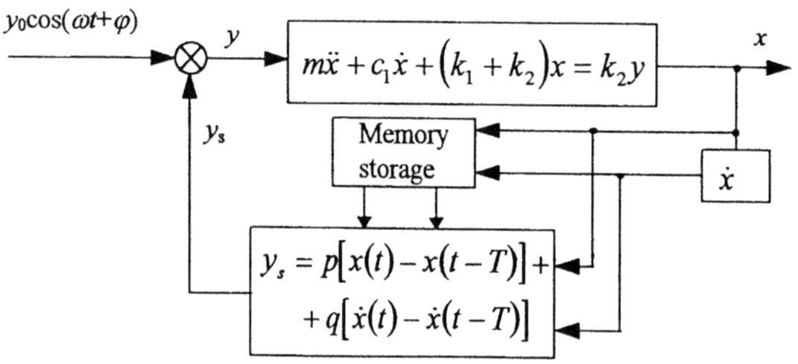

Fig. 1 - One-degree-of-freedom vibro-impact system kinematically excited (a) and the control diagram (b).

It is assumed (Figure 1) that the stabilised periodic orbit $x_0(t) = x_0(t-T)$ has the period of the excitation the same as in the system without that loop and that $x_0(t)$ is the particular solution for the system with and without the delay loop [8-10]. The delay loop starts to operate when the disturbances occur and is going to act on the dynamics in order to achieve the vibro-impact periodic motion quicker than in the case without a loop.

3. Control far from the resonance

In order to find the analytical solution to equation (1) the approximate analytical method has been applied assuming that:

- the difference $x(t) - x(t-T)$ is small;
- the damping c is of the same order as the introduced formally perturbation parameter ε.

In order to find the complete solution of the equation (1) in the far from the resonance motion the Krylov-Bogoliubov-Mitropolskij method (KBM) was applied.

A new variable "z" is introduced to the equation (1), defined as:

$$x = z + \frac{P_0}{\alpha^2 - \omega^2}\cos\omega t . \tag{2}$$

We get:

$$\ddot{x} + \alpha^2 x = \varepsilon f(a, \eta, \psi), \tag{3}$$

where:

$$\varepsilon f(a,\eta,\psi) = \varepsilon S\left[z + \frac{P_0}{\alpha^2 - \omega^2}\cos\omega t - z(t-T) - \frac{P_0}{\alpha^2 - \omega^2}\cos\omega(t-T)\right] +$$
$$+ \varepsilon Q\left[\left(1 - \frac{c}{Q}\right)\left(\dot{z} - \frac{P_0\omega}{\alpha^2 - \omega^2}\sin\omega t\right) - \dot{z}(t-T) + \frac{P_0\omega}{\alpha^2 - \omega^2}\sin\omega(t-T)\right],$$

$$\eta - \omega t, \qquad \psi - \alpha t.$$

Using the KBM method we have truncated the ε series up to the order $O(\varepsilon)$ and we have obtained:

$$z(t) = a(t)\cos\psi(t), \tag{4}$$

where:

$$\frac{\mathrm{d}a}{\mathrm{d}t}=\frac{\varepsilon Sa}{2\alpha}\sin\alpha T+\frac{\varepsilon}{2}(Q-c)a-\frac{\varepsilon Qa}{2}\cos\alpha T,$$
$$\frac{\mathrm{d}\psi}{\mathrm{d}t}=\alpha-\frac{\varepsilon S}{2\alpha}+\frac{\varepsilon S}{2\alpha}\cos\alpha T+\frac{\varepsilon Q}{2}\sin\alpha T \tag{5}$$

For $S=Q=0$ we get the uncontrolled solution, which testifies the validity of our approach.

After integration of equations (5) one obtains:

$$a(t)=C_0e^{Vt},\qquad \psi(t)=\alpha_0 t+\theta_0,$$

$$V=\frac{1}{2\alpha}\{S\sin\alpha T-\alpha[c-Q(1-\cos\alpha T)]\}, \tag{6}$$

$$\alpha_0=\alpha-\frac{S}{2\alpha}(1-\cos\alpha T)+\frac{Q}{2}\sin\alpha T.$$

Therefore, we analyse the following equivalent solution (for vibro-impact motion):

$$x(t)=\frac{P_0}{\alpha^2-\omega^2}\cos(\omega t+\theta)+\mathrm{e}^{Vt}\left(C\cos\alpha_0 t+D\sin\alpha_0 t\right)\quad \text{for}\quad x<s, \tag{7}$$

$$x_+=x_-,\quad \dot{x}_+=-R\dot{x}_-\qquad \text{for}\quad x=s,$$

where:

$$C=C_0\cos\theta_0,\quad D=-C_0\sin\theta_0,\quad C_0=\sqrt{C^2+D^2},$$

and the values of the parameters C, D and θ are defined according to the formulas obtained for the system without a delay loop [6].

In order to investigate the stability we use the following approach [6]. If we perturb the periodic solution (12) with a small value δx_l, this perturbance will cause the change of the parameters C and D and the phase shift θ with the value of δC_l, δD_l and $\delta\theta_l$, correspondingly. The perturbed solution will be then:

$$x+\delta x_l=\mathrm{e}^{Vt}\left[\left(C+\delta C_l\right)\cos\alpha_0 t_l+\left(D+\delta D_l\right)\sin\alpha_0 t_l\right]+a\cos\left(\omega t_l+\theta+\delta\theta_l\right), \tag{8}$$

where: $a=P_0/\left(\alpha^2-\omega^2\right)$.

In the above formula the time t is measured beginning from the l-th impact, and the change of the constants C and D were taken into account. The impact l+1 will occur in the time moment $t_{l+1} = 2\pi l/\omega + \delta T_l$, where δT_l denotes the period change T. Taking into account the fact that t concerns also the unperturbed equation, then $t_l = t + \delta t_l$, where $\delta t_l = 0$ for the l-th impact, and $\delta t_l = \delta T_l$ for the (l+1)-th impact.

After a few transformations we get:

$$\begin{aligned}\delta x_l = \mathrm{e}^{Vt}\Big[\big(\delta C_l + \alpha_0 D\delta t_l + VC\delta t_l\big)\cos\alpha_0 t + \big(\delta D_l - \alpha_0 C\delta t_l + VD\delta t_l\big)\sin\lambda t\Big] + \\ - a\big(\omega\delta t_l + \delta\theta_l\big)\sin(\omega t + \theta).\end{aligned} \tag{9}$$

Introducing the following boundary conditions for l-th and (l+1)-th impact:

$$l: \qquad t = 0, \quad \delta t_l = 0, \quad \delta x_l = 0, \quad \delta\dot{x}_l = \delta\dot{x}_{l+}, \tag{10}$$

$$l+1: \quad t = \frac{2\pi k}{\omega} + \delta t_l = 2\beta + \delta t_l, \quad \delta t_l = \delta T_l, \quad \delta x_l = 0, \quad \delta\dot{x}_l = \delta x_{(l+1)-},$$

we receive six equations.

Since $\delta\theta_{i+1} = \delta\theta_0 + \sum_{i=0}^{l}\omega\delta T_i$, we finally get the following three equations:

$$\delta C_1 - a\delta\theta_1 \sin\theta = 0,$$

$$\mathrm{e}^{2\beta V}\Big\{\delta C_l \cos 2\beta\alpha_0 + \delta D_l \sin 2\beta\alpha_0 + \frac{1}{\omega}\big(\delta\theta_{l+1} - \delta\theta_l\big)[(\alpha_0 D + VC)\cos 2\beta\alpha_0 +$$

$$(VD - \alpha_0 C)\sin 2\beta\alpha_0]\Big\} - \delta C_{l+1} = 0,$$

$$\begin{aligned}V\mathrm{e}^{2\beta V}\Big\{\big(\alpha_0\delta D_l + V\delta C_l\big)\cos 2\beta\lambda - \big(\alpha_0\delta C_l - V\delta D_l\big)\sin 2\beta\lambda + \\ + \frac{1}{\omega}\big(\delta\theta_{l+1} - \delta\theta_l\big)\Big[\big(2V\alpha_0 D + C\big(V^2 - \alpha_0^2\big)\big)\cos 2\beta\alpha_0 + \\ + \big(-2V\alpha_0 C + D\big(V^2 - \alpha_0^2\big)\big)\sin 2\beta\alpha_0\Big]\Big\} + V\delta C_{l+1} + \alpha_0\delta D_{l+1} - (R+1)\delta\theta_{l+1} a\omega\cos\theta = 0.\end{aligned} \tag{11}$$

The solutions are sought in the following form:

$$\delta C_l = a_1\gamma^l, \qquad \delta D_l = a_2\gamma^l, \qquad \delta\theta_l = a_3\gamma^l, \tag{12}$$

where γ is a constant. After substituting (12) to (11) we get the following characteristic equation:

$$b_2\gamma^2 + b_1\gamma + b_0 = 0, \tag{13}$$

where:

$$b_2 = \alpha_0\left\{a\sin\theta - \frac{1}{\omega}e^{-2\beta V}\left[\left(\alpha_0 D + VC\right)\cos 2\beta\alpha_0 + \left(VD - \alpha_0 C\right)\sin 2\beta\lambda\right]\right\},$$

$$b_1 = e^{2\beta V}\left\{a\sin\theta\left[(R-1)\alpha_0\cos 2\beta\alpha_0 + (R+1)V\sin 2\beta\alpha\right] - (R+1)a\omega\cos\theta\sin 2\beta\alpha_0 + \right.$$
$$\left. + \frac{1}{\omega}\alpha_0\left[\left(VC + \alpha_0 D\right)\left(\cos 2\beta\alpha_0 - Re^{2\beta V}\right) + \left(VD + \alpha_0 C\right)\sin 2\beta\alpha_0\right]\right\}, \tag{14}$$

$$b_0 = R\alpha_0 e^{4\beta V}\left[\frac{1}{\omega}\left(\alpha_0 D + VC\right) - a\sin\theta\right].$$

The problem of the investigation of this equation stability is reduced to the analysis of the roots of the second power equation (13). If these roots fulfil the following inequality:

$$\left|\gamma_{1,2}\right| < 1, \tag{15}$$

then, in accordance with the expressions (12), the solutions δC_l, δD_l and $\delta\theta_l$ approach zero at $l \to +\infty$, and the solution will be called asymptotically stable. The above inequality is equivalent with the placement of the roots inside the unit circle of the complex plane.

4. Simulation results

The simulation model of a one-degree-of-freedom system and the control of the system is constructed with a use of the MATLAB-Simulink package.

The following system parameters were adopted for the simulation: $m = 2\,kg$, $c_1 = 0.01\,Ns/m$, $k_1 = 7\,N/m$, $k_2 = 1\,N/m$, $\omega/\alpha = 2.2$, $y_0 = 1.5\,m$, $R = 0.65$, and $s = 0.001\,m$.

The analytical method presented in the previous section has been used to detect the delay loop coefficients. Figure 2 presents moduli of the roots of algebraic equation (13) versus the delay loop coefficients p and q (for the considered parameters, the $|\gamma_{1,2}|$ roots are complex conjugate numbers). The p and q parameters have been taken from the interval (-0.5, 0.5).

In order to check the analytical predictions, $p.=0$and $q=-0.4Ns/m$ have been assumed and then the simulation results have been compared with the case when $p = q = 0$.

For these parameters the delay loop control coefficients p and q allow us to obtain quicker damping of oscillations in the solution (7) than without control. Additionally, for the given parameters we have found from equation (13) that $|\gamma_{1,2}|$ are lying closer to the origin for the system with the control coefficients than without control (with the delay loop $|\gamma_{1,2}| = 0.539$, whereas without the loop $|\gamma_{1,2}| = 0.648$).

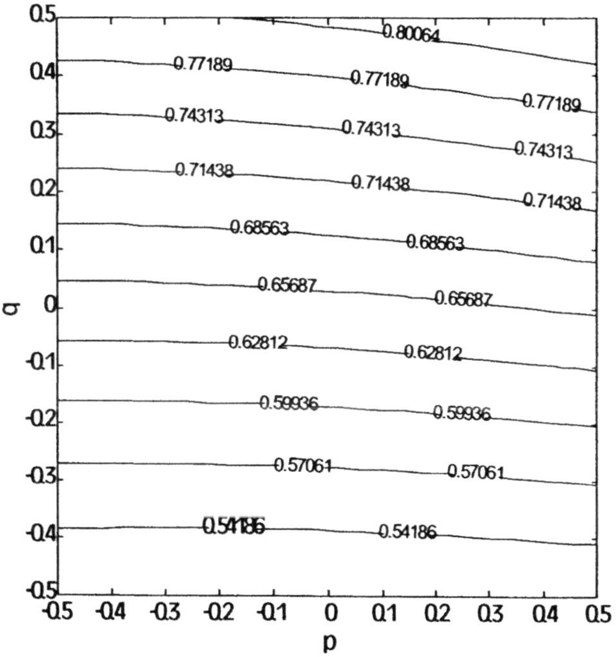

Fig. 2 - The moduli of $|\gamma_{1,2}|$ versus the delay loop coefficients p and q.

Figure 3 presents the simulation results in the form of phase planes and the transients of the difference $x(t) - x(t-T)$. To compare these transients an additional μ parameter has been adopted. That parameter defines the time interval where the signal $|x(t) - x(t-T)| < \mu$. (during the simulation $\mu = 10^{-3}$, has been used).

It can be seen in Figure 3 that with control the transients vanish quicker than in the case without control. In the case presented above the periodic orbit is achieved after about 17.3 seconds for the system analysed without the delay loop and after 14.9 seconds for the system analysed with the delay loop ($\mu = 10^{-3}$), respectively.

(a)

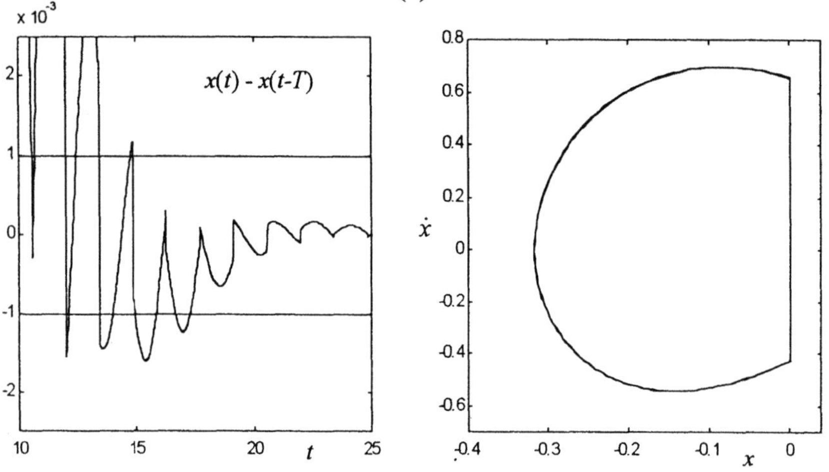

(b)

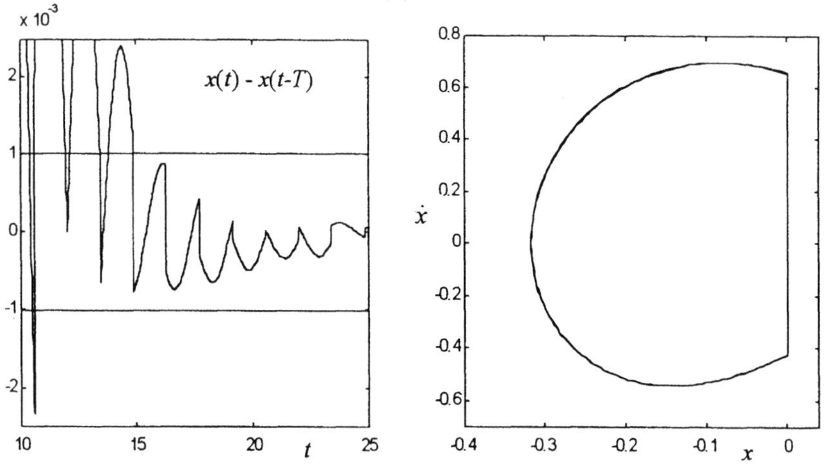

Fig. 3 - Difference between two transients $x(t)$-$x(t-T)$ and phase plane approaching periodic orbit for the system: (a) without control ($p = 0$, $q = 0$) and (b) with control ($p = 0$, $q = -0.4$).

5. Conclusions

In this paper we have presented an analytical approach to estimate the delay control coefficients for efficient stabilisation or destabilisation of the periodic orbit under consideration. Although the efficiency of the method is presented for $k = 1$ (periodic orbit with the same period as the excitation period) but our considerations are also valid for subharmonics (for arbitrarily taken $k > 1$). The validity of our analytical approach has been testified by numerical simulations.

To date, in the literature available to the authors, in order to achieve the mentioned objective, the feedback loop coefficients have been adopted in a random way, using the numerical observation. In this paper this problem was solved analytically.

6. References

1. Rusakov, I.G. and Kharkevich, A.A.: Forced vibration of systems impacting against a step, *Zhurnal Tekh. Phys.*, **12**, Vol. XII (1942), in Russian.
2. Bespalova, L.V.: To the theory of vibro-impacting mechanisms, *Izv. AN SSSR OTN*, **5** (1957), in Russian.
3. Fedosenko, Yu.S. and Feygin, M.L.: Periodic motion of vibrating hammer including the presence of sliding regime, *Prikl. Math. i Mekh.*, **5**, Vol. 35 (1974), 892-898, in Russian.
4. Goldsmith, W.: *Theory and physical aspects of impacting systems*, Moscow, 1965, in Russian.
5. Kolovskij, M.Z.: *Nonlinear theory of impact systems*, Nauka, Moscow, 1966, in Russian.
6. Babitskij, W.I.: *Theory of impact systems*, Nauka, Moscow, 1978, in Russian.
7. Peterka, F.: Laws of impact motions of mechanical systems with one degree of Freedom, Part I, II, *Acta Technica ČSAV*, **4**, (1974), 462-473, **5** (1974), 569-580.
8. Awrejcewicz, J., Tomczak, K. and Lamarque, C.-H.: Controlling Systems with Impacts. In Awrejcewicz, J. and Lamarque, C.-H.:

Proceedings of the International Conference on Nonlinearity, Bifurcation and Chaos: the Doors to the Future, Łódź-Dobieszków (1996), 79-82.

9. Awrejcewicz, J., Tomczak, K. and Lamarque, C.-H.: Multibody Vibro-Impact Dynamics, *Proceedings of the 2nd European Nonlinear Oscillations Conference*, Prague, **2** (1996), 9-12.
10. Awrejcewicz, J., Andrianov, I.V. and Manevitch, L.I.: *Asymptotic approach in nonlinear dynamics: new trends and applications*, Springer-Verlag, Berlin 1998.

THEORETICAL AND EXPERIMENTAL EVIDENCE OF SYMMETRIC RESPONSE INSTABILITY IN THE FINITE, PLANAR DYNAMICS OF A CIRCULAR ARCH

F. BENEDETTINI
Dipartimento di Ingegneria delle Strutture, Acque e Terreno, Universita' dell'Aquila, Monteluco Roio, 67040 L'Aquila, Italy

1. Introduction

The role of experimental tests as a necessary step in the analysis and the design of slender structures undergoing large displacements is assessed in this paper. In particular, the finite, forced dynamics of elastic structures having initial curvature show a series of interesting phenomena due to the presence of both symmetric and non symmetric nonlinearities. In the case of a circular arch excited by an harmonic vertical load applied on the tip and having the dynamics confined to the plane of initial configuration, the simple unimodal symmetric solution, stable for low excitation levels, loses stability around meaningful resonance conditions, and coupled symmetric anti-symmetric solutions appear. The nonlinear modal coupling is furthermore strengthened by possible internal resonance conditions depending on elasto-geometrical structural parameters and/or on the location of possibly applied concentrated masses.

The problem already known for a long time (Bolotin [1]) has been recently revisited in the case of non shallow (Thomsen [2], Benedettini [3]) and shallow (Win-Min Tien et al. [4]) arches. In the cited papers, planar models of arches are analyzed with the objective of describing the overall dynamics and the bifurcation scenarios leading eventually to non regular motions occurring in some areas of the frequency-amplitude excitation parameter plane. Concerning the case of non shallow arches, the analytical model obtained by Thomsen [2], contains a mixed continuous-discrete formulation: the symmetric oscillations are, in fact, modeled as a discrete degree of freedom parametrically forcing the antisymmetric vibrations, modeled, on the contrary, with a correct continuum mechanics approach. In ref. [3], after revisiting the analytical results obtained in [2] and discussing a minor improvement proposed by Lakrad et al. [5], an analysis of the results obtained by using a first companion experimental model was done. In this work, after discussing the results obtained with systematic tests on a new experimental model, a new planar analytical model obtained with a monodimensional nonlinear elastic approach is considered (Sheinman [6], Alwar and Narasimhan [7]). To realize an arch having an actually planar dynamics, a cross section of the arch having an out of plane moment of inertia well above the planar one, has been considered (out of plane frequencies well above the planar ones). Within the aforementioned hypothesis and owing the assumption of non-shallow arch, the dynamic deformation is practically

N. Van Dao and E.J. Kreuzer (eds.),
IUTAM Symposium on Recent Developments in Non-linear Oscillations of Mechanical Systems, 11–22.

inextensional and free of shear effects. Coherently the new analytical model taking into account the nonlinear change in curvature, the tangential inertia forces and disregarding axial (Nayfeh and Raouf [8,9]) and shear deformations is derived by using the extended Hamilton's principle. The *pde's* of the motion are then discretized by using a two-mode Galerkin approximation validated by experimental tests. The Karhunen-Loeve decomposition [10] was applied on experimental time-series contemporaneously acquired on eight different positions on the arch. During both periodic and non-periodic evolutions, the spatial shape of the system remains *close* to the eigenfunctions of the first two, symmetric and anti-symmetric modes, justifying in such a way the strong reduction adopted in the Galerkin discretization. The analytical model obtained with the preceding assumptions appears therefore correct to describe both the regular and non-regular dynamics of the arch. The experimental and analytical models were analyzed around the primary external resonance condition of the symmetric mode in the case of a nearly 2:1 internal resonance condition between the directly excited and the anti-symmetric mode. The nonlinear modal coupling produces an instability of the simple symmetric solution and the extension and the nature of this instability zone are analyzed in detail for the given detuning. When the unimodal solution loses stability, two-mode periodic, quasi-periodic and chaotic motions have been observed. Each solution belongs to a sub-region of the main instability region: furthermore, entering the region from its boundary the complexity of the motion increases. The extention and shape of such sub-areas depending on the internal detuning have been observed and sample time histories representative of the class of responses have been analyzed. Classical global complexity indicators like the *maximum Lyapunov exponent* and the *correlation dimension* have been used to quantify the chaoticity of the system and the possible fractal nature of underlying attractors. In the case of experimental tests the preceding analysis has been conducted on the basis of the delay map technique [11, 12, 13]. Furthermore the spatial complexity observed during the strong chaotic evolutions has been unfolded with the aid of the Karhunen-Loeve decomposition.

2. Preliminary analytical and experimental models

The analytical model proposed by Thomsen [2] is based on the following assumption:

- the tangential inertia forces are neglected,
- the shear and axial deformations are neglected,
- the mass of the arch is concentrated on its crown.

With the preceding assumption and using a mixed continuous-discrete formulation, the following *ode's* of the motion were derived; in this case, α=0 and the eqs. read:

$$\begin{aligned} &\ddot{f} + 2\beta\dot{f} + (1 - m\omega^2 u)f + \alpha \cdot \left[-m\omega^2 \frac{k}{8} f^3 \right] = 0 \\ &\ddot{u} + 2\beta\omega\dot{u} + \omega^2 u + \kappa(f\ddot{f} + \dot{f}^2) + \alpha \cdot \left[+\omega^2 \frac{k}{8} f^2 \right] = (q/m)\cos(\Omega\tau) \end{aligned} \tag{1}$$

In eqs. (1), with the symbology present in [2], f and u are respectively the anti-symmetric and symmetric amplitudes, β denotes the damping coefficient, m and κ are the coefficients of non-linearities depending on the mass and the opening angle of the arch, ω is the ratio between the symmetric and anti-symmetric natural frequencies and, eventually, q and Ω express the non-dimensional amplitude and frequency of a concentrated vertical force at the tip of the arch. A minor improvement of the model (suggested in [2] and proposed in [5]) is obtained taking into account the 2nd order vertical displacement of the tip of the arch due to anti-symmetric vibrations; accordingly eqs. (1) are modified by the presence of the terms in brackets (α=1).
With the aim of comparing and validating the previous models, a double hinged, steel circular arch having a radius $R=90\ cm$, a cross sectional area $A=3 \times 0.4\ cm^2$ and an opening angle $2\varphi_0=160°$ was constructed in the Nonlinear Dynamics Lab of University of L'Aquila. The choice of the cross section was done with the aim of confining the dynamics *practically* in the plane of initial configuration. In Figure 1 the analytical and the experimental models are shown.

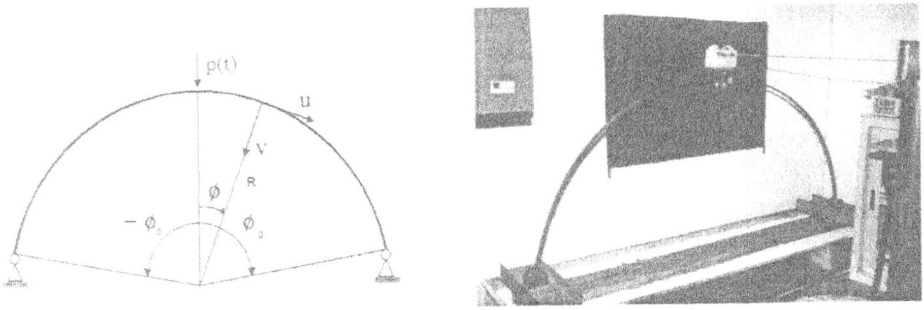

Figure 1—The analytical and experimental models

Interesting similitudes were discussed in [3] both for the regular and non regular dynamics. The corresponding analytical and experimental behavior charts explaining, the bifurcation phenomenon are reported in Figure 2.

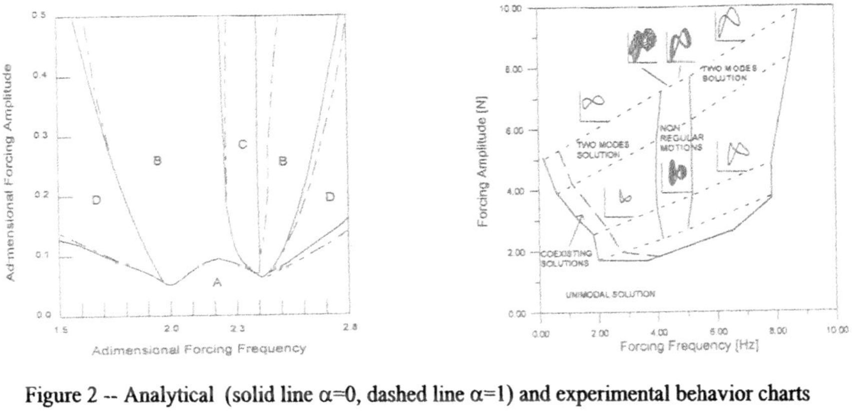

Figure 2 -- Analytical (solid line α=0, dashed line α=1) and experimental behavior charts

A small dynamic exciter to be applied on the tip of the arch was then constructed: it mainly consists of two opposite counter-rotating non balanced masses mounted on two coupled cog-wheels moved by a circular thin shaft transmitting a couple and driven by a step controlled, asynchronous, electrical engine.
The two models, synchronized in a 2:1 internal resonance condition by applying on the tip of the arch a concentrated mass, furnish a qualitatively correct prediction of the symmetric response instability by evidencing the occurrence of strong subharmonic antisymmetric vibrations in a large frequency-range even at low excitation amplitudes. In the analytical chart shown in Figure 2 (left), several zones are observable: in the A-zone the unimodal symmetric solution is the only stable solution; the B-zone, is the region of stability of the coupled, periodic, symm.-anti-symm. solution; the D-zone is the region where the two kind of solution belonging to the previous classes are both stable and coexisting and, eventually, in the C-zone, no periodic solutions are stable, and, after Hopf bifurcations, quasi-periodic solutions arise, leading eventually, to chaotic motions. In Figure 2 (right), the same region for the experimental model is reported. The shape of the plot qualitatively agrees with the preceding one: within the main instability region an inner core and two sub-regions are present: the latter are zones of stability of the coupled periodic motion, while the former, coherently with the analytical prediction, is the zone of non-periodic, complex motions.
Even if the analyzed models give comparable results, four main points could be observed.
Concerning the analytical model (proposed in [2] and [5]):

- in the case of an high opening angle (160° in the considered case) the tangential inertia forces cannot be disregarded owing also the presence of a concentrated mass at the tip of the arch subjected, in the case of coupled solution, to an high anti-symmetric (tangential) displacement component,
- the distributed mass of the arch is of the same order of magnitude of the dead load applied to the crown and to concentrate it could produce imperfect results.

Concerning the experimental model (proposed in [3]):

- the behavior chart, even qualitatively in agreement with the analytical one, shows a quite bigger region of dynamic instability occupying the whole range of analyzed frequencies, circumstance probably related to the model imperfections and to the driving shaft transmitting extra-actions to the model,
- during the observed complex (QP and chaotic) evolutions no measures where done on the spatial coherence of the measured displacements, circumstance not *a priori* justifying the proposed truncation on the companion analytical model.

To go over the preceding points a new experimental rig (model, dynamic exciter and driving shaft) and a new analytical models are proposed and analyzed in the following.

3. New experimental set-up

A new experimental model with same characteristics of the preliminary one, improved in the assembling of the pin-end constraints, was realized with the aim of verifying the possible accuracy of a reduced analytical two degree-of-freedom model even during non regular regimes observed in a large area of the forcing-frequency excitation control parameter plane. A parallel improvement in the assembling of the dynamic exciter was realized as well by reducing an asymmetry present in the first exciter and constructing a new shaft, linking, with a double cardan-joint, the exciter to the step-controlled engine and able to transmit, in the limit of a real mounting, only the required couple. The new experimental set-up, because of the cited improvements, gave more reliable results permitting to draw some useful observations that oriented the derivation of a new analytical model described in paragraph 3.

The observed motions, cleaner in each test than those exhibited by the first model, again indicated a large area in which coupled symmetric-anti-symmetric motions arise with strong anti-symmetric components accompanying the directly driven symmetric ones. In the coupled case the *mean amplitude* of the motion considerably increase indicating the need to understand and possibly control the relevant bifurcation between the two different, somewhere coexisting, solutions. Figure 3 (left and middle) has been obtained by using a strobo-flash camera, and they permit to estimate the amplitudes of the motion during the dynamic evolution. In particular the two pictures were taken in a coexistence zone *i.e.* when at the same values of excitation parameters, both unimodal-symmetric (left) and coupled (middle) solutions are competing in function of i.c..

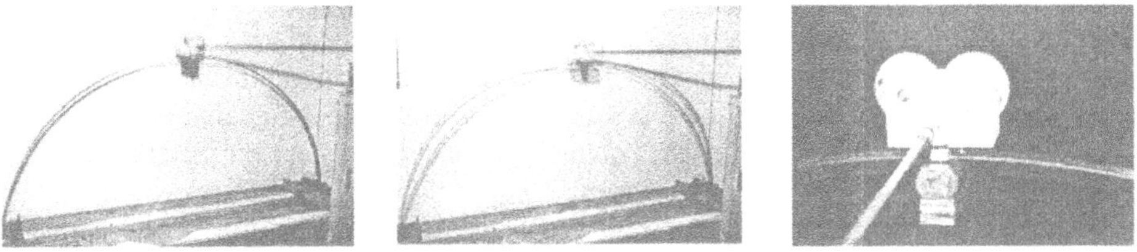

Figure 3 – New exp. setup: symm. motion (left), anti-symm. motion (center) and the exciter (right)

From the picture in the middle, keeping in mind the observation on the necessity to include the tangential inertia forces in the analytical modeling, is also easy to estimate the order of magnitude of the horizontal component of the tip motion. In Figure 3 (right) the exciter with the shaft and the cardan-joint is shown as well.

The first step of the experimental analysis was to realize a nearly 2:1 internal resonance condition between the frequencies of the first symmetric and anti-symmetric modes by adding an extra mass on the tip of the arch (Figure 3 right). An identification of a mass value of 2.211 Kg (set-up A) gave experimental frequencies very close to the analytical ones obtained with the new model described in the following. In Table 1 these frequencies are reported (first and second columns) together with those evaluated with a f.e. code both in the cases of undeformed (third column) and deformed (fourth column) initial configuration: The agreement between experimental and analytical frequencies is considered very good with a relative error well below the 4%.

Natural frequencies (setup A)				
	Experimental	Analytical	Numerical	Numerical (def)
$\omega_{anti\text{-}symm}$	2.65	2.685	2.667	2.582
ω_{symm}	5.42	5.607	5.498	5.325

Table 1 – Experimental, analytical and numerical natural frequencies

After measuring the natural frequencies an identification of the dampings of symmetric and anti-symmetric vibrations was done acquiring free-decaying oscillations and interpolating the relevant maxima by means of an exponential law (see Figure 4):

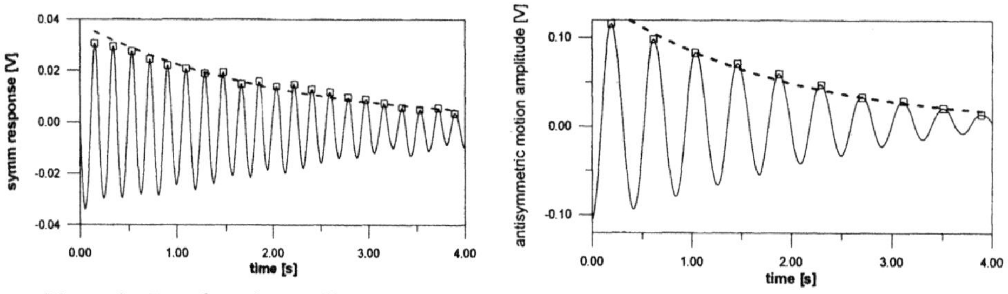

Figure 4—Free decaying oscillations for symmetric (left) and anti-symmetric (right) oscillations

The interpolated values to be used in the companion analytical model (with adimensionalized time on the respect of the anti-symmetric natural frequency) are respectively: $2\xi_s\omega_s=0.071$ (symmetric component) and $2\xi_a\omega_a=0.048$ (anti-symmetric). After evaluating the modal parameters to be used in the analytical model, having observed in the experimental tests strong non-periodic oscillations inside the coupling zone, was necessary to validate the *a priori* choice concerning the spatial truncation (2 modes) assumed in the first model. To this end, mounting accelerometers on eight points of the arch (in the radial direction first and in the tangential after), synchronous time series were acquired during the motion. Repeating the acquisition during periodic (simple and coupled) and a chaotic (always coupled) evolutions, it was possible to evaluate the spatial coherence by means of the Karhunen-Loeve decomposition [10]. In the periodic cases the proper orthogonal modes *(pom's)* correspond to spatial shapes *similar* to the normal modes. In Figure 5, such simple cases (left and middle) were compared with the *pom's* corresponding to a full developed chaotic regime (right).

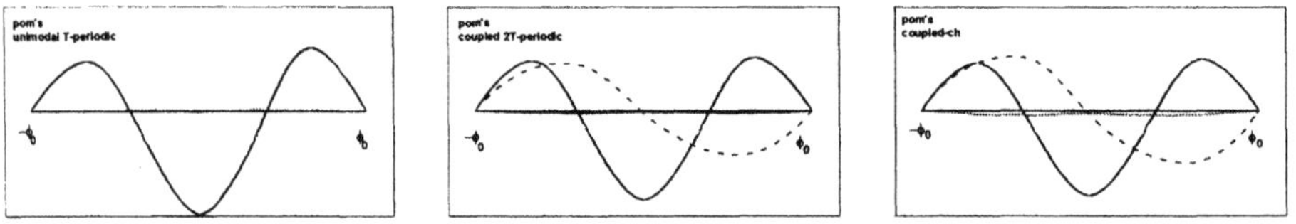

Figure 5 – Proper orth. modes (radial comp.): periodic (left and middle) and chaotic (right) evolutions

From Figure 5 it was possible to try the following conclusions:

- the spatial shapes visited in average during a chaotic evolution *looks like* the ones pertaining the periodic response hence justifying, at a first level of approximation, the use of a simple, reduced 2 d.o.f. model also to predict chaotic motions,
- no more than two shapes are involved in the planar dynamics of the arch being, further contribution, practically absent.

With the validation of preliminary tests it was therefore possible to formulate a new analytical model coherent with the experimental evidence.

4. New analytical model

Following a nonlinear elastic mono-dimensional formulation the equations of the motion of the circular arch were derived by applying the extended Hamilton's principle. In the case of negligible shear effects, the planar deformation of the arch is described when the relations of finite extensional strain and change in curvature are expressed in terms of $u(\varphi,t)$ and $v(\varphi,t)$, the radial and tangential components of the displacement field (see Figure 1). They are adimensionalized in the following on the respect of the radius R of the circular arch. Considering two points P and Q on the arch axis, infinitely close in the undeformed configuration, calling ds their initial distance and ds' their corresponding distance in the deformed configuration, expressing the distances in function of the displacement gradient and adopting the *Lagrangian strain* ε as the measure of the axial deformation, the following expression is obtained:

$$\varepsilon = u' - v + \frac{u'^2 + v'^2}{2} + \frac{u^2 + v^2}{2} + uv' - vu' \tag{2}$$

Analogously, calling $\Delta\varphi$ the difference between the angles formed by two infinitely close tangent vectors in P and Q (in the deformed and initial configurations) and dividing by ds the following relation for the change of curvature $\Delta\chi$ is obtained:

$$\Delta\chi = \left(\frac{1}{R}\right)\cdot\left(v'' + v + \frac{v'^2 - 2u'^2}{2} + \frac{2v^2 - u^2}{2} + 2u'v'' - v'u'' + 2vv'' - uu''\right) \tag{3}$$

Expressing the potential energy $\mathcal{V}$ by means of ε and $\Delta\chi$, the kinetic energy $\mathcal{T}$ by means of time-derivatives of the displacement field, taking into account the work $\mathcal{W}_{nc}$ done during the motion by damping and external forces, the extended Hamilton's principle reads:

$$\delta \int_{t_1}^{t_2} \int_{-\varphi_0}^{\varphi_0} (\mathcal{T} - \mathcal{V})\, d\varphi\, dt + \int_{t_1}^{t_2} \int_{-\varphi_0}^{\varphi_0} \delta W_{nc}\, d\varphi\, dt = 0 \tag{4}$$

From eq. (4) the *pde's* of the motion and the relevant boundary conditions are then obtained. A reduced 2 d.o.f. model is obtained by assuming a truncated Galerkin expansion of the displacement field using, as shape functions, the first symmetric (f_1, g_1) and anti-symmetric (f_2, g_2) eigenfunctions of the linear problem [14]:

$$\begin{aligned} u(\varphi,t) &= f_1(\varphi)\cdot q_1(t) + f_2(\varphi)\cdot q_2(t) \\ v(\varphi,t) &= g_1(\varphi)\cdot q_1(t) + g_2(\varphi)\cdot q_2(t) \end{aligned} \tag{5}$$

Carrying out the space integrations, two *ode's* of the motion were obtained: the relevant coefficients are expressed in terms of space integration of the functions f and g. Eqs. (6) contain both quadratic and cubic nonlinearities and are accurate to describe moderately large oscillations. With high opening angles, like in the examined case, the dynamic deformation is practically inextensional: modifying the Hamiltonian (4) by adding, via a Lagrange multiplier, the inextensibility condition, the final equations read:

$$\begin{aligned} &\ddot{q}_1 + \mu_1\dot{q}_1 + c_{10}q_1 + c_{20}q_1^2 + c_{02}q_2^2 + c_{30}q_1^3 + c_{12}q_1q_2^2 = Psin(\Omega\tau) \\ &\ddot{q}_2 + \mu_2\dot{q}_2 + q_2 + c_{11}q_1q_2 + c_{02}q_2^2 + c_{03}q_2^3 + c_{21}q_1^2q_2 = 0 \end{aligned} \tag{6}$$

5. Analytical and experimental results

Analytical frequency-responses at two different forcing levels are reported in Figure 6: at the lower forcing amplitude (left) the unimodal solution loses stability entering the resonance region and a coupled, periodic, two-mode solution bifurcates from it. As noted in the preliminary experimental tests, the coupled motion has an higher *mean amplitudes* of oscillations. At an higher forcing level, the preceding scenario is further complicated by the loose of stability, entering the resonance region, of the coupled periodic solution: the latter modification corresponds to an Hopf bifurcation activating solutions having complex time evolutions. Forcing-response curves for two different frequencies, are reported in Figure 7 and show the classical saturation phenomenon (see [8,9]): the pitch-fork bifurcation activating the two mode solution, supercritical at the lower frequency (left), becomes sub-critical at the higher one.

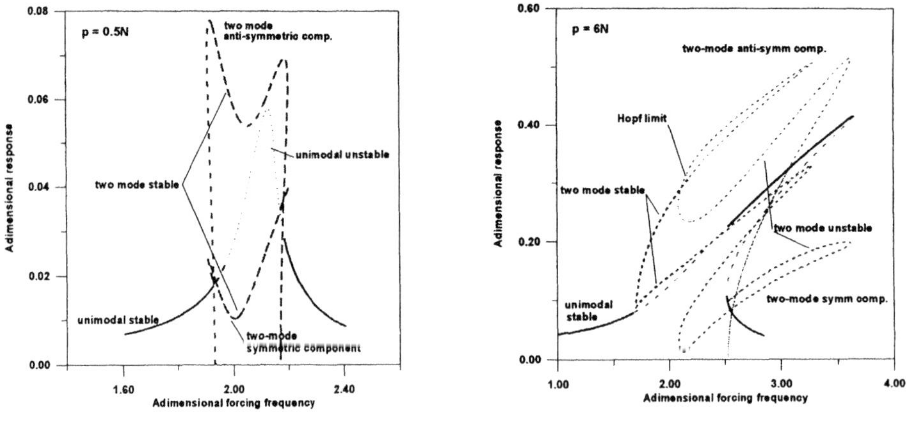

Figure 6—Analytical model: frequency-response plots

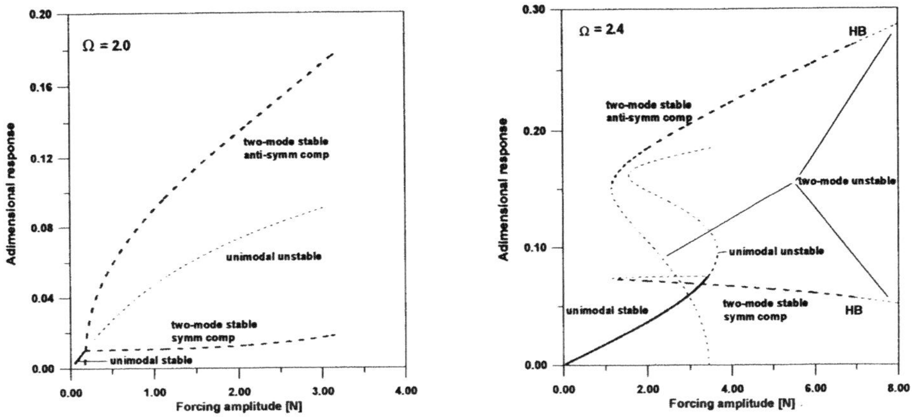

Figure 7—Analytical model: forcing amplitude-response plots

Similar plots are not directly obtainable with the experimental model: in fact, coherently with the nature of the exciter (the harmonic vertical force is obtained with two counter-rotating unbalanced masses), varying the frequency of the excitation (the angular velocity of the rotors), an even wider change in amplitude, linked to the power two of the angular frequency Ω, is realized as well. This circumstance makes actually impossible to perform experimental tests at fixed values of excitation amplitudes and every test corresponds to a parabolic path in the frequency-amplitude excitation plane. A response curve, following the cited path, is reported in Figure 8.

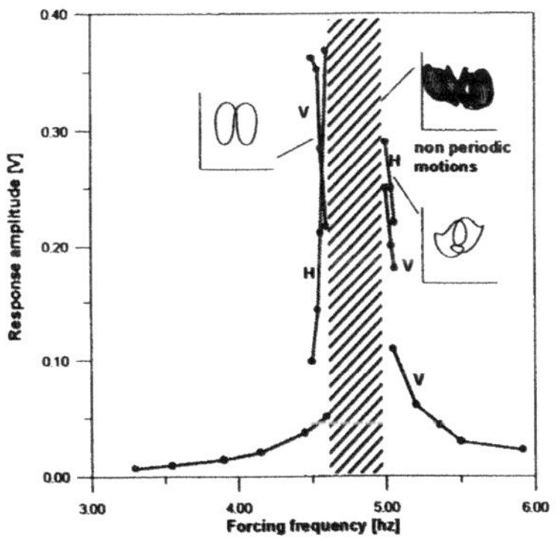

Figure 8—Experimental model: response plot

Within the limits linked to the previous observation, the agreement between the analytical and experimental plots appears good and encouraging the use of the new analytical model in predicting both the regular and non regular dynamics of the examined arch. A more general comparison between the two models can be observed on Figure 9 reporting the behavior charts of analytical and experimental models:

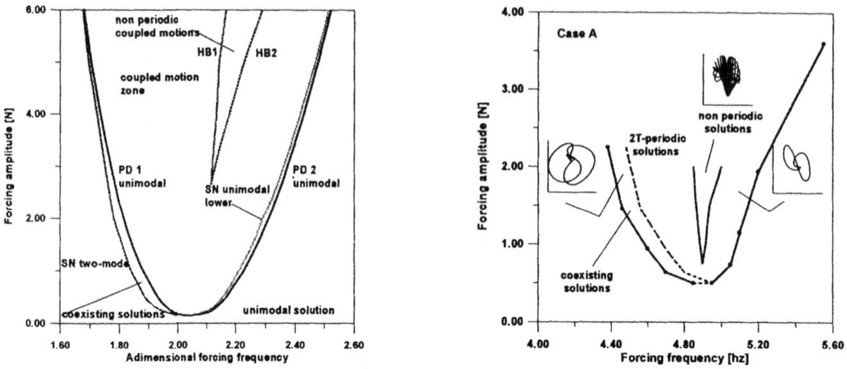

Figure 9—Analytical and experimental behavior charts

The two charts were obtained reporting on the forcing-frequency plane the bifurcation points read on various response-curves. The agreement between the two charts appears good both qualitatively and quantitatively, again confirming the accuracy of the new analytical model in a wide range of excitation amplitudes and frequencies. Concerning the main characteristics of the response a good agreement is found not only in the periodic zones: inside the sub-region of complex motions, higher increasing complexity is revealed approaching the core of the sub-zone both in the analytical and in the experimental case. In particular, increasing the forcing frequency and entering the instability region from it's boundary qualitatively different motions have been detected: when the periodic coupled solutions loses stability via the Hopf bifurcation, quasi-periodic, chaotic and then again quasi-periodic motions are observed, both in the analytical (see Figure 10) and experimental case.

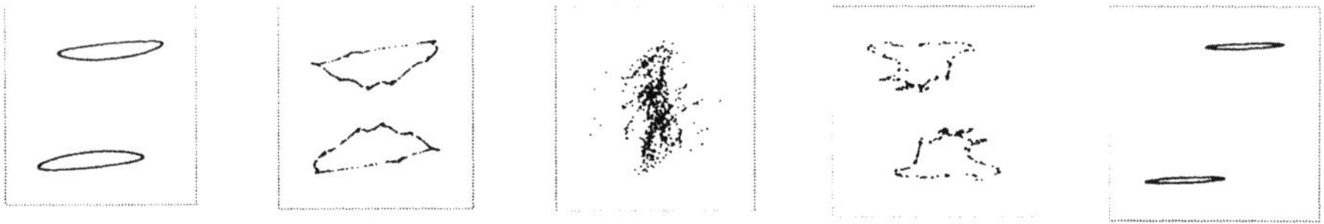

$\Omega = 2.113, \lambda_1 = 0$ $\Omega = 2.157, \lambda_1 = 1.14$ $\Omega = 2.185, \lambda_1 = 3.36$ $\Omega = 2.215, \lambda_1 = 0.18$ $\Omega = 2.293, \lambda_1 = 0$

Figure 10—Analytical model: complex transition inside the sub-area of non-periodic responses

In Figure 10 a transition of complex response for a forcing amplitude of *6N* is reported. Just after the Hopf bifurcation observable in the corresponding chart (Figure 9, left) the attractor is the two-torus whose Poincaré section is reported in the first column. Evolving this solution from a 2T-periodic motion (the anti-symmetric component always is subharmonic), the section is coherently constituted of two disjointed closed curves. The evaluation of the maximum Lyapunov exponent, correctly gives a *zero* value. Increasing the forcing frequency, the surface of the torus becomes fuzzy, the Lyapunov exponent increases (second column), then, after the torus breakdown, a fully developed chaotic regime is observable in column 3. Further increasing the forcing frequency, passing again for a fuzzy torus (column 4), the motion becomes again quasi-periodic (column 5).

A corresponding analysis was conducted on experimental time series using the delay map technique allowing the estimation of global complexity indicators like the *correlation dimension* and the *maximum Lyapunov exponent*: following this way it was possible to detect, also in the case of exeprimental tests, the increasing complexity entering the core of sub-region pertaining the complex-motions.

At a fixed forcing amplitude of $p=2N$ and frequencies respectively $\Omega_{qp}=4.87$ and $\Omega_{ch}=4.92$, two sample reconstruction are reported in Figure 11; they correspond to a quasi-periodic (first line) and a chaotic (second line) case. Respectively for the two cases, on the left, middle and right part of the plot are reported:

- a 2D-projection of the reconstructed phase space (delay coordinates) using the right delay coming from the *mutual average information* analysis,
- the classical *log[C(r)]-log(r) correlation curves* corresponding to embedding dimension ranging from $n=2$ to $n=12$,
- the slope estimation giving, respectively a *correlation dimension* $d_c=2$ for the quasi-periodic case, and $d_c=3.1$ for the chaotic one.

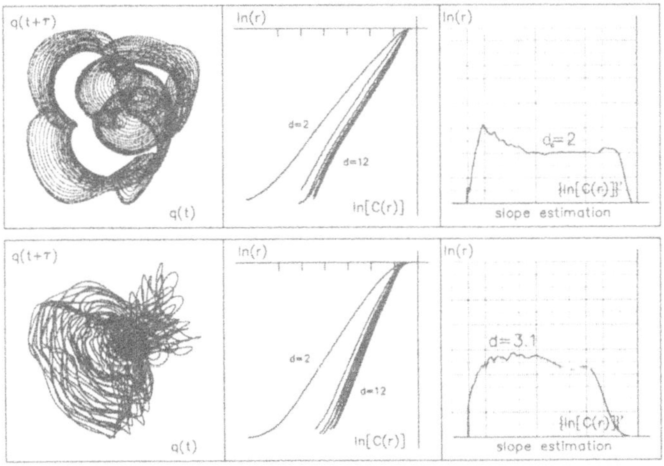

Figure 11—Experimental model: complex transition inside the sub-area of non-periodic responses

6. Conclusions

The contemporaneous analysis of an experimental and an analytical models of a double hinged circular arch gave a complete understanding of the dynamical behavior concerning the problem of the dynamic instability of the simple unimodal symmetric solution under the action of a concentrated, vertical load applied on its tip. Preliminary experimental tests furnished hints for a correct analytical modeling. Then, a systematic analysis conducted on the two models on wide ranges of the forcing parameters showed a good qualitative and quantitative agreement between the two models: while in the cases of periodic evolutions the comparison was done on the basis of the modal components, when complex motion arise, the comparison was done on the basis of synthetic complexity indicators like the *correlation dimension* and *the maximum Lyapunov exponent* (delay map technique). Eventually, because theoretical results concerning strangeness and chaoticity could be related more to the analytical modeling than to the real behavior of structures, in the case of complex time evolutions the help of a companion experimental model is crucial for furnishing an actual mechanical framework to the understanding and interpretation of strange phenomena [13].

7. Acknowledgments

This work was partially supported by the 40% 1997 funds.
The author tanks Dr. R. Alaggio for the help during the experimental analysis.

6. References

1. Bolotin VV: *The Dynamic Stability of Elastic Systems* (Translated by V. I. Weingarten et al.), Holden-Day, San Francisco, 1964.
2. Thomsen, J.J.: Chaotic Vibrations of Non-Shallow Arches, *Journal of Sound and Vibration* 153(2), (1992), 239-258.
3. Benedettini, F.: Planar Finite Forced Dynamics of a Double Hinged Circular Arch: Theory and Experiments, *Proceedings of ASME Design Engineering Technical Conferences,* Sacramento, (1997), VIB4092.
4. Win-Min Tien, Sri Namachchivaya, N. and Bajaj, A.: Non-Linear Dynamics of a Shallow Arch Under Periodic Excitation—1. 1:2 internal resonance. *Int. J. Non-Linear Mech.* 29 (1994), 349-366.
5. Alaggio, R., Belhaq, M., Benedettini, F., Lakrad, F. and Rega, G. Comparison and validation of different analytical models of the finite, forced, planar dynamics of a double hinged circular arch by means of experimental tests, *Proceedings of 3th National Congress of Mechanics,* Tetouan, Marocco, (1997).
6. Sheinman I.: Dynamic Large-Displacemet Analysis of Curved Beams Involving Shear Deformation, *Int. J. Solids and Struct.*, 16 (1980), 1037-1049.
7. Alwar R.S. and Narasimhan M.C.: Non-Linear Analysis of Laminated Axisymmetric Spherical Shells, *Int. J. Solids and Struct.*, 30,6 (1993), 857-872.
8. Nayfeh A.H. and Raouf A.R.: Non-linear Oscillations of Circular Cylindrical Shells, *Int. J. Solids and Struct.*, 23,12 (1987), 1625-1638.
9. Nayfeh A.H. and Raouf A.R.: Non-Linear Forced Response of Infinitely Long Circular Cylindrical Shells, *J. Appl. Mech.*, 54 (1987), 571-577.
10. Cusumano, J.P. and Bai, B.Y.: Period-infinity Periodic Motions, Chaos, and Spatial Coherence in a 10 Degree of Freedom Impact Oscillator, *Chaos Solitons & Fractals* 3,5 (1993), 515-535.
11. Takens, F. (1981) Detecting Strange Attractors in Turbulence, *Dynamical Systems and Turbulence,* D.A. Rand and L.S. Yang, eds., Springer Lecture Notes in Mathematics, Springer Verlag, New York, 898, 266-281.
12. Rosenstein, M.T., Collins, J.J. and De Luca, C.J. (1993) A Practical Method for Calculating Largest Lyapunov Exponents from Small Data Sets, *Physica D* 65, 117-134.
13. Benedettini, F.: An experimental time series analysis approach in the classification of non periodic motions in nonlinear structural dynamics, *Proceedings of the Third European Conference on Structural Dynamics: EURODYN 96*, Augusti, Borri, Spinelli eds., A.A. Baklema, Rotterdam (1996), 415-421.
14. Henrych, J: *The Dynamics of Arches and Frames,* Helsevier, Amsterdam, 1981.

DYNAMICAL ANALYSIS OF AN OFFSHORE PLATFORM WITH VIBRATION ABSORBERS

L.BEVILACQUA, R.B.BATTISTA and N. F. F. EBECKEN
Graduate School of Engineering (COPPE)
Federal University of Rio de Janeiro (UFRJ)
PO Box 68506
Rio de Janeiro, 21945-970 RJ, Brazil

1. Introduction

Recent studies on alternatives for structural systems designed to support deepwater offshore platforms have introduced the *compliant structure* concept, encompassing a wide class of structures into which may be incorporated, as sub-classes, *compliant towers* such as guyed towers [1], Gamma tower [2], Roseau and others [3]; and *buoyant systems* such as semi-submersibles [4] and tension -leg platforms [5]. The common characteristic of these structures is their ability to undergo large displacements under the action of environmental loads. A particular concept of compliant tower is known as guyed tower, whose behavior has been extensively studied, leading to the design, fabrication and installation of Lena Guyed Tower at the Gulf of Mexico [6]. The guylines are connected to the tower at a position near to the level of the resultant environmental loads and inertia forces. These guylines present highly variable stiffness, provided by a "clump weight" mechanism near the seabed touchdown point. The stiffness increases as these weights are being suspended; however, in storm condition the greater values of the displacement causes the total uplift of some guylines, thus
increasing the catenary length, reducing the guy line stiffness and limiting the maximum stress on the cable. It is the purpose of this paper to study the effect produced by coupling a vibration absorber to the tower

N. Van Dao and E.J. Kreuzer (eds.),
IUTAM Symposium on Recent Developments in Non-linear Oscillations of Mechanical Systems, 23–34.

in order to reduce the displacements of the tower for storm conditions and excitation near the first natural frequency.

2. The Finite Element Model for a Guyed Tower

The present paper studies the global behavior of a 330 m guyed tower, considering environmental loading typical of the south-eastern Brazilian coast. The analysis employed a "complete" finite element model, shown in the figure 1, with 468 three-dimensional nonlinear beam elements. The behavior of the guylines is simulated by nonlinear springs with associated forcing functions determined by previous static analysis. The forcing function vector consists of environmental loading generated by the ocean waves, sea currents and wind effects.

The results presented here correspond to the analysis for environmental storm conditions, consisting of deterministic Airy waves with periods 10s, 12 s and 14 s and height values close to 9m; current velocity at sea surface is 1.45 m/s; at the sea bottom is 0.25 m/s; the reference wind velocity is 55 m/s.

The load distribution profiles corresponding respectively to the three forcing functions are depicted in the figure 1. The tower is free to move along the three directions in space and the time history related to each mode has correspondingly the three components in x, y and z.

The dynamical analysis has been performed using a computer code developed at COPPE. The details could be found in reference [9].

For the present discussion, in order to build up a simplified model, it is necessary to know the nodal forces and the nodal displacements. Some typical time histories for these two variables are presented in the figure 2 a-c. The load has clearly a nonlinear periodic behavior, with period T = 14 s equal to the wave exciting force. The displacement displays also some non-linearity, but much less markedly as compared with the nodal forces. The period is again the same as that of the wave.

It is important to point out that the fluid-structure interaction was taken into account in the calculations. The effect of the variable relative velocities of the fluid and the structure at the different members are therefore already computed in the time histories.

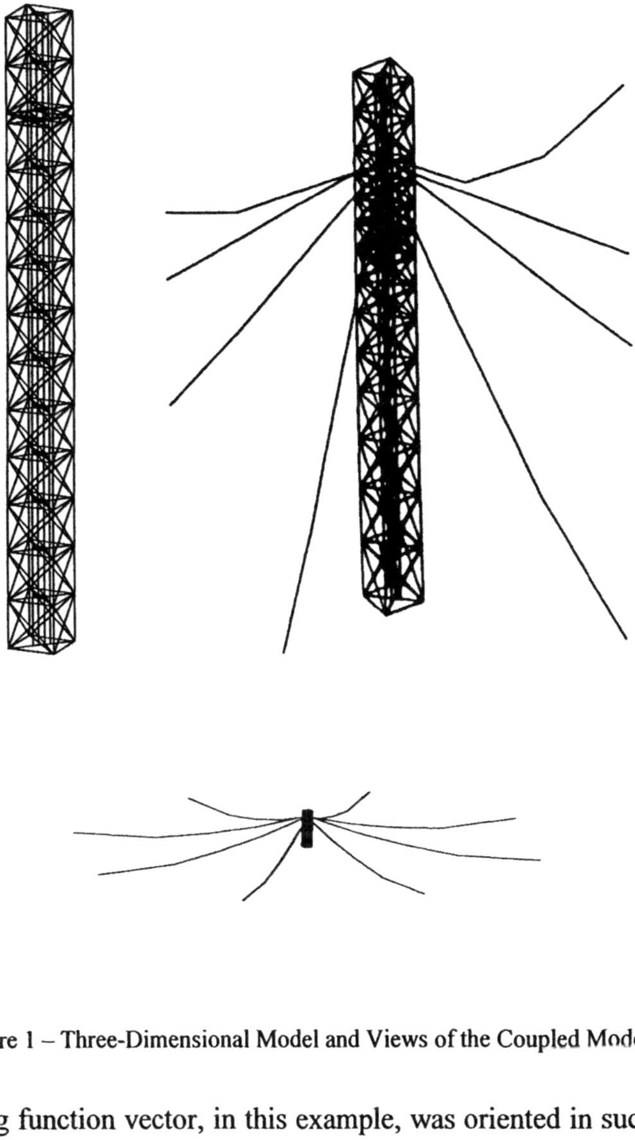

Figure 1 – Three-Dimensional Model and Views of the Coupled Model

The forcing function vector, in this example, was oriented in such a way as to preserve the motion essentially in the same plane, namely the x-z plane. This results in an almost plane motion, with displacements in the y and z directions practically equal zero.

This complete analysis has the advantage of giving reliable results, very close to the real physical model. It is however expensive. In order to reduce the computational costs a simplified model can be used initially to perform a parametric analysis, leading to the best configuration that minimizes the displacement amplitudes. Afterwards this solution can be checked with a more sophisticated model.

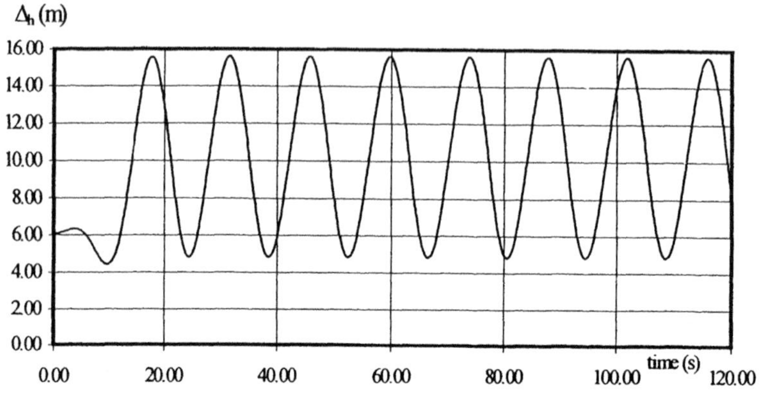

Figure 2.a - Time response for horizontal displacement (x direction) at 3D model tower top (z = 330.0 m)

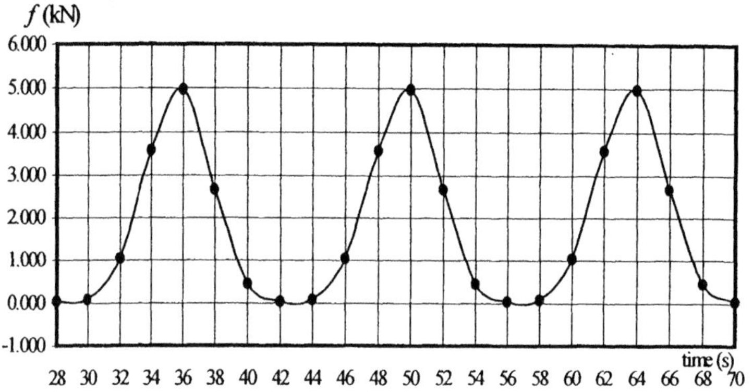

Figure 2.b - Time history of nodal load vector in x direction applied on the axis of the 3D tower model at level z = 75.0 m

3. The Simplified Model

The simplified model consists of two steps. First we search for a lumped mass system that approximately reproduces the displacement time history of the tower, as calculated by the finite element model, under equivalent loading conditions. This is a kind of inverse problem that could be very complicated, unless some simplifying assumptions are made, within the expected approximation range.

Consider the lumped mass system of the figure 3. From the FEM model it is possible to evaluate the load vector associated to this system, consisting of equivalent forces acting on each virtual mass. The motion will be considered plane, consistent with the previous result. The rod is hinged at the basis and the guylines are substituted by springs as in the

FEM model. Now, since the displacement amplitude of the points connected to the guylines are moderate, the variation of the spring force with the displacement amplitude
can be taken linear, with a spring constant K equal to the tangent to the

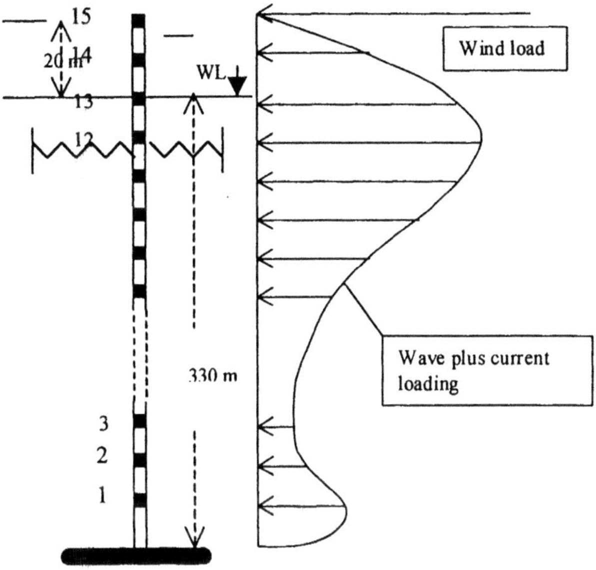

Figure3. Wave and current load distribution

Table 1. Nodal forces

Node	Load
15	8144.00
14	6601.36
13	3428.22
12	1778.80
11	949.66
10	260.49
9	66.15
8	16.73
7	5.26
6	2.95
5	3.91
4	8.07
3	16.85
2	31.27
1	0.24

curve force-displacement at the point corresponding to the average force on the guyline (Fig. 4).
As can be seen from the FEM analysis, both the displacements and the nodal forces are periodic functions of time with period equal to the period of the exciting wave. We will consider here, the displacement and nodal force vectors as harmonic functions with period coincident with the exciting external force. This is not strictly true for non-linear behavior, but it is a good approximation for practical purposes and within the frame of this simplified analysis. Considering

TABLE 2. Natural spectrum of the guyed tower. First 10 natural modes. Period in seconds.

Mode	1	2	3	4	5	6	7	8	9	10
Period	9.998	2.479	2.437	1.258	1.106	1.086	0.787	0.634	0.623	0.584

the spring displacement within the linear range as shown in the figure 4, and using the complete FEM model analysis the natural spectrum of the tower is found (Table 2).The problem we have now is to find the appropriate lumped mass oscillator, such that it will reproduce approximately the same displacements at the corresponding levels measured - along tethe z axis - under the action of the same forcing functions, with the additional requirement of having the same frequency spectrum. The tower is reduced to a bar with concentrated masses located at the main horizontal planes. Figure 3 shows the distribution of the external forces which will be approximated by a harmonic variation with time, with period consistent with the time history shown in the figure 2 Following the standard procedures of modal analysis, it can be shown that the above requirements are met if the masses are calculated as follows:

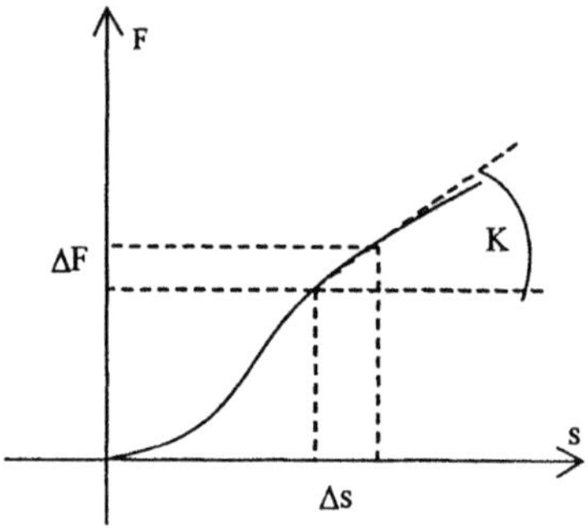

*Figure 4.*Typical guyline force-displacement function

$$M = P^{-T} M^{*} P^{-1} \tag{1}$$

$$M_k^{*} = \frac{1}{\omega_k^2 + 2\xi_k \omega_k - \omega_e^2} \frac{\phi_k^T p}{\phi_k^T x} \|\phi\|^2 \tag{2}$$

where M and M^* are the mass matrix and the generalized -diagonal- mass matrix respectively. The vectors ϕ_k , p and x are the k-th modal shape vector, the load vector and the displacement vector respectively. The frequencies ω_k and ω_e correspond to the k-th mode and the exciting external force - wave - respectively. The parameter ξ_k is the generalized damping coefficient related to the k-th mode. The matrix P is the mode-shape matrix.

Although the analysis has been performed for the wave, current and wind data specified in the previous section, it is expected that for small variations of the exciting forces the displacement and load distribution will be kept essentially the same. The amplitude varying according to the denominator of the first term in (2).

Summarizing, the present method is aimed to determine equivalent masses that would give for the simplified model, approximately the same time history for the forcing functions and the displacements at the corresponding z-coordinates of the complete model. Note that we are interested in the dynamical contribution of the load and displacement vectors, that is, the total deviation from the average value. The average value of the load is only important, in this case, to determine the constants of the equivalent springs.

Figure 5 shows the mode shape for the first mode. Clearly the first mode is a bending free mode and for the frequency range appearing in practical

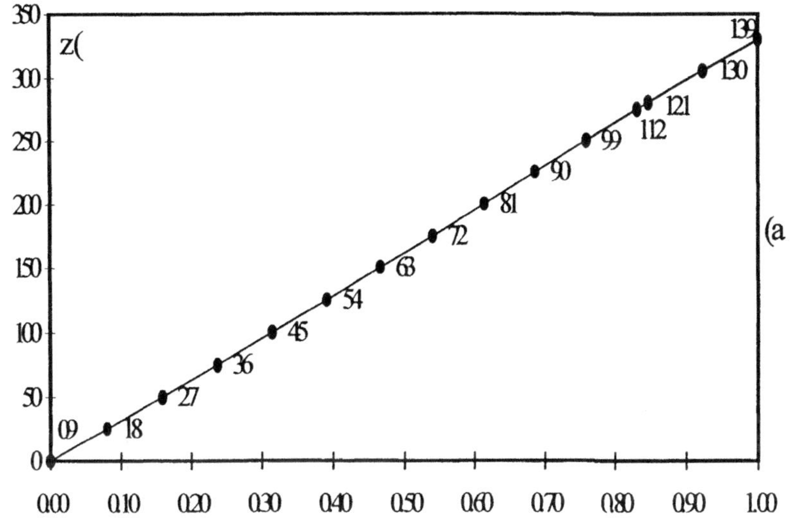

Figure 5 - Vibration mode shapes of tower axis in xz plane. (a) pendular mode

application this mode will store practically all the energy transferred to the structure. So it is reasonable to represent the tower structure by a single mass attached to a spring such that the natural frequency will coincide with the first natural frequency of the tower or of the lumped mass system, which is the same. The parametric study of a pendulum introduced as a vibration absorber can therefore be performed much more easily, and the results obtained with this two degree of freedom, non linear system, reproduces satisfactorily the most important aspects of the amplification factors for the tower, as far as the basic design variables are concerned.

4. The Vibration Absorber Model

Let us now turn our attention to the two-degree of freedom model shown in the figure 6. The mass M represents the equivalent mass of the tower corresponding to the position where the pendulum is connected. In this paper the pendulum will be considered hinged at the top of the tower. The mass M is taken equal to M_1^*. The fundamental frequency for this spring-mass single system is chosen to be ω_1 and consequently the spring constant is $K = \omega_1^2 M_1^*$. The load vector is given by $F = F_0 \operatorname{sen}(\omega_e t + \alpha)$ where $F_0 = \phi_1^T p$. Clearly ω_e is the frequency of the exciting external force. Note that the eigenvectors are normalized such that the displacement at the level where the pendulum is hinged equals 1. The governing equations for this system are classical. They read:

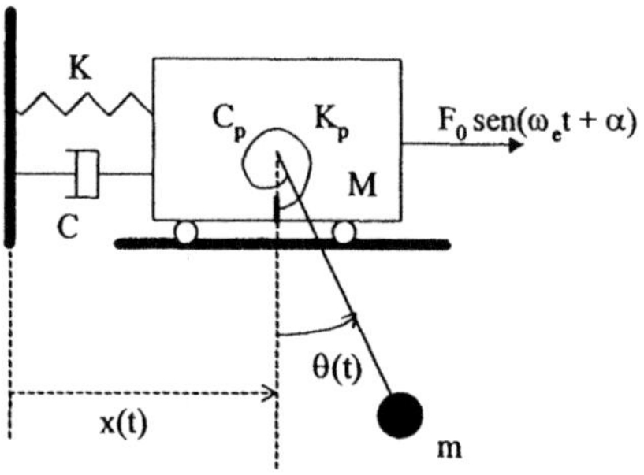

Figure 6. Simplified two degrees of freedom model

$$(M+m)\frac{d^2x}{dt^2} + C\frac{dx}{dt} + Kx + \frac{d^2}{dt^2}(\theta \operatorname{sen}\theta) = F_0 \operatorname{sen}(\omega_e + \alpha) \tag{3}$$

$$ml^2\frac{d^2\theta}{dt^2} + C_p\frac{d\theta}{dt} + K_p\theta + mgl\operatorname{sen}\theta + (ml\cos\theta)\frac{d^2x}{dt^2} = 0 \tag{4}$$

A previous parametric analysis [7] for a similar case lead to the following indications for the choice of the pendulum characteristics - m and l - for large values of θ, that is in the non linear region:

For $\omega_e < \omega_1$ best choice $\omega_p < \omega_e$
For $\omega_e = \omega_1$ ω_p arbitrary
For $\omega_e > \omega_1$ best choice $\omega_p > \omega_e$

Of course for the linear behavior the classical result holds, that is $\omega_p = \omega_e$. For all cases the amplitude of the pendulum should be limited to the stability region of this system and to the conditions imposed by the design specifications.

The following combinations have been test to obtain the time response of the displacements θ(t) and x(t) :

Table 3. Selected configurations. K_p has been put equal zero.

CASE	I	II	III	IV	V
ω_1 (rad/sec)	0.628	0.628	0.628	0.628	0.628
T_1 (sec)	10	10	10	10	10
ω_e (rad/sec)	0.628	0.628	0.524	0.524	0.449
T_e (sec)	10	10	12	12	14
M (pendulum mass kg)	$3x10^5$	$6x10^5$	$3x10^5$	$6x10^5$	$6x10^5$

L (pendulum length m)	25	25	36	36	50
ω_p (rad/sec)	0.626	0.626	0.521	0.521	0.4429

Case V represents the introduction of the vibration absorber for the case studied with the FEM without the pendulum. There is an effective reduction of the amplitude of about 33% (Fig. 7), for a pendulum with mass equal to 600 metric tons and a damping ratio ξ_p=0.02. Although high, if we consider the absolute value, the mass of the pendulum is about 10% of the tower total mass, which falls within an acceptable design value. The pendulum length is equal to 50 m in this case which depending on the tower design specifications could be a limiting factor for the use of this device.

Cases III and IV compare the effectiveness of two pendulums as energy absorbers, both with the same natural frequency but different masses, when the structure is under an exciting force with period equal to 12s. Clearly from the figures 8 a and 8 c, the pendulum with mass equal to 600 metric tons is superior. The amplitude reduction reaches about 38.5% for this case, while is only about 8.2% for a pendulum with half of that mass. The maximum pendulum angular amplitude is about the same for both cases, circa 1.0 rad. for the pendulum with mass equal to 300 metric tons and 0.9 rad. for 600 tons. For both cases the damping ratio was taken equal to .02.

For an exciting force with frequency coincident with the natural frequency of the structure, ω_e=ω_1 it is seen from the time histories in figures 9 a and 9 c that, although the heavier pendulum introduces a higher reduction factor of around 48%, nevertheless the pendulum with mass equal to 300 tons provides a reduction in the amplitude of circa 29%. So it can be said that for the resonant frequency both pendulums are effective in the reduction of the structure amplitude. Again the angular amplitude for the lighter pendulum is slightly greater than that for the heavier one. Both fall within the range {+1 rad. , -1 rad. } which is reasonable for practical purposes.

The system (3)-(4) was integrated numerically using the Runge -Kutta method. For the present analysis the structural damping C was assumed to vanish.

Although the model used here introduces considerable simplifications, the results are satisfactory for practical applications and platform design.

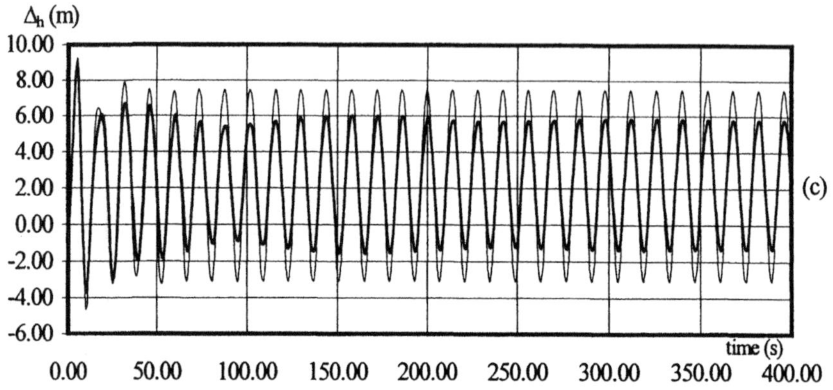

Figure 7 - Uncontrolled and controlled tower top displacement (Δ_h) responses of 2D model under wave (T_o = **14** s) for pendular absorbers (l = 50 m; m_p = 600 t) with damping factors $\xi_p = 0.02$

5. Conclusions

The use of a pendulum as energy absorber, in order to reduce the amplitudes of the tower, is effective provided that the proper combination of frequency and mass is used. The frequency of the pendulum should be tuned close to values of the frequency of the external force. This is however not enough, it is also necessary to adopt a mass sufficient large as to store an energy amount that would allow for substantial reduction in the tower displacement. This can be tested with a simple model as described above. The damping effect is effective in the reduction of amplification factor but less important than the mass.

Another item that should be considered in the design is the maximum angular displacement of the pendulum and its length as well. Both should be limited to an acceptable range specified in the design. The introduction of a torsional spring K_p can reduce the angular amplitude. In that case however it is necessary to change the length of the pendulum in order to keep the frequency at a convenient value necessary to reduce the displacements. In general it would be necessary to increase the length up to a value given by:

$$l = \frac{1}{2} l_i \left(1 + \sqrt{1 + \frac{4K_p}{g l_i m}} \right)$$

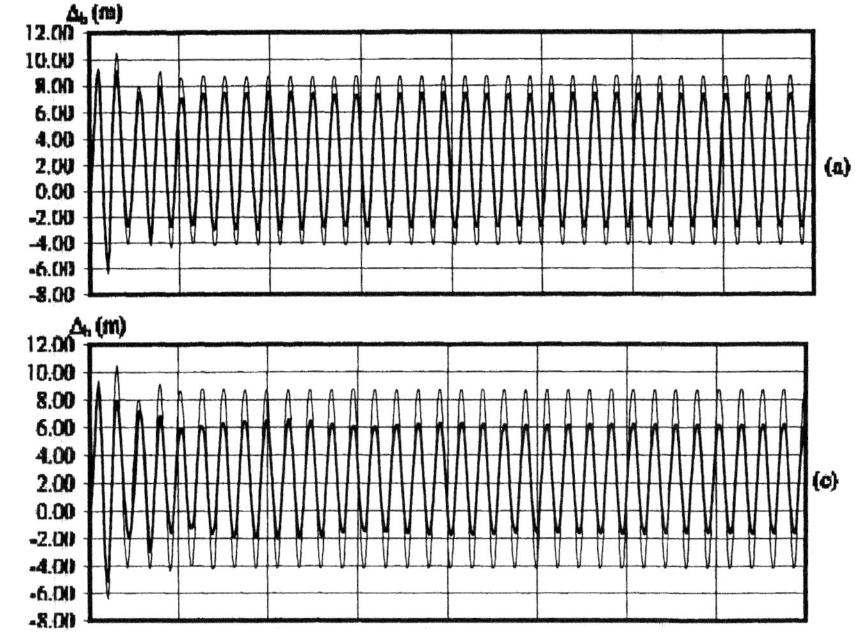

Figure 8 - Uncontrolled and controlled tower top horizontal displacement (Δ_h) responses of 2D model under wave ($T_s = 12$ s), ($l = 36.0$ m; $\xi_p = 0.02$) with $m_p = 300.0$ t (figs. a) and $m_p = 600$ t (figs. c)

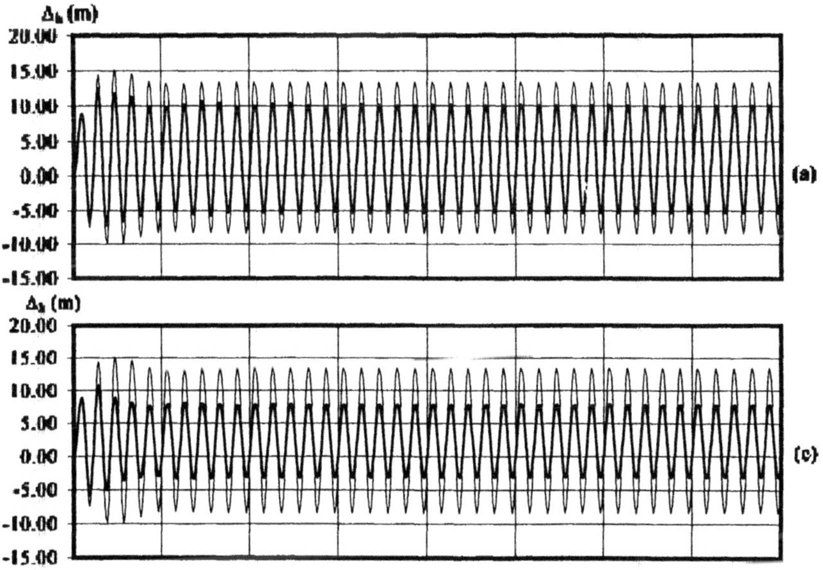

Figure 9 - Uncontrolled and controlled tower top horizontal displacement (Δ_h) responses of 2D model under wave ($T_s = 10$ s), ($l = 25$ m; $\xi_p = 0.05$) with $m_p = 300$ t (figs. a) and $m_p = 600$ t (figs. c)

where l_i is the length of the pendulum to provide for the "optimal" frequency without the torsional spring. For $\omega_p/\omega_1 > 1$, ω_p should be always greater or equal to ω_e, in which case it will be not necessary to increase the length of the pendulum.

One of the most important advantages in diminishing the displacement amplitudes of the tower is the reduction of the danger of collapse due to the fatigue of the guyed lines.
The use of this type of energy absorber is becoming more and more frequent in towers of the type described in this paper and in tension-leg platforms as well. Other possibilities including active control and more complex mechanisms are under investigation now. The use of variable mass devices deserve also the attention of engineers and investigators, controlling for instance the water contents of reservoirs placed in strategic points of the structure.

6. Acknowledgements

The work contained in this paper has been partly supported by the CAPES-PROBRAL program under the number 47/96 and by the CNPq through fellowships and scholarship granted to students s working under the supervision of the authors.

7. References

1. L. D. Finn. A New Deepwater Offshore Platform - The Guyed Tower, *Proceedings of the Offshore Technological Conference. OTC'76.*(1976). **Paper 2688**
2. F. Sedillot and L. A. Boston. The Technological Development of Two Deep Water Compliant Structures. V. M. Mourelle, B. P. Jacob and N. F. F. Ebecken. Nonlinear Dynamic Behavior of Semisubmersible Offshore Platforms. *Offshore Engineering. Vol.3* (eds. F.L.L.B. Carneiro, R. C. Batista and A. J. Ferrante), Pentech Press, London (1988) 385-401
3. V. M. Buslov and D. I. Karsan. Deepwater Platform Designs: an Illustrated Review. *Ocean Industry.* (1986) 53-62
4. V. M. Mourelle, B. P. Jacob and N. F. F. Ebecken. Nonlinear Dynamic Behavior of Semisubmersible Offshore Platforms. *Offshore Engineering. Vol.3* (eds. F.L.L.B. Carneiro, R. C. Batista and A. J. Ferrante), Pentech Press, London (1988) 385-401
5. D. C. Angelides, S. A. Will and R. F. Figgers. Design and Analysis Framework of Tension Leg Platforms. *Offshore Engineering. Vol.3* (ed. F.L.L.B. Carneiro), Pentech Press, London (1982)
6. D. E. Boening and E. R. Howell. Lena Guyed Tower Project Overview. *Proceedings of the Offshore Technological Conference. OTC'84.*(1984). **Paper 4649**
7. R. C. Batista and M. • Silva Pinheiro. Absorsor Pendular Não Linear Para Redução de Vibrações em Torres Esbeltas. *V Congresso de Engenharia Mecânica Norte-Nordeste.* (1998) Fortaleza, Ceará, Brasil
8. M. S. Glasscock, J. W. Turner, L. D. Finn and P. J. Pike. Design of Lena Guyed Tower. *Proceedings of the Offshore Technological Conference. OTC'84.*(1984). **Paper 4650**
9. B. P. Jacob and N. F. F. Ebecken. A computational System for the Nonlinear Dynamic Analysis of Deepwater Offshore Structures. *Proceedings of the 8th International Conference on Offshore Marine and Artic Engineering - OMAE'89.* (1989) The Hague

COMPLEX DYNAMICS OF A 'SIMPLE' MECHANICAL SYSTEM: THE PARAMETRICALLY EXCITED PENDULUM

S. R. BISHOP
Centre for Nonlinear Dynamics and its Applications
University College London
Gower Street, London WC1E 6BT, UK

1. Introduction

A planar pendulum is perhaps the simplest and most quoted example of a dynamical system, yet when driven its simplicity of description belies a range of complex dynamical motions. Historically interest was focused on, and indeed restricted to, small displacements from the vertical, but more recently numerical simulations may now incorporate full nonlinear effects and large amplitudes without restrictions to small parameters. If driven vertically at the pivot, a pendulum which can freely move in the plane exhibits equilibrium states (corresponding to the hanging and inverted positions), periodic solutions (oscillations and rotations), as well as chaotic motions, all of which can easily be seen in a mechanical experiment. Miles [11] produced an excellent overview of the basic, qualitative dynamics but research interest is not yet exhausted with some recent results focusing on a purely vertical forcing investigating the topological structure of phase space [6], chaos [2] and the stability of the inverted state [1,8].

We consider here the so-called parametrically excited pendulum idealised as a mass on a light, rigid, inextensible rod, moving in a plane and driven by a periodic vertical force. For mathematical convenience the model is often scaled [4], leading to the fundamental equation

$$\ddot{\vartheta} + c\dot{\vartheta} + \left(1 + p\cos(\omega t)\right)\sin\vartheta = 0 \qquad (1)$$

where ϑ measures the anti-clockwise angle from the downward hanging state. The terms p and ω correspond to the scaled amplitude and frequency of the driving force used generally as control parameters. In practice the damping may have a nonlinear velocity dependence and for small, low velocity oscillations, be governed by friction in the pivot,

N. Van Dao and E.J. Kreuzer (eds.),
IUTAM Symposium on Recent Developments in Non-linear Oscillations of Mechanical Systems, 35–43.

but is taken here as a constant ($c = 0.1$) times the velocity in line with earlier experimental and numerical investigations. Writing

$$\frac{dV}{d\vartheta} = \sin\vartheta \qquad V = -\cos\vartheta \tag{2}$$

allows us to view the dynamics as a particle moving in a potential energy function (V).

Small oscillations correspond to periodic motions within the well while rotations lead to escape from the local potential well between ($+\pi$, $-\pi$), see figure 1. In this way comparisons can be made with general results regarding escape from a potential well which underpins a wide range of problems including ship capsize and buckling [14].

For small driving amplitudes the downward hanging state forms a stable equilibrium of the model system (1). However, in line with the behaviour of linear counterpart known as the Mathieu equation [10], this hanging state becomes unstable in a series of zones, the most predominant of which is located about $\omega = 2$, on which we focus our attention. A linear analysis can reveal some insight but for the full nonlinear problem numerical simulations are preferable.

Numerically we can fix one parameter, ω say, and follow the stable solutions which emanate from the hanging state as we vary the other parameter p using numerical path following techniques [9]. For example the schematic diagram of figure 2 shows that if ω is fixed at 2.1 and p increased, then the hanging solution remains stable until a pitchfork bifurcation occurs (denoted at PF). Thereafter the only local stable solution is an oscillation whose period is twice the period of the forcing. As p increases further this solution remains stable until the system breaks symmetry (at S), after which two anti-symmetric oscillating solutions exist of which only one is drawn on the figure. This solution then undergoes a cascade of period doubling bifurcations at F culminating in a chaotic oscillating motion, which is stable over only a very small range of the parameter. For a very small increase in p the chaotic motion loses stability via a global bifurcation and thereafter a trajectory seeks an alternative steady state which almost always corresponds to escape from the local potential well. Such solutions can be rotating periodic solutions, but more typically this escape leads to a chaotic motion which can be thought of as a global cross-well motion involving oscillations, in some well, followed by an irregular series of left and right rotations. Experimentally such motion produces a series of clockwise and anti-clockwise rotations and oscillations of the pendulum in a random-like manner where almost any sequence of left and right swings is possible. This motion is referred to as a tumbling chaotic solution.

Standard techniques of path following can also be extended to locate and follow bifurcations as both parameters are varied [9], so that we can plot, in figure 3a, the zones in parameter space in which the various solutions described above exist.

The same routine can also be carried out to locate and follow rotating solutions. If we consider the most basic rotating solution which appears at a saddle node bifurcation at A and performs one rotation in one period of the forcing, then this also undergoes a period doubling bifurcation (which we shall denote B) and subsequent cascade (at F) to a rotating chaotic motion shown in figure 3b. Again this chaotic solution is stable only over an extremely small range of the parameter space, giving way typically to the tumbling chaotic motion. The zone in parameter space where this basic period-1 rotating solution exists and is stable is shown between the lines A and B on figure 3b. This periodic motion restabilises at U creating another zone of stable motion.

Stable subharmonic solutions also exist, both of an oscillatory and rotatory nature. These may be classified according to the period (n) and, in the case of rotations (r), the number of complete rotations, i.e. (n,r). These subharmonic solutions may coexist with the simpler, harmonic motions and a full examination requires investigation of their basins of attraction [5].

If we superimpose all of these solutions on the parameter space then a rich structure exists, with coexisting solutions the norm rather than the exception. The solution onto which the system decays will depend crucially on the initial conditions given, which leads to complicated basins of attraction. Furthermore, a Melnikov analysis [5] shows that as we increase p the basin boundaries become fractal [12,15]. As a consequence the resulting dynamics is complicated even for this 'simple' system. Many of these motions can easily be viewed experimentally, and worthy of note is the fact that the tumbling chaos is stable over a reasonably broad range of the parameter space. Whereas the oscillating and rotating chaotic motions exist over very narrow windows of parameter space and even then are almost indistinguishable from a periodic motion with noise superimposed.

2. Flexible Control

So far we have restricted our attention to stable solutions, but it is also known that embedded within the chaotic attractor is a large (possibly infinite) number of unstable periodic solutions. Precisely locating these unstable solutions can be carried out purely numerically (using a Newton-type scheme), by using a topological analysis to guide numerical studies [7] or via a direct examination of a time series using the method of close returns.

In a seminal paper by Ott, Grebogi and Yorke [13] proposed a method of control to utilise these embedded solutions. Their method as proposed was brilliantly simple, using the chaotic dynamics of the system to approach as close as one wished to any desired solution. The control method forced the system onto a stable manifold of the unstable orbit (or more precisely a linear approximation to it) and so thereafter no control is theoretically needed; once again the system does all the work. Since that time many other methods have been proposed to carry out the same task, though to some extent these disguise the simplicity of the original concept.

Combining a mixture of methods to locate unstable orbits, and developing robust methods for control even in the presence of noise, means that now we are able to select a desired solution from a large selection of unstable motions onto which the system can be controlled [4]. Furthermore, a chaotic solution can be controlled on to a combination of these solutions without globally changing the system parameters, as shown in the time history of angular velocity of figure 4; 4 different desired orbits have been stabilised and each time control is turned off the trajectory is attracted to the chaotic motion.

3. Inverted Solution

It has been known for some time that if the pendulum system is driven hard enough, the inverted state can stabilise [8]. Recently interest in this phenomenon has been rejuvenated considering multiple pendulums [1]. An examination of an effective potential energy function (V_{eff}) by separating the rapidly oscillating motions from the smooth components for $\omega >> 1$ yields

$$V_{eff}\,(\vartheta) = -\cos\vartheta + \frac{p^2}{4\omega^2}\sin^2\vartheta \tag{4}$$

plotted in figure 4. From this relationship we may analytically determine the parameter values for which the inverted solution stabilises, p_A. For $\omega > 2$ results from numerical simulation closely match this analytical result [3] though this analysis gives no information of the subsequent dynamics within the well.

If, instead of a purely vertical forcing, a small tilt is given to the system (1°, say) then perhaps it is not surprising that the subsequent stable inverted solution is no longer perfectly upright ($\vartheta = \pi$). What is remarkable is that, for fixed ω, as p is increased, this solution first stabilises in a solution whose mean variation (ψ) of angular displacement from the inverted state ($\vartheta=\pi$) attains reasonably large values, as shown in figure 6, and only approaches $\psi = 1$ as ω increases.

4. Conclusions

A driven pendulum forms an archetypal dynamical system which can exhibit a wide range of dynamical response. With parametric forcing the system displays three types of chaotic motion; rotations, oscillations and equilibrium states. Unstable periodic orbits embedded within the tumbling chaos motion can also be viewed in conjunction with a suitable scheme. In addition such a pendulum has the interesting property that rapid excitation leads to stabilisation of the inverted state.

5. References

[1] Acheson, D.J. and Mullin T. (1993) Upside down pendulums, *Nature* **366**, 215-216.

[2] Bishop, S.R. and Clifford, M.J. 1996 Zones of chaotic behaviour in the parametrically excited pendulum, *J. Sound & Vibration* **189, No.1**, 142-147.

[3] Bishop, S.R. and Sudor, D.J. (1999) The "not quite" inverted pendulum, Int. J. Bif & Chaos **9** to appear.

[4] Bishop, S.R., Xu, D. and Clifford, M.J. 1996 Flexible control of the parametric pendulum, *Proc. Roy. Soc. Lond.* **A452**, 1-18.

[5] Capecchi, D. and Bishop, S.R. 1994 Periodic oscillations and attracting basins for a parametrically excited pendulum, *Dynamics and Stability of Systems* **9, No.2**, 123-143.

[6] Clifford, M.J. and Bishop, S.R. 1994 Bifurcational precedences for parametric escape from a symmetric potential well, *International J. Bifurcation & Chaos* **4, No.3**, 623-630.

[7] Clifford, M.J. and Bishop, S.R. 1996 Locating oscillatory orbits of the parametrically excited pendulum, *J. Austral. Math. Soc. Series **B*** **37**, 309-319.

[8] Clifford, M.J. and Bishop, S.R. 1998 Inverted oscillations of a driven pendulum, *Proc. Roy. Soc. Lond.* A **454**, 2811-2817.

[9] Foale, S. and Thompson, J.M.T. (1991) Geometrical concepts and computational techniques of nonlinear dynamics, *Comput. Meth. Appl. Mech. Engng.* **89**, 380-394.

[10] Jordan, D.W. and Smith, P. (1987) *Nonlinear Ordinary Differential Equations*, Clarendon Press: Oxford.

[11] Miles, J. (1989) The pendulum from Huygens' Horologium to symmetry breaking and chaos, in *Theoretical and Applied Mechanics*, edited by P. Germain, M. Piau and D. Caillerie, Elsevier Science: North Holland, pp 193-215.

[12] Moon, F.C. and Li G.X. (1985) Fractal basin boundaries and homoclinic orbits for periodic motion in a two-potential well, *Phys..Rev. Lett.* **55**, 1439-1442.

[13] Ott, E., Grebogi, C. and Yorke, J.A. (1990) Controlling chaos, *Phys. Rev. Lett.* **64**, 1196-1199.

[14] Thompson, J.M.T. (1989) Chaotic phenomena triggering escape from a potential well, *Proc. Roy. Soc. Lond.* A **421**, 195-225.

[15] Thompson, J.M.T., Bishop, S.R., and Leung, L.M. (1987) Fractal basins and chaotic bifurcations prior to escape from a potential well, *Phys. Lett.* A **121**, 116-120.

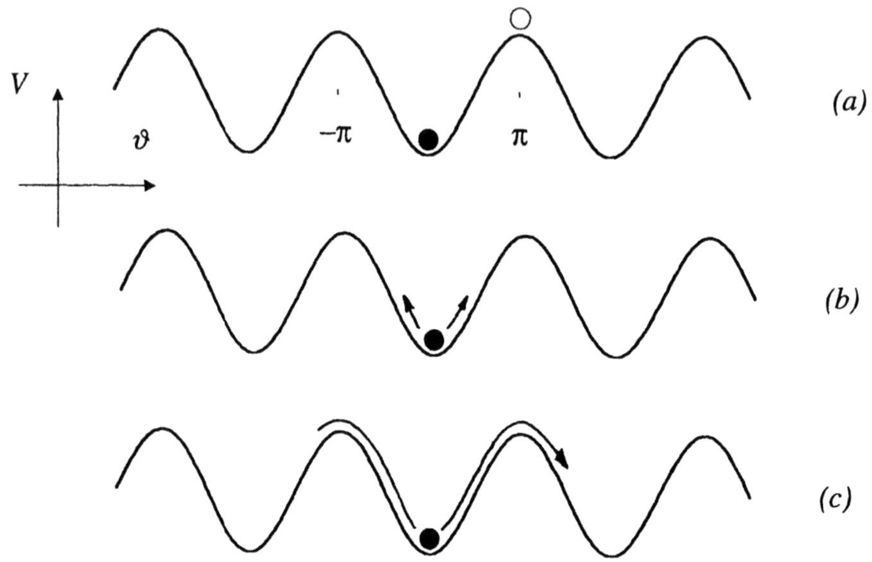

Figure 1. Solutions visualised as motions of a particle in the governing potential energy function: (a) stable (filled circle) and unstable (hollow circle) equilibrium, (b) oscillations and (c) rotations.

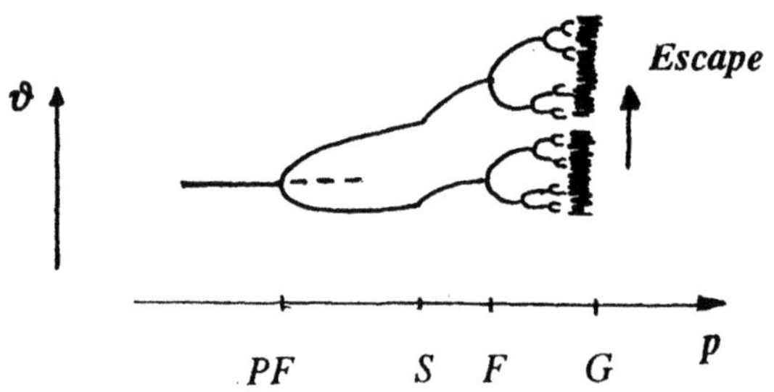

Figure 2. Schematic representation of solution paths for fixed forcing frequency ω=2.1. *PF* is where the equilibrium state becomes unstable (the dotted line indicates the unstable path). *S* corresponds to a symmetry-breaking bifurcation, while *F* represents the complete Feigenbaum cascade of period doubling bifurcations. G corresponds to a global bifurcation.

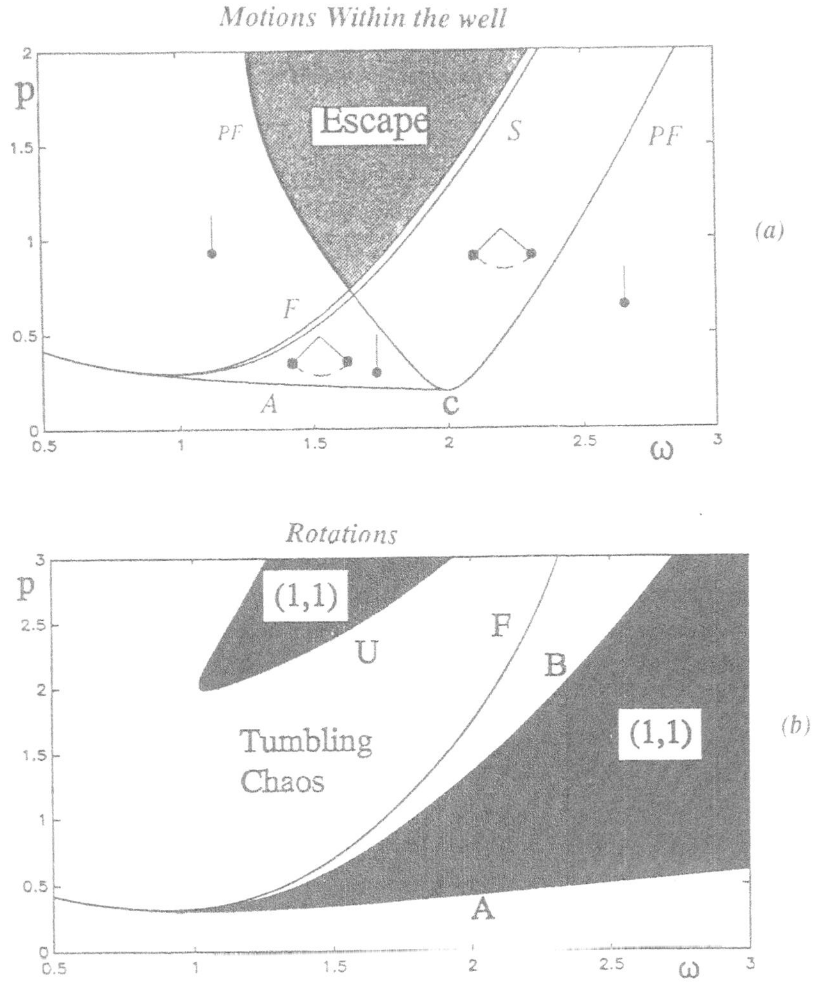

Figure 3. Zones of stable solution of (a) hanging and oscillating motions, and (b) rotating motions in the space of parameters. In (a) *PF* denotes the line of pitchfork bifurcation which is subcritical to the left of the point c, *S*, the symmetry breaking, *F* the Feigenbaum cascade of period doubling, *A* the period two fold (saddle-node) line. Escape denotes the zone for which no major stable solutions exist so that almost all trajectories leave the local potential well. In (b) A is the saddle-node, B the first period doubling to a R(2,2) motion, F the subsequent cascade to rotating chaos.

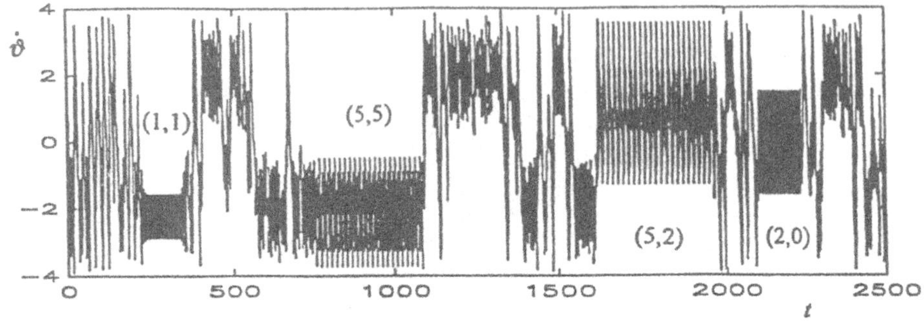

Figure 4. Flexible control. The chaotic tumbling motion is controlled on to various solutions. Once control is turned off the system once again behaves chaotically and may approach another chosen solution. The controlled unstable solutions are a rotating (1,1), a subharmonic rotating (5,5), a tumbling (5,2) and an oscillating (2,0) motion.

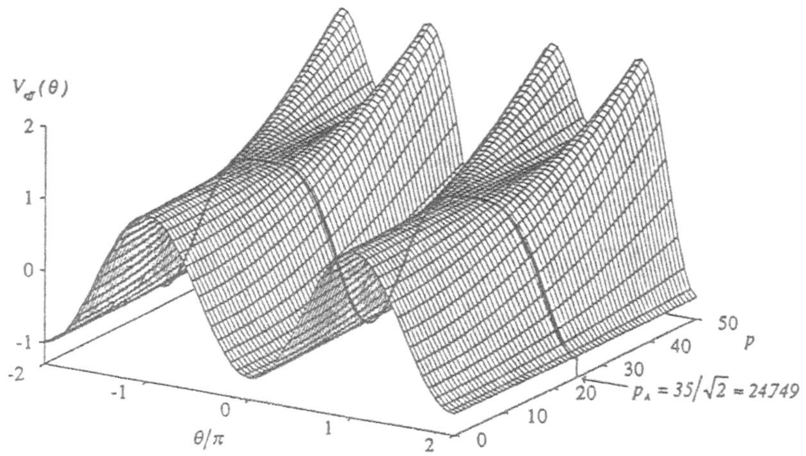

Figure 5. Effective potential function for the inverted state

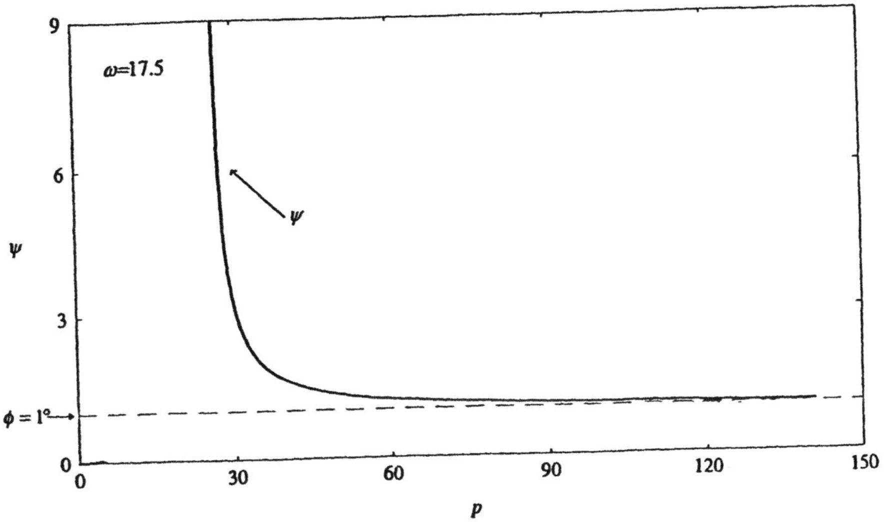

Figure 6. Variation of mean oscillation from the upward vertical ψ versus p.

CONTROL OF OSCILLATIONS IN SYSTEMS WITH MANY DEGREES OF FREEDOM

F. L. CHERNOUSKO
Institute for Problems in Mechanics
of the Russian Academy of Sciences
pr. Vernadskogo 101-1, Moscow 117526, Russia

Abstract. Two procedures for the control design in dynamical systems subjected to the control and state constraints are described and applied to the control of oscillations. The first approach is based on the decomposition of the original system with many degrees of freedom into simple oscillators; the second one is an extension of Kalman's method. Both approaches produce explicit control laws satisfying the constraints imposed and driving the system from a given initial state to the terminal state in finite time. Several examples are presented.

1. Introduction

There exist several approaches to the control design in dynamical systems with many degrees of freedom. In the classical linear methods of automatic control, the control u is represented as a linear operator of the state x: $u = Lx$. This approach has the following shortcomings. In the vicinity of the terminal state $x = 0$, where x is small, the control u is also small. Thus, the control possibilities are not used in full here; as a result the time of motion T is infinite, and $x \to 0$ as $t \to \infty$. On the other hand, if x is large enough, then u also becomes large, and the constraints imposed on control can be violated. Besides, the application of linear methods to nonlinear systems is often questionable. The methods of optimal control [1] are applicable to nonlinear systems and can drive the system to the terminal state in minimal time, taking into account various constraints. However, it is very difficult to obtain closed-loop optimal controls for systems with many degrees of freedom. Such well-known methods for the control design as the method of variable structure systems and feedback linearization do not explicitly take into account the control and state constraints.

In this paper, we consider two approaches to the control design for dynamical systems in the presence of constraints. The first one is based on the decomposition of the system with many degrees of freedom into linear oscillators controlled by bounded forces. The second approach is an extension of well-known Kalman's method [2] (originally developed for linear systems in the absence of constraints) to the case of constraints imposed on the control and state variables. Note that, even for linear systems, the control problems are themselves essentially nonlinear, if the control constraints are taken into account. As examples of our approaches, we consider the control for systems of oscillators, distributed-parameter elastic systems, and a system driven by an electric DC motor. More details and examples related to the approaches described below can be found in [3-6].

N. Van Dao and E.J. Kreuzer (eds.),
IUTAM Symposium on Recent Developments in Non-linear Oscillations of Mechanical Systems, 45–54.

2. Decomposition

We consider a dynamical system with n degrees of freedom described by equations

$$A\ddot{x} + Cx = Bv + f(x, \dot{x}, t); \quad v(t) \in V, \quad f(x, \dot{x}, t) \in F. \tag{1}$$

Here, $x \in R^n$ is a vector of generalized coordinates, A and C are constant symmetric positive definite $n \times n$ matrices of the kinetic and potential energy, respectively, B is a constant $n \times m$ matrix, f is a given n-vector of nonlinear terms, and $v \in R^m$ is a vector of controls. The values of v and f are bounded by the given sets $V \subset R^m$ and $F \subset R^n$, respectively. We seek for a feedback control law $v(x, \dot{x})$ which satisfies the imposed constraints and drives the system (1) from any given initial state

$$x(0) = x^0, \quad \dot{x}(0) = \dot{x}^0 \tag{2}$$

to the zero terminal state $x(T) = \dot{x}(T) = 0$ in finite time T (not fixed a priori). Let us introduce normal coordinates $q = (q_1, \ldots, q_n)$ defined by the transformation

$$x = Hq. \tag{3}$$

Here, the $n \times n$ invertible matrix H consists of n columns $h_1, \ldots, h_n$ which are the eigenvectors of the eigenvalue problem

$$(C - \lambda_i A)h_i = 0, \quad i = 1, ..., n, \tag{4}$$

For the sake of simplicity, we assume that the problem (4) has n different eigenvalues $\lambda_1, \ldots, \lambda_n$. Then the transformation (3) reduces equations (1) to the system of linear oscillators

$$\ddot{q}_i + \omega_i^2 q_i = w_i + z_i, \quad \omega_i = \lambda_i^{1/2}, \quad i = 1, ..., n. \tag{5}$$

Here, ω_i is the eigenfrequency of the ith oscillator, whereas w_i and z_i are the components of the n-vectors w and z defined by

$$w = H^{-1}A^{-1}Bv, \quad z = H^{-1}A^{-1}f. \tag{6}$$

The oscillators in (5) are coupled only through the control and nonlinear terms. On the strength of (1), the vectors w and z belong to the following sets in R^n:

$$w \in W = H^{-1}A^{-1}BV, \quad z \in Z = H^{-1}A^{-1}F. \tag{7}$$

Let us consider w_i and z_i in each equation (5) as controls of two independent players. The first player which chooses w_i tends to bring the ith equation (5) to the zero terminal state $q_i = \dot{q}_i = 0$ in finite time, whereas the second player choosing z_i counteracts. The first player can succeed, if his control possibilities exceed those of the second player. Thus, we come to the following conditions. Let the n-dimensional parallelepiped P defined by

$$P: \ |w_i| \leq w_i^0, \quad i = 1, ..., n \tag{8}$$

exist such that the following inclusions hold for some $\varepsilon > 0$ (see Fig. 1)

$$Z + S_\varepsilon \subset P \subset W. \tag{9}$$

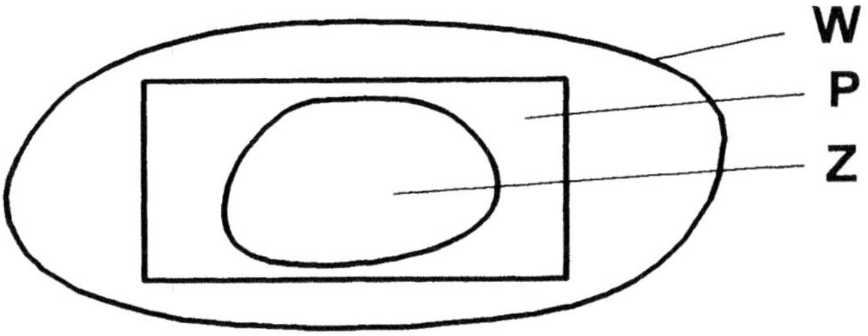

Figure 1. Inclusions (9).

Here, S_ε is an n-dimensional ball of the radius ε. Under the condition (9), we take

$$w = -z + u \tag{10}$$

where u is a new n-dimensional vector of control. Substituting (10) into (5), we obtain

$$\ddot{q}_i + \omega_i^2 q_i = u_i. \tag{11}$$

The inclusions (9) ensure that there exists an n-dimensional rectangular parallelepiped

$$U: \quad |u_i| \leq u_i^0, \quad i = 1, ..., n, \tag{12}$$

such that any values $u \in U$ are admissible. It means that, for any $u \in U$ and any $z \in Z$, the vector w from (10) satisfies the constraints (8). In other words, for any $u \in U$ and any $f \in F$, there exists $v \in V$ such that the corresponding w given by (6) satisfies (8) and is presented in the form (10).

Thus, the inclusions (9) can be regarded as sufficient controllability conditions for the system (1). Under these conditions, the control design for the system (1) is reduced to the control of simple subsystems (11) with one degree of freedom each by means of independent control forces u_i bounded by constraints (12). To minimize the time of control, let us choose the time-optimal control for each subsystem (11). This feedback control is given by [1]

$$\begin{gathered}
u_i(q_i, \dot{q}_i) = u_i^0 \operatorname{sign}[\psi(x) - y], \quad \psi \neq 0 \\
u_i(q_i, \dot{q}_i) = u_i^0 \operatorname{sign} x = -u_i^0 \operatorname{sign} y, \quad \psi = 0 \\
\psi(x) = (-x^2 - 2x)^{1/2}, \quad -2 \leq x \leq 0 \\
\psi(x) = \psi(x+2), \quad x < -2; \quad \psi(x) = -\psi(-x), \quad x > 0 \\
x = (u_i^0)^{-1}\omega_i^2 q_i, \quad y = (u_i^0)^{-1}\omega_i \dot{q}_i, \quad \omega_i > 0.
\end{gathered} \tag{13}$$

Here, x and y are, respectively, the non-dimensional coordinate and velocity of the ith oscillator. The switching curve $y = \psi(x)$ for the control (13) possesses central symmetry and consists of semicircles of unit radii with centres at the points $y = 0$, $x = \pm(2k+1)$, $k = 1, \ldots$. In Fig. 2 the solid line gives the switching curve, and the thin line is one of the time-optimal trajectories. The arrows show the direction of motion along the trajectory.

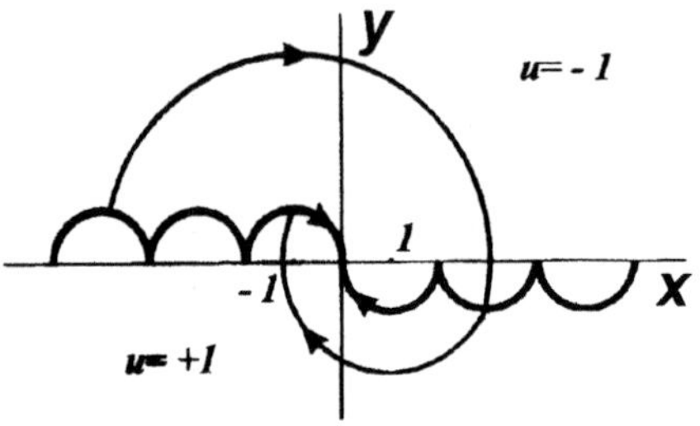

Figure 2. Time-optimal feedback control of the linear oscillator.

The total time of control is given by

$$T = \max_i T_i, \quad i = 1, \ldots, n, \tag{14}$$

where T_i is the time of control for the ith oscillator. The following upper estimate on T_i is obtained in [3]

$$T_i(q_i^0, \dot{q}_i^0) \le \pi (u_i^0)^{-1} \left[\rho_i/2 + (2u_i^0 \omega_i^{-1} \rho_i)^{1/2}\right], \tag{15}$$

$$\rho_i = \left[\omega_i^2 (q_i^0)^2 + (\dot{q}_i^0)^2\right]^{1/2}, \quad \omega_i > 0, \quad i = 1, \ldots, n.$$

Here, q_i^0 and $\dot{q}_i^0$ are the initial data for the ith oscillator related to the initial data (2) by equation (3).

The procedure described above makes it possible to obtain the control satisfying the imposed constraints and driving the system (1) to the zero terminal state $x(0) = \dot{x}(0) = 0$ in finite time which is estimated from above by means of formulas (14) and (15). The choice of parallelepiped P satisfying the conditions (9), in other words, the choice of the constraints w_i^0 in (8) and u_i^0 in (12), can be used in order to minimize the total time of control (14).

3. Distributed-Parameter Elastic System

Let us apply the decomposition approach to an elastic distributed-parameter system described by the equation

$$w_{tt} = Aw + v. \tag{16}$$

Here, $w(x,t)$ is a scalar elastic displacement depending on the position vector $x \in R^n$ and time t, where x belongs to the domain $\Omega \subset R^n$ and $t \ge 0$. In (16), the distributed control force v is bounded by the constraint

$$|v(x,t)| \leq v_0, \quad x \in \Omega, \quad t \geq 0, \tag{17}$$

where v_0 is a given constant, and A is a linear elliptic differential operator containing partial derivatives with respect to x_i, $i = 1, ..., n$. The coefficients of A are independent of t, and its order is even: ord $A = 2m$. For example, if $m = 1$ and $A = \Delta$, equation (16) becomes the wave equation, and if $m = 2$ and $A = -\Delta^2$, it describes the vibration of beams (for $n = 1$) or plates (for $n = 2$). Homogeneous boundary conditions are imposed at the boundary Γ of the domain D:

$$Mw = 0, \quad M = (M_1, ..., M_m), \quad x \in \Gamma, \quad t \geq 0. \tag{18}$$

Here, M_j is a linear differential operator of order ord $M_j < 2m$ with coefficients independent of t. In particular, if $Mw = w$, we have the Dirichlet condition. The initial conditions are

$$w(x,0) = w_0(x), \quad w_t(x,0) = w_{t0}(x). \tag{19}$$

We seek for a control $v(x,t)$ satisfying (17) and such that the corresponding solution of (16) under the boundary conditions (18) and initial conditions (19) satisfies the zero terminal conditions $w(x,T) = w_t(x,T) = 0$ at some finite (unspecified) instant T. Let us introduce the eigenvalue problem related to our initial-boundary-value problem

$$A\varphi_i = -\lambda_i \varphi_i, \quad x \in \Omega; \quad M\varphi_i = 0, \quad x \in \Gamma \tag{20}$$

$$(\varphi_i, \varphi_k) = \int_\Omega \varphi_i(x)\varphi_k(x)dx = \delta_{ik}.$$

Here, δ_{ik} is the Kronecker delta. Under well-known conditions, the eigenvalue problem (20) has a discrete denumerable spectrum of nonnegative eigenvalues which can be numbered in non-decreasing order: $\lambda_1 \leq \lambda_2 \leq ..., \lambda_i \to \infty$, and the corresponding eigenfunctions $\varphi_k(x)$ form an orthonormal system complete in Ω. Using the Fourier method, we seek w and v as the series

$$w(x,t) = \sum q_i(t)\varphi_i(x), \quad v(x,t) = \sum u_i(t)\varphi_i(x). \tag{21}$$

Summation over i is performed from 1 to ∞. We assume that $\lambda_1 > 0$; the case $\lambda_1 = 0$ was considered in [3]. Substituting (21) into (16), (19) and using (20), we obtain for q_i, u_i equations (11) with $\omega_i = \lambda_i^{1/2}$. We impose the constraints (12) on u_i, where u_i^0 are to be chosen later, and denote

$$\Phi_i = \max_{x \in \Omega} |\varphi_i(x)|, \quad i = 1, 2, ... \tag{22}$$

To ensure the control constraint (17), it is sufficient, on the strength of (21) and (22), to require:

$$\sum u_i^0 \Phi_i \leq v^0. \tag{23}$$

We choose the control u_i for each subsystem (11) according to (13) and take

$$u_i^0 = \varkappa \rho_i, \quad i = 1, 2, ..., \quad \varkappa > 0 \tag{24}$$

where $\varkappa$ is a constant and ρ_i is defined in (15). Substituting (24) into (15) and (14) and taking into account that $\omega_i \geq \omega_1$, we obtain the upper estimate on the control time

$$T \leq \pi \left[(2\varkappa)^{-1} + 2^{1/2}(\omega_1 \varkappa)^{-1/2} \right], \quad i = 1, 2, ... \tag{25}$$

It remains to choose the constant $\varkappa$ so that the constraint (17) holds. Substituting (24) into (23), we obtain

$$\varkappa = v^0 \sum \rho_i \Phi_i \tag{26}$$

where ρ_i and Φ_i are specified in (15) and (22), respectively. As follows from (15), if the following two series converge

$$\sum \omega_i |q_i^0| \Phi_i < \infty, \quad \sum |\dot{q}_i^0| \Phi_i < \infty, \tag{27}$$

then the series (26) converges too.

Let us summarize the obtained results. Suppose the Fourier coefficients q_i^0, $\dot{q}_i^0$ for the initial functions w_0, w_{t0} in (19) are such that the both series in (27) converge. Then the constants $\varkappa$ and u_i^0 can be taken according to (26) and (24). The control v defined by (21) where u_i are given by (13) satisfies the constraint (17) and brings the system (16) from the given initial state (19) to the zero terminal state in finite time T which is estimated from above by (25). The sufficient controllability conditions (27) can be reduced [3] to simple differentiability conditions imposed on the initial functions (19).

For example, in the case of a string ($m = 1$, $A = \Delta$, $n = 1$), these conditions are $w_0 \in C^3$, $w_{t0} \in C^2$ for both Dirichlet and Neumann boundary conditions. For an elastic beam ($m = 2$, $A = -\Delta^2$, $n = 1$), the sufficient controllability conditions are $w_0 \in C^4$, $w_{t0} \in C^2$ for various boundary conditions at the ends of the beam.

4. Application of Kalman's Method

We consider now a general linear control system

$$\dot{x} = A(t)x + B(t)u + f(t) \tag{28}$$

under mixed constraints imposed on the state $x \in R^n$ and control $u \in R^m$

$$|C^i(t)x(t) + D^i(t)u(t)| \leq 1, \quad i = 1, ..., l. \tag{29}$$

Here, A, B, f, C^i, and D^i are given $n \times n$, $n \times m$, $n \times 1$, $l \times n$, and $l \times m$ matrices, respectively. We seek for the control $u(t)$ satisfying the constraints (29) and driving the system (28) from the given initial state at $t = 0$ to the zero terminal state in finite (non-fixed) time T:

$$x(0) = x^0, \quad x(T) = 0. \tag{30}$$

Denote by $\Phi(t)$ the fundamental matrix of the system (28):

$$\dot{\Phi} = A(t)\Phi, \quad \Phi(0) = E. \tag{31}$$

Here, E is the $n \times n$ unit matrix. The solution of the system (28) under the initial condition (30) is

$$x(t) = \Phi(t)\{x^0 + \int_0^t \Phi^{-1}(\tau)[B(\tau)u(\tau) + f(\tau)]d\tau\}. \tag{32}$$

Inserting (32) into the terminal condition (30), we obtain

$$\int_0^T \Phi^{-1}(t)B(t)u(t)dt = x^*, \quad x^* = -x^0 - \int_0^T \Phi^{-1}(t)f(t)dt \tag{33}$$

Kalman [2] proposed to seek the control in the absence of constraints (29) in the form

$$u(t) = Q^T(t)c, \quad Q(t) = \Phi^{-1}(t)B(t), \tag{34}$$

where c is a constant n-vector, and the superscript ^{T} denotes the transpose. Substituting (34) into (33), we obtain the equation for c

$$R(T)c = x^*, \quad R(t) = \int_0^t Q(\tau)Q^T(\tau)d\tau. \tag{35}$$

Here, $R(t)$ is a symmetric nonnegative definite matrix for $t \geq 0$. We assume that $R(t)$ is positive definite for $t \geq 0$. This condition implies that the linear system (28) is controllable. In this case we have $c = R^{-1}(T)x^*$, and our control $u(t)$ is completely defined by (34) for any T.

Let us extend this approach to the case of the constraints (29) by choosing the appropriate T. We substitute the control u from (34) into (32) and then insert both x and u into (29). We obtain

$$|F^i(t)c + \sigma^i(t)| \leq 1, \quad c = R^{-1}(T)x^*, \quad i = 1, ..., l \tag{36}$$

$$F^i(t) = C^i(t)\Phi(t)R(t) + D^i(t)Q^T(t),$$

$$\sigma^i(t) = C^i(t)\left[\Phi(t)x^0 + \int_0^t \Phi^{-1}(\tau)f(\tau)d\tau\right]$$

Estimating the left-hand sides of the inequalities in (36) where x^* is given by (33), one can deduce sufficient controllability conditions imposed on the time T and initial state x^0 under which the control (34) satisfies the imposed constraints (29). This approach was implemented in [4-6].

5. System of Oscillators

Let us consider the system of oscillators similar to (11) but controlled by one scalar bounded control:

$$\ddot{q}_i + \omega_i^2 q_i = u, \quad |u| \leq u^0, \quad i = 1, ..., n. \tag{37}$$

This system is a model for n pendulums (or elastic oscillators moving horizontally) attached to a trolley which can move horizontally with a bounded acceleration u (see Fig. 3).

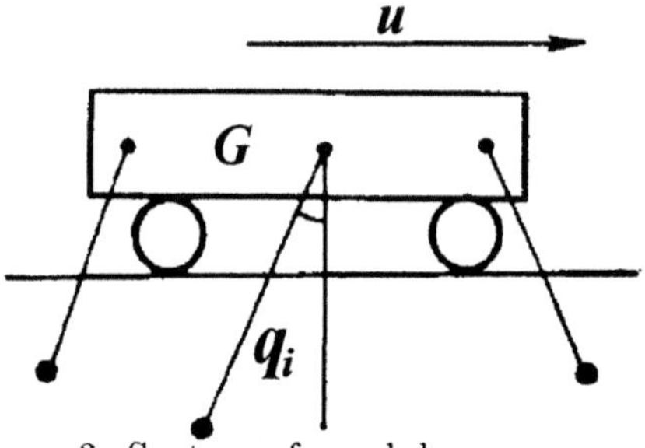

Figure 3. System of pendulums.

We assume that all ω_i in (37) are different; otherwise the system is not controllable. Let

$$\omega_0 = 0 < \omega_1 < \omega_2 < ... < \omega_n, \quad \Omega = \min_{0 \leq i \leq n-1} (\omega_{i+1} - \omega_i) > 0.$$

The control bringing the system (37) to the state of rest ($q_i(T) = \dot{q}_i(T) = 0, \quad i = 1, ..., n$) can be found by the approach of Section 4 and is given [4] by

$$u(t) = \sum_{i=1}^{n} [a_i \cos(\omega_i t) + b_i \sin(\omega_i t)] \tag{38}$$

where the constants a_i, b_i are the components of the vector c depending on T and determined according to (35). The sufficient controllability condition can be [4] expressed as follows

$$T \geq 2(u^0)^{-1}(2nE_0)^{1/2} + 2k_n \Omega^{-1} \tag{39}$$

$$E_0 = \frac{1}{2} \sum_{i=1}^{n} [\dot{q}_i^2(0) + \omega_i^2 q_i^2(0)], \quad k_n = [5n(64n - 55)/72]^{1/2}$$

It is quite natural that the required time T of the control increases, if the number of oscillators n and their initial energy E_0 increase, and also if the control bound u^0 and the minimal difference of eigenfrequencies Ω decrease. For any T satisfying the inequality (39), the control $u(t)$ is given in an explicit form (38).

6. Electromechanical System

Let us consider a system of two masses m_1 and m_2 connected by an elastic spring of stiffness c_0 and driven by a DC electric motor whose driving force F applied to the mass m_1 is proportional to the current I in the circuit of the rotor. The motion of the system and the balance of the electric voltages in the rotor circuit are described by equations

$$m_i \ddot{\xi}_1 = c_0(\xi_2 - \xi_1) + F, \quad m_2 \ddot{\xi}_2 = c_0(\xi_1 - \xi_2) \tag{40}$$

$$F = k_1 I, \quad RI + k_2 \dot{\xi}_1 = U, \quad k_1 > 0, \quad k_2 > 0.$$

Here, ξ_1 and ξ_2 are the coordinates of the masses m_1 and m_2, respectively, k_1 and k_2 are constant coefficients, R is the electric resistance, U is the controlling voltage, and the term with the inductance is neglected. Equations (40) hold also for other oscillatory systems with two degrees of freedom driven by an electric motor such as a pendulum attached to a moving trolley or a rotating beam when only the lowest eigenfrequency of elastic vibrations is taken into account. By introducing non-dimensional variables and parameters

$$t' = \omega t, \quad x_1 = \frac{m_1\xi_1 + m_2\xi_2}{(m_1+m_2)l}, \quad x_2 = \frac{m_1\dot{\xi}_1 + m_2\dot{\xi}_2}{(m_1+m_2)l\omega}, \quad x_3 = \frac{m_1(\xi_1 - \xi_2)}{(m_1+m_2)l}, \tag{41}$$

$$x_4 = \frac{m_1(\dot{\xi}_1 - \dot{\xi}_2)}{(m_1+m_2)l\omega}, \quad u = \frac{k_1 I}{(m_1+m_2)l\omega^2}, \quad \omega^2 = \frac{c_0(m_1+m_2)}{m_1 m_2}, \quad \mu = \frac{m_1}{m_2}$$

where l is some constant length, we reduce equations (40) to the normalized form

$$\dot{x}_1 = x_2, \quad \dot{x}_2 = u, \quad \dot{x}_3 = x_4, \quad \dot{x}_4 = -x_3 + u \tag{42}$$

Here, dots denote derivatives with respect to the non-dimensional time. The constraints imposed on the angular velocity of the rotor, electric voltage, and current in the rotor circuit of the motor can be presented as a system of inequalities imposed on the non-dimensional variables (41). A typical example of these constraints is shown in Fig. 4. Applying the approach of Section 4, we obtain the control $u(t)$ satisfying the mixed constraints of Fig. 4 and driving the system (42) from any initial state $x_i(0) = x_i^0$ to the zero terminal state $x_i(T) = 0$, $i = 1,2,3,4$ in finite time T. The required time T depending on the initial state is determined by a special numerical procedure. Typical numerical results are illustrated by Fig. 5. Here, the projections of the four-dimensional phase trajectory of the system (42) in the (x_1, x_2)- and (x_3, x_4)-planes are shown by the respective curves 1 and 2. The corresponding trajectory of the control and state variables bounded by the constraints is shown in Fig. 4. Here $p = 0.4$, $\mu = 0.5$.

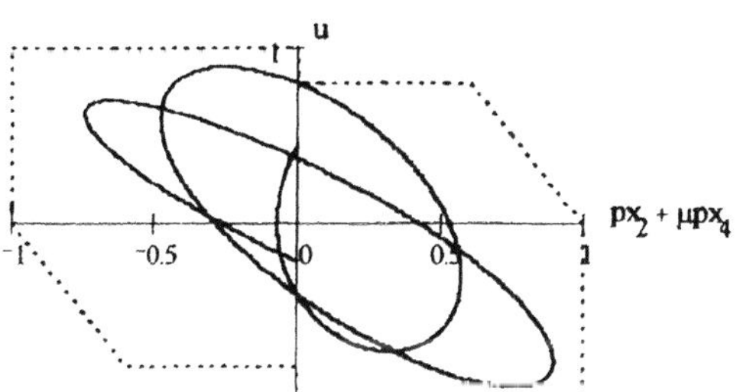

Figure 4. Constraints for the electromechanical system.

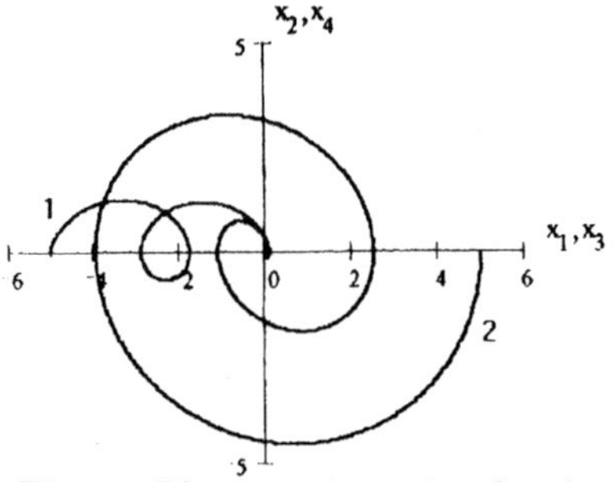

Fig. 5. Phase trajectories for the electromechanical system.

7. Conclusion

Two possible approaches are described which can be used for the control of oscillations in the presence of constraints imposed on the control and state variables. The control laws are obtained in an explicit form, satisfy all constraints, and drive the system from any initial state to the prescribed terminal state in finite time which is estimated from above. Sufficient controllability conditions in the presence of the imposed constraints are derived.

8. Acknowledgments

The research was carried out with the support of the Russian Foundation for Basic Research (Project 99-01-00258) and the A. von Humboldt Research Award.

9. References

1. L.S.Pontryagin, V.G. Boltyanski, R.V. Gamkrelidze, and E.F. Mishchenko. *Mathematical Theory of Optimal Processes.* Wiley-Interscience, New York, 1962.

2. R.E.Kalman. On the general theory of control systems. In *Proc. 1st IFAC Congress.* Butterworth, London (1960) 481-500.

3. F.L. Chernousko. Bounded controls in distributed-parameter systems. *J.Applied Mathematics and Mechanics* **56**, 5 (1992) 707-723.

4. F.L. Chernousko. On the construction of a bounded control in oscillatory systems. *J.Applied Mathematics and Mechanics* **52**, 4 (1988) 426-433.

5. F.L. Chernousko and I.S. Dobrynina. Constrained control in a mechanical system with two degrees of freedom. In *Proc. IUTAM Symposium on Optimization of Mechanical Systems* (Eds. D.Bestle and W.Schiehlen). Kluwer Academic Publishers, Dordrecht (1996) 57-64.

6. F.L. Chernousko. Design of control under mixed constraints. In *Proc. IUTAM Symposium on Interaction between Dynamics and Control in Advanced Mechanical Systems* (Ed. D.H. van Campen). Kluwer Academic Publishers, Dordrecht (1997) 67-74.

ELIMINATION OF INSTABILITIES IN ROTORS SUPPORTED IN GAS BEARINGS

K. CZOLCZYNSKI[1], T. KAPITANIAK[1] and J. BRINDLEY[2]
[1] Division of Dynamics
Technical University of Lodz
Stefanowskiego 1/15, 90-924 Lodz, Poland
[2] Dept. of Applied Mathematical Studies
University of Leeds
Leeds LS2 9JT, U.K.

Abstract

When during the operation of rotors supported in gas bearings their rotational velocity reaches a sufficiently high value, loss of steady-state stability occurs. This instability is caused by the loss of damping properties of the gas film, which leads to self-excited vibrations. These vibrations are the basic obstacle to a widespread application of gas bearings. The phenomenon of self-excited vibrations can be avoided by introducing an elastic supporting structure between the bearing bushes and the casing, characterized by properly selected stiffness and damping coefficients. In practice such a structure can have the form of an externally pressurized gas ring with a chamber feeding system, which ensures the required values of stiffness and damping coefficients (with regard to the stability). The investigations have been carried out by means of a numerical simulation method with the use of a mathematical model of the gas bearing, verified already many times.

1. Introduction

Gas bearings in comparison with oil bearings and rolling bearings exhibit numerous indisputable advantages: they operate without noise, they have a low coefficient of friction, they do not generate heat and are not subjected to wear. These advantages of gas bearings are due to the fact that the surfaces of the journal and bush are separated by a gas (mainly air) layer characterized by a very low (when compared with oil) viscosity. Gas bearings retain their advantages at high rotational velocities which exceed significantly the maximum rotational

N. Van Dao and E.J. Kreuzer (eds.),
IUTAM Symposium on Recent Developments in Non-linear Oscillations of Mechanical Systems, 55–64.

velocities admissible for oil bearings and rolling bearings.

The main disadvantage of gas bearings, which prevents their widespread applications, are the self-excited vibrations occurring when a sufficiently high rotational velocity is achieved. The phenomenon of self-excited vibrations is manifested by the fact that, at a critical value of the rotational velocity, the steady-state stability is lost and the bearing journal begins to move along a trajectory whose radius increases until the journal reaches a stable boundary cycle. When the boundary rotational velocity is exceeded even by a few per cent, the radius of the boundary cycle is bigger than the radial clearance of the bearing, and thus the phenomenon of self-excited vibrations leads rapidly to journal-bush contact and, as a result, to the destruction of the bearing.

Czolczynski's numerical experiments [Czolczynski 1994a,b,c] have shown that the introduction of an isotropic system of linear springs and viscous dampers between the bearing bushes and the casing leads to a limitation of the range of rotational velocities at which self-excited vibrations occur. The main outcome of his work was to demonstrate that a proper selection of the values of stiffness and damping coefficients of the elastic bush support leads to a vanishing of the unstable regions, that is, to an elimination of the phenomenon of self-excited vibrations. Further investigations carried out by Czolczynski and Marynowski have provided data on the ranges of stiffness and damping coefficients, which make it possible to avoid the loss of the steady-state stability of symmetrical rotors supported in self-acting bearings [Czolczynski and Marynowski 1996a] and in externally pressurized bearings [Czolczynski and Marynowski 1996b]. The object of their considerations was a symmetrical rigid rotor supported in two gas bearings with flexibly mounted bushes. As the practical design of the elastic support an externally pressurized air ring supporting the bearing bush with a chamber feeding system is proposed, which stiffness and damping coefficients are presented in this paper. To compute coefficient values an original method described in detail in [Czolczynski 1996] has been used.

2. Gas Bearings

Gas bearings have been used to support rotors since the early 1960s. They have been designed for such applications as gyros, supports for magnetic heads in hard discs of computers, dental drills, or grinding machines. Though they have, as described above, many advantages in comparison with oil bearings or rolling bearings, gas bearings have two main disadvantages: their load capacity is comparatively small, but the major problem in gas bearings application is the phenomenon of self-excited vibrations. This phenomenon is the reason of their low stability.

Because of the self-excited vibration, the rotor supported in gas bearings is

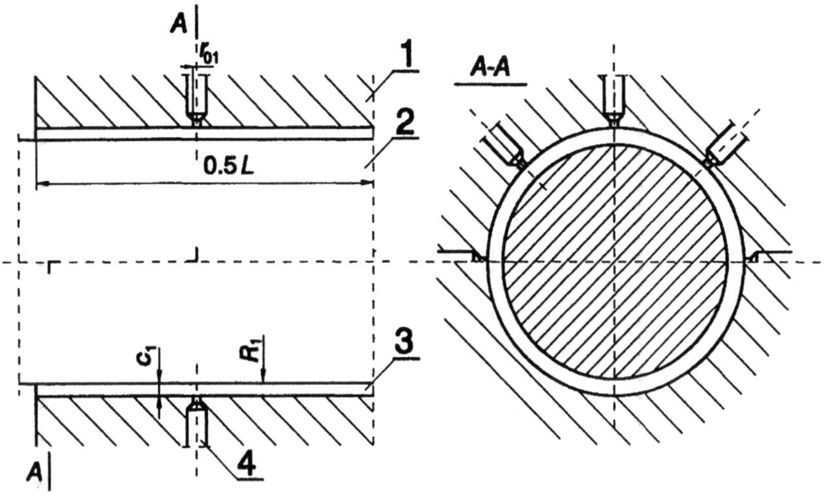

Figure 1.

stable only when the rotational velocity is lower than a certain stability threshold, after which the amplitude of self-excited vibrations exceeds quickly the maximum value determined by the bearing clearance and shaft-bush eccentricity. This leads to journal-bush contact and to the destruction of both the rotor and the bearings.

Figure 1 shows the typical gas journal bearing, which consists of two parts: 1 is the bearing bush, and 2 - the rotating journal. 3 denotes the air gap between the bush and the journal. The average thickness of this gap (radial clearance of the bearing) is about 20-60 micrometers. The gas bearing may be selfactig or externally pressurized. In Figure 1, 4 denotes the feeding system which consists of 16 feedholes, located in two rows. The radius of each feedhole varies from 0.15 to 1 milimeter. Through these feedholes the air is transported into the bearing gap from the compressor. The pressure of the air from the compressor is about 0.4-0.7 Mpa. Other bearings have a chamber feeding system in which the air from the compressor goes first to a chamber of comparatively big volume, and then from this chamber it flows through the feedhole into the bearing gap. The mathematical model of such bearing consists of the Reynolds equation describing the pressure distribution in the bearing gap

$$-\frac{\partial}{\partial\theta}\left(PH_1^3\frac{\partial P}{\partial\theta}\right)-\frac{\partial}{\partial\xi}\left(PH_1^3\frac{\partial P}{\partial\xi}\right)+\Lambda\frac{\partial}{\partial\theta}(PH_1)+\frac{\partial}{\partial\tau}(PH_1)=0 \tag{1}$$

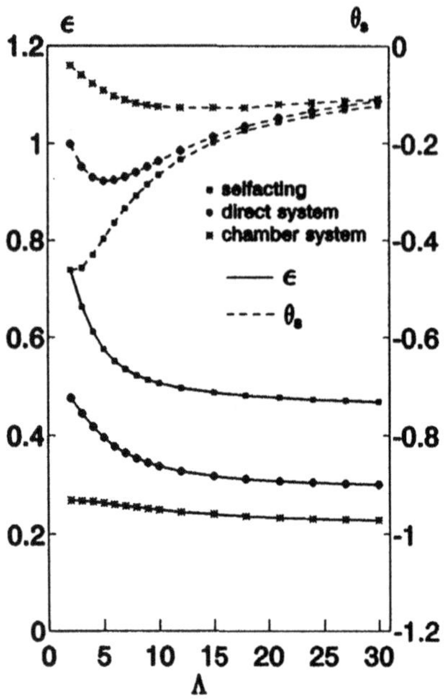

Figure 2.

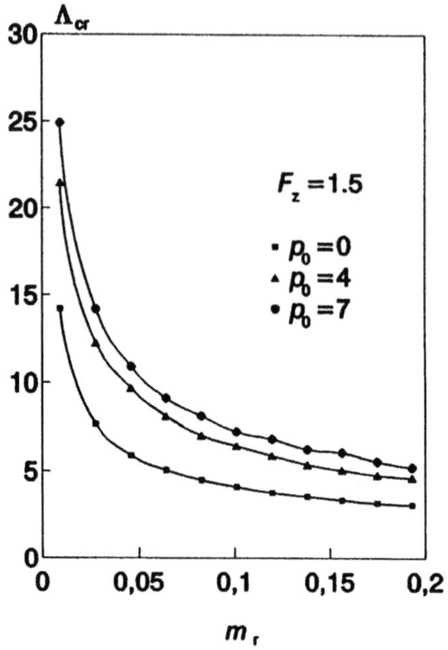

Figure 3.

(P - pressure, H - bearing gap, Λ - velocity, ξ, θ - coordinates, τ - time) and the equations describing the mass flow through the feeding system. This model is described in details in [Czolczynski el al, 1996], and was the basis of calculations of the stiffness and damping coefficients by means of the method described in [Czolczynski, 1996]. In this method the coefficients are calculated from the dynamical response (force) of the bearing on the kinematically forced harmonical motion of the journal. In the described below results of the numerical experiments, the bearings (and the air rings) were represented by sets of the stiffness and damping coefficients.

Figure 2 shows the static characteristics of three different bearings: self-acting, externally pressurized with a direct feeding system and externally pressurized with a chamber feeding system. In this Figure we can see the relative eccentricity ratio ϵ between the journal and the casing, and the angle θ_s between the direction of the journal displacement and the direction in which the loading force acts. On the horizontal axis is the dimensionless rotational velocity of the rotor Λ.

From this figure it follows that at the same loading the eccentricity is the biggest and the most strongly dependent on Λ in the self-acting bearing. The bearing with a chamber feeding system has the biggest load capacity so its eccentricity ratio is the smallest and hardly depends on Λ. This means that this bearing is almost gas static.

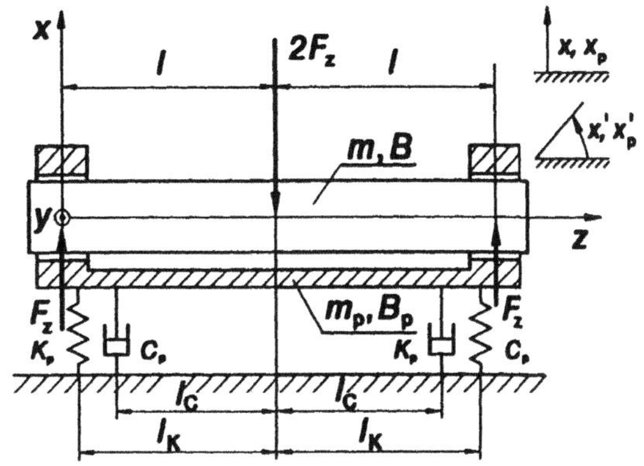

Figure 4.

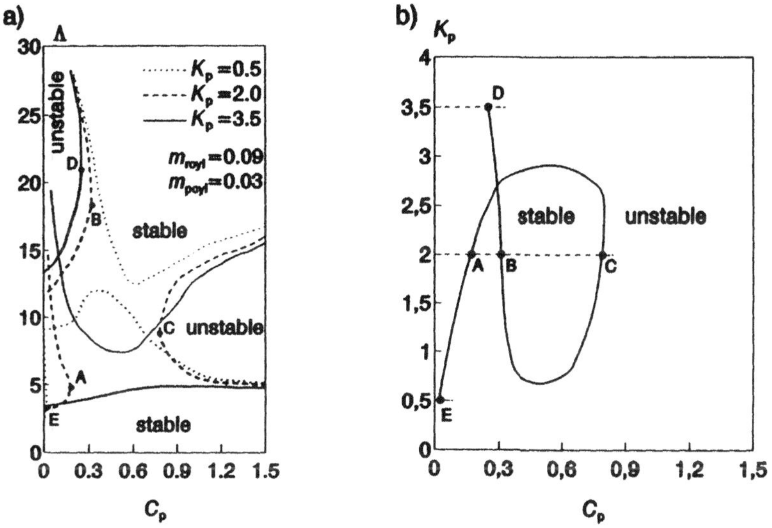

Figure 5.

3. Rotor Stability - Rigidly Mounted Bushes

Figure 3 shows the comparison between the stability thresholds for the selected value of the load capacity $F_z = 1.5$ and various supply pressures. As can be easily seen, the thresholds for the systems with externally pressurized bearings are located above the threshold for the system with self-acting bearings. The increasing of the supply pressure from 4 to 7 ($p_0 = 7$ is the maximum value from the practical point of view) does not cause any significant increase of the stability threshold. The most important fact is that the region of self-excited vibrations is unlimited above, so there is no possibility to operate above this region.

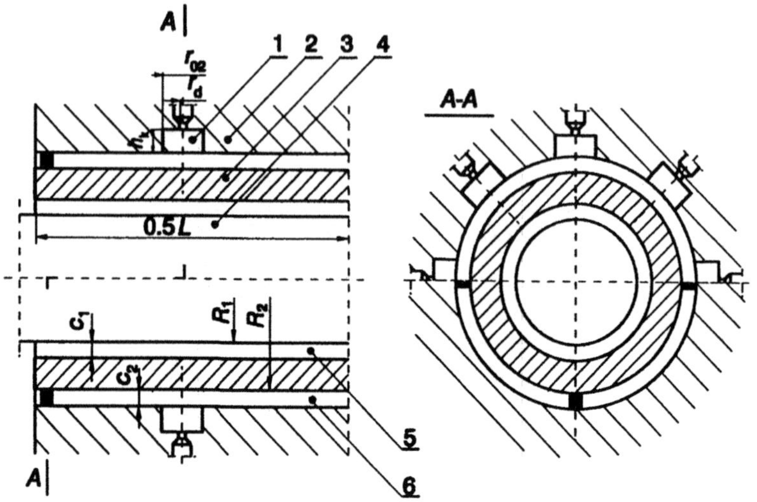

Figure 6.

4. Rotor Stability - Flexibly Mounted Bushes

The situation changes when we introduce an elastic support, consisting of the linear springs K_p and the viscous dampers C_p, between the journal bushes and the casing (Figure 4). Figure 5a shows a sample of the stability thresholds of cylindrical vibrations of the rotor with flexibly mounted bushes for three selected values of the stiffness coefficient K_p, and various values of the damping coefficient C_p. As can be seen, for each stiffness coefficient K_p the unstable regions in which self-excited vibrations appear have a limited size unlike in the case of the rotor with rigidly mounted bushes. For a small (0.5) and for a big (3.5) values of K_p, the unstable regions exist for any value of the damping coefficient C_p and, in order to operate above the unstable regions, the system has to pass through these dangerous regions. This is not possible because, if the rotational velocity of the rotor is only a few percent bigger than the critical one, the journals hit the bearing bushes, which leads to the destruction damage.

For a properly chosen value of K_p, for example $K_p = 2$, the main unstable region is divided into two sub-regions, bounded at points **A** and **C** and the second (upper) unstable region ends at the point B. This means that, when C_p is equal for example to 0.6 no self-excited vibrations appear during the system operation for any value of the rotational velocity Λ. Figure 5b shows the ends of unstable regions for various values of the stiffness and damping coefficient of the elastic support. As we see, these points derive the so called "always stable" loop. When the values C_p and K_p are from the inside of this loop, the static equilibrium position of the rotor will always be stable, so no selfexcited vibrations will appear.

5. Air Ring

The introduction of the elastic support between the bushes and the casing allows us to eliminate self excited vibrations from the numerical experiment of the rotor operation, but it is difficult to realize it in prctice as a system of massless linear springs and viscous dampers. What is proposed here is the air ring with the chamber feeding system, which is shown in Figure 6 (1-chamber, 2-casing, 3-movable bush, 4-rotor, 5-bearing gap, 6-air ring). A mathematical model of such a bearing is the same as the model of the bearing with the chamber feeding system.

Figure 7 compares the main stiffness (K_{11}) and damping (C_{11}) coefficients of the air ring with the direct and with the chamber feeding system. As we can see, the stiffness coefficient of the chamber ring is smaller, and the damping coefficient is bigger than in the ring with the direct feeding system. This means that it is more easy to design a ring with a stiffness and damping coefficient from the inside of the always stable loop for the chamber feeding system. For sufficiently small values of the frequency of vibrations ν, the damping coefficient of the chamber ring is negative. This means that for these frequencies the system might undergo the phenomenon of air-hammer, which is another form of self-excited vibrations. It appeared from our investigations, that these frequencies are lower than the natural frequencies of the rotor investigated.

A parameter of the feeding system which influences the coefficients C_{11} and K_{11} is the radius r_d of the orifice through which the air enters the chamber (Figure 8). For $r_d = r_{02} = 1.0 \times 10^{-3}$ m, the ring has the same damping and stiffness coefficients as the ring with the direct feeding system and $r_{02} = 1.0 \times 10^{-3}$ m. A decrease in the value of r_d causes a decrease in the value of the damping coefficient, especially in the region of the air hammer, but outside of this region the changes of C_{11} are very small. What is important is that the decrease in the radius r_d brings causes a significant (advantageous!) decrease of the stiffness coefficient K_{11} and "introduces" us into the "always stable" loop.

6. Example

As an example of the influence of the air ring coefficients on the stability of the rotor, the stability map of the rotor with the selfacting bearings supported in viscous dampers and linear springs for two selected values of the stiffness coefficient K_p=5 and 16, and for various values of the damping coefficient C_p is shown (Figure 9). In this example, parameters of the rotor have been selected in

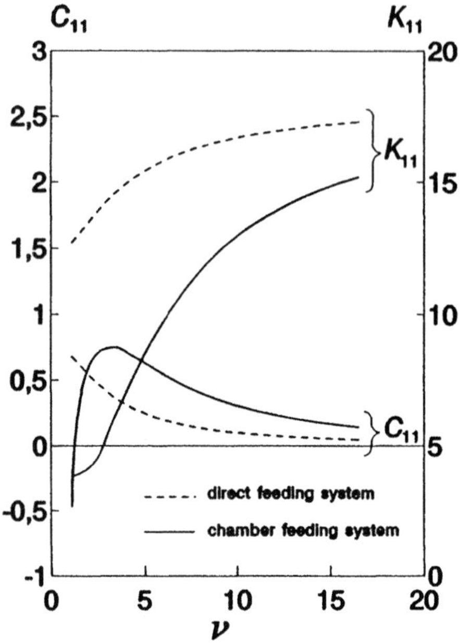

Figure 7.

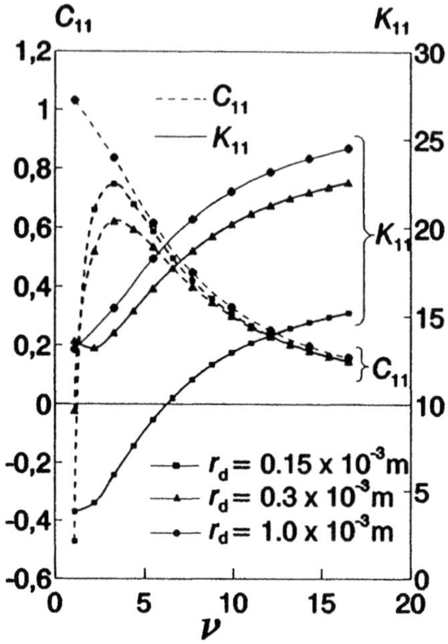

Figure 8.

such a way that the system has the same eigenvalues of cylindrical and conical modes. As may be seen, for big value of K_p the unstable region exists for any damping coefficient C_p - and this situation repeats for the rotor with air rings with big orifices (practically with the direct feeding system).

As may be seen from Figure 9, for $K_p=16$ and $C_p=0.9$ at $\Lambda \approx 5$ the system undergoes a Hopf bifurcation, and at $\Lambda \approx 14$ - a reversed Hopf bifurcation. We may eliminate self-excited vibrations changing the radius of the orifice from $r_d=1.0\times10^{-3}$ m to $r_d=0.15\times10^{-3}$ m. After this change, K_p diminishes from 16 to 5, and C_p from 0.9 to 0.75. For such values of K_p and C_p, there are no unstable regions on the stability map.

Figure 10 shows amplitudes of vibrations of the journal (x_c - solid lines) and of the bush (x_p - broken lines) in the plane in which the force F_z acts as functions of the rotational velocity Λ. As can seen, when $r_d=1.0\times10^{-3}$ m (Figure 10a), in the range $7{,}5<\Lambda<12$ the amplitude of self-excited vibrations of the journal exceeds the value which is permitted by the radial clearance of the bearing and the value of the journal-bush eccentricity ratio. When $r_d=0.15\times10^{-3}$ m (Figure 10b), only the unbalanced vibrations and the small resonance of the bush but no self-excited vibrations can be observed.

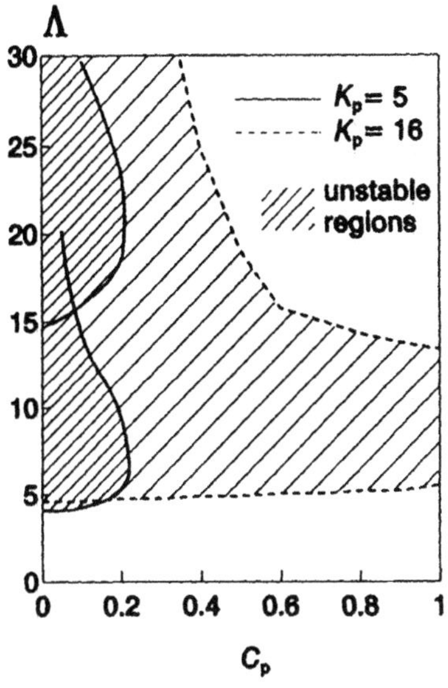

Figure 9.

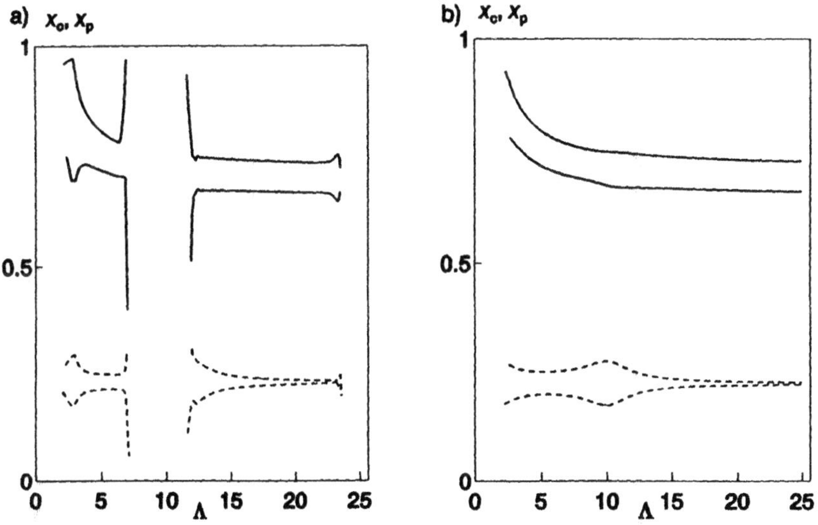

Figure 10.

7. Conclusions

We may avoid self excited vibrations of the rotor supported in gas journal bearings by introducing an air ring with chamber feeding system between the bearing bushes and the casing. The air hammer phenomenon does not appear in the range of the natural frequencies of the system.

8. References

Czolczynski K. (1994a) Stability of Flexibly Mounted Self-Acting Gas Journal Bearings. Nonlinear Science **B** 7, 286-299.

Czolczynski K. (1994b) Stability of High Stiffness Gas Journal Bearing. Wear 172, 175-183.

Czolczynski K. (1994c) Stability and self-excited vibrations of a rotor supported in gas bearings. Zeszyty Naukowe PL 694, 2-132.

Czolczynski K.; Marynowski K. (1996a) Stability of symmetrical rotor supported in flexibly mounted, self-acting gas journal bearings. Wear 194, 190-197.

Czolczynski K.; Marynowski K. (1996b) How to avoid self-excited vibrations in symmetrical rotors supported in gas journal bearings. Machine Dynamics Problems 15, 7-20.

Czolczynski K. (1996) How to obtain stiffness and damping coefficients of gas bearing. Wear 201, 265-275.

Czolczynski K.; Kapitaniak T. and Marynowski K. (1996) Stability of ritirs with bushes mounted in air rings. Wear 199, 100-112.

INTEGRATED ELECTROMECHANICAL MULTI-COORDINATE DRIVE WITH AN ELASTIC DEFORMABLE SLIDE

H. FREUDENBERG, P. MAISSER and P. A. TUAN
Institute of Mechatronics
Chemnitz University of Technology
Reichenhainer Str. 88, 09126 Chemnitz, Germany

1. Introduction

The aim of the paper is to show the use of the LAGRANGIAN approach to electromechanical driving systems also containing flexible mechanical substructures. This approach is based on the dynamics of Hybrid Multibody Systems (HMBS) and Discrete Electromechanical Systems (EMS), respectively. The procedure will be demonstrated by an example of a planar motor with an elastic deformable slide, using the software tool **alaska**.

The planar motor consists of a stator and an aerostatical supported slide. It works as a hybrid stepper drive where the stator is passively. The slide is a plate (circular or rectangular) which is assumed to be an elastic deformable body.
The dynamics of Multibody Systems with flexible substructures can be regarded as a point dynamics in a Riemannian space V^n using a (global) RITZ-approach to the displacement field of the elastic deformation. An elastic deformable body is regarded as a manifold endowed with a curvilinear coordinate system. The equations of motion are LAGRANGE's equations of second kind.
For modelling of the elastic deformable slide eigenfunctions are used as the shape-functions of the plates. The so-called shape-numbers in metric coefficients, Christoffel-symbols and generalized forces of the LAGRANGE's equations of motion are calculated using symbolic computation.

2. Hybrid Multibody Systems

Using well-known concepts and definitions from continuum mechanics the transition from a reference configuration $\mathfrak{B}^*$ to the actual configuration $\mathfrak{B}_t$ can be described by the displacement field

$$\mathfrak{u}(\xi, t) = u^i(\xi, t)\mathfrak{g}_i(\xi, q, t) \tag{1}$$

where

$$\mathfrak{x} \rightarrow \bar{\mathfrak{x}} : \bar{\mathfrak{x}} := \mathfrak{x} + \mathfrak{u} \equiv \xi^i \mathfrak{E}_i(q, t) + u^i(\xi, t)\mathfrak{g}_i(\xi, t)\,, \tag{2}$$

$\{\mathfrak{E}_i\}$: affine frame fixed on $\mathfrak{B}^*$,
$\{\mathfrak{g}_i\}$: local frame on $\mathfrak{B}^*$.

N. Van Dao and E.J. Kreuzer (eds.),
IUTAM Symposium on Recent Developments in Non-linear Oscillations of Mechanical Systems, 65–73.

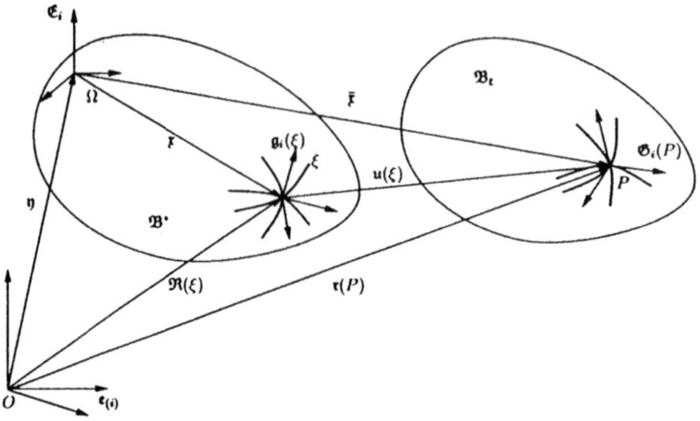

Figure 1: Kinematics of an elastic deformable body

The local frames $\{\mathfrak{g}_i\}$ and $\{\mathfrak{G}_i\}$ on $\mathfrak{B}^*$, $\mathfrak{B}_t$, respectively, are defined by

$$\mathfrak{g}_i(\xi) := \partial_i \mathfrak{r}(\xi) \quad , \quad \mathfrak{G}_i(P) := \partial_i[\mathfrak{r}(\xi) + \mathfrak{u}(\xi, t)]\,, \tag{3}$$

$$\mathfrak{G}_i := \mathfrak{g}_i + \nabla_i u^k \mathfrak{g}_k = (\delta_i^k + \nabla_i u^k)\mathfrak{g}_k \equiv E_i{}^k(u)\mathfrak{g}_k\,, \tag{4}$$

where

$E_i{}^j(u) \equiv \delta_i^j + \nabla_i u^j$	: motion-tensor of elastic deformation,
$\nabla_i u^k \equiv \partial_i u^k + \Gamma^k_{ij} u^j$	: covariant derivative in $\mathfrak{B}^*$,
Γ^k_{ij}	: Christoffel-symbols of second kind in $\mathfrak{B}^*$.

Then, the metrices on $\mathfrak{B}^*$,$\mathfrak{B}_t$ respectively, are defined by $g_{ij} := \mathfrak{g}_i \cdot \mathfrak{g}_j$, $G_{ij} := \mathfrak{G}_i \cdot \mathfrak{G}_j \equiv g_{ij} + \varepsilon_{ij}$. G_{ij} denotes the CAUCHY-GREEN's metric tensor on $\mathfrak{B}_t$ and $\varepsilon_{ij} := 2\nabla_{(i}u_{j)} + \nabla_i u^r \nabla_j u_r$ is the GREEN's strain tensor. In general, g_{ij}, Γ^k_{ij} and ε_{ij} have to be computed in curvilinear coordinates.

The kinetics of an elastic deformable body is based on the principle of virtual work in LAGRANGE's form. Starting from the kinetic energy and generalized forces (inclusive the elastic potential $U(x)$) and using the RITZ-approach

$$u^i(\xi, t) = x^{i\nu_i}(t)\varphi_{\nu_i}(\xi) = x^\nu(t)\varphi_\nu(\xi), \tag{5}$$

the LAGRANGE's equations of motion of an elastic body explicitly read:

$$\begin{aligned} g_{ab}\ddot{q}^b + g_{a\nu}\ddot{x}^\nu + \Gamma_{abc}\dot{q}^b\dot{q}^c + 2\Gamma_{ab\nu}\dot{q}^b\dot{x}^\nu &= Q_a\,, \\ g_{\nu b}\ddot{q}^b + g_{\nu\mu}\ddot{x}^\mu + \Gamma_{\nu bc}\dot{q}^b\dot{q}^c + 2\Gamma_{\nu b\mu}\dot{q}^b\dot{x}^\mu &= Q_\nu\,. \end{aligned} \tag{6}$$

q^a denote the generalized rigid body coordinates and x^ν denote the generalized coordinates of the elastic deformations. $(q^a, x^\nu) \equiv (\eta^\rho) = \eta$ denotes the representing point of the elastic body. $\varphi_\nu(\xi)$ are shape-functions which have to satisfy the kinematic constraints imposed on the boundary of the elastic body.

The metric coefficients $g_{ab}, g_{a\nu}, g_{\mu\nu}$, the Christoffel-symbols of first kind $\Gamma_{abc}, \Gamma_{ab\nu}, \Gamma_{\nu bc}, \Gamma_{\nu b\mu}$ and the generalized forces Q_a, Q_ν in the LAGRANGE's equation (6) have to be generated using kinematic basic functions [1] algebraically due to the RITZ-approach presented in [2].

In order to describe geometric constraints between elastic deformable bodies it is necessary to introduce local body-fixed frames in $\mathfrak{B}^*$ and in $\mathfrak{B}_t$, respectively. In general, geometric, kinematic and dynamic constraints between two elastic deformable bodies in HMBS can be written, as

$$f^r(\underset{r}{q}, \underset{s}{q}, \underset{r}{u}^i(\underset{r}{\xi}), \underset{s}{u}^i(\underset{s}{\xi}), \nabla_i \underset{r}{u}^j, \nabla_i \underset{s}{u}^j; \xi, t) = 0\,, \tag{7}$$

$$\dot{f}^r \equiv \partial_\rho f^r \dot{\eta}^\rho = 0 \tag{8}$$

and

$$\ddot{f}^r \equiv \partial_\rho f^r \ddot{\eta}^\rho + \partial_\rho \partial_\sigma f^r \dot{\eta}^\rho \dot{\eta}^\sigma = 0\,, \tag{9}$$

respectively. These equations will be generated in a derivative-free manner for several ordinary kinematic joints (spherical, universal, cylindrical, revolute and translational joints) to get a complete set of model equations for a HMBS.

3. Discrete Electromechanical Systems

Electromechanical Systems (EMS) are physical structures characterized by interactions between electromagnetic fields and inertial bodies [3], [4]. The interaction can be expressed by constitutive equations (generalized force laws) describing the coupling of Maxwell's theory and mechanics. Constitutive equations describing the coupling between the dynamics of Multibody Systems (MBS) with a finite degree of freedom and Kirchhoff's theory (as quasi stationary approximation of Maxwell's theory) define discrete EMS. A mathematical description following the classical analytical mechanics and completed by some basic concepts and methods of graph theory to characterize topological properties of electrical networks plays a fundamental role for a unified modelling and simulation of discrete EMS.

The general motion equations of an EMS read

$$\begin{aligned} g_{\mu\nu}\ddot{q}^\nu \qquad\qquad\qquad + \partial_\lambda g_{\mu\nu}\dot{q}^\lambda \dot{q}^\nu + (2\Gamma_{\mu b0} + s_{\mu b})\dot{q}^b + \Gamma_{\mu 00} + s_{\mu 0} &= 0\,, \\ g_{\kappa\lambda}\ddot{q}^\lambda + \Gamma_{\kappa\lambda\varrho}\dot{q}^\lambda \dot{q}^\varrho - \tfrac{1}{2}\partial_\kappa g_{\nu\omega}\dot{q}^\nu \dot{q}^\omega + (2\Gamma_{\kappa b0} + s_{\kappa b})\dot{q}^b + \Gamma_{\kappa 00} + s_{\kappa 0} &= 0\,, \end{aligned} \tag{10}$$

where

$$\begin{aligned} g_{\mu\nu} &= \dot{\partial}_\mu \dot{\partial}_\nu \Psi = l_{\mu\nu} = A^i{}_\mu A^j{}_\nu L_{ij}\,, \\ g_{\kappa\lambda} &= \dot{\partial}_\kappa \dot{\partial}_\lambda T = \sum_k [\underset{k}{m} \underset{k}{u}_{i\kappa} \underset{k}{u}_{i\lambda} + (1 - s_\kappa)(1 - s_\lambda) \underset{k}{\Theta}^{ij} \underset{k}{\Omega}_{i\kappa} \underset{k}{\Omega}_{j\lambda}] \end{aligned} \tag{11}$$

denote the generalized inductivities and masses, respectively.

Using the topology of the representing electrical network and the constitutive equations, the LAGRANGE's equations (10) become

$$
\begin{aligned}
&A^i{}_\mu A^j{}_\nu [L_{ij}\ddot{q}^\nu + \partial_\lambda L_{ij}\dot{q}^\lambda \dot{q}^\nu] + A^i{}_\mu \partial_\kappa [L_{ij}\dot{q}_0^j(t) + \Psi_{i0}]\dot{q}^\kappa + A^i{}_\mu V_{i0}(t) \\
&\qquad = -A^i{}_\mu A^j{}_\nu R_{ij}\dot{q}^\nu - A^i{}_\mu R_{ij}\dot{q}_0^j(t) \; , \\
&g_{\kappa\lambda}\ddot{q}^\lambda + \Gamma_{\kappa\lambda\varrho}\dot{q}^\lambda \dot{q}^\varrho - \tfrac{1}{2}A^i{}_\mu A^j{}_\nu \partial_\kappa L_{ij}\dot{q}^\mu \dot{q}^\nu - A^i{}_\mu \partial_\kappa [L_{ij}\dot{q}_0^j(t) + \Psi_{i0}]\dot{q}^\mu \\
&\qquad = \sum_{k=\kappa}^{K} [\underset{k}{K}{}^i \underset{k}{u}_{i\kappa} + \underset{k}{M}{}^i \underset{k}{\Omega}_{i\kappa}] + \tfrac{1}{2}\partial_\kappa L_{ij}\dot{q}_0^i(t)\dot{q}_0^j(t) + \partial_\kappa \Psi_{i0}\dot{q}_0^i(t) \; .
\end{aligned}
\tag{12}
$$

These equations will be generated automatically by the simulation tool **alaska**.

Under the following assumptions

A1:	quasistatic approximation of the Maxwell-Theory
A2:	no hysteresis
A3:	no saturation
A4:	electrically linear constitutive equations

the following steps for getting inductivities have to be carried out:

1. A^i_μ — Fundamental loop matrix (Topology)
2. $L_{ij}(x)$ — Inductivities depending on mechanical coordinates
3. $\Psi_{i_0}(x)$ — Permanent magnetic flux (in a coil generated flux by permanent magnets)
4. $W'_m := \int \psi_\mu d\dot{q}^\mu = W'_m(\dot{q}, x)$ — Magneto-mechanical copotential (Coenergy)

   $\psi_\mu := A^i_\mu \Psi_i$ — linkage magnetic flux in the fundamental loop μ

   $\Psi_i := L_{ij}(x) I^i + \Psi_{i_0}(x)$ — linkage magnetic flux in branch i

   $\sim W'_m = \frac{1}{2} L_{ij}(x) I^i I^j + \Psi_{i_0}(x) I^i = W'_m(I, x)$

   (Maxwell: $w'_m = \int \mathfrak{H}\, d\mathfrak{B} = \int H^i\, dB_i = \frac{1}{2}\mu_{ij} H^i H^j$ coenergy density)
   $B_i := \mu_{ij} H^j$

5. $L_{ij} := \frac{\partial^2}{\partial I^i \partial I^j} W'_m, \Psi_{i_0} := \frac{\partial}{\partial I^i} W'_m|_{I=0}$

   $(l_{\mu\nu} \equiv A^i_\mu A^j_\nu L_{ij} := \dot{\partial}_\mu \dot{\partial}_\nu W'_m(\dot{q}, x))$
6. $\psi_i = n\Phi_i$ n coil windings, $\Phi_i := \int_{F_i} \mathfrak{B}\, d\vec{f}$, $\mathfrak{B} : div\, \mathfrak{B} = 0$, $rot\, \mathfrak{B} = j$

4. Integrated Electromechanical Multi-Coordinate Drive (Planar Motor) with an Elastic Deformable Slide

Modelling of Integrated Electromechanical Multi-Coordinate Drive

For the research project "Modelling, simulation and validation of integrated electromechanical Multi-Coordiante Drives" supported by the DFG (German Research Council), several electromechanical simulation models of linear and planar drives are developed. An detailed description of these models can be found in [5]. These models include beside the mechanical and electrical substructure also the measurement and control systems, and the slide bearing (aerostatical air bearing). Using such drives a high positioning accuracy should be achieved. Therefore, the dynamics of the slide during positioning movements is an essential investigation subject. Elastic slide deformations are possibly one of the reasons for undesired oscillations. The used simulation tool allows the modelling of hybrid multibody systems by various ways. In this case the global RITZ-approach is used. The driving system works as a hybrid stepping one. The driving principle is based on the superposition of magnetic flux caused by permanent magnets and coils. Hence, the magnetic flux is increasing or decreasing, respectively.

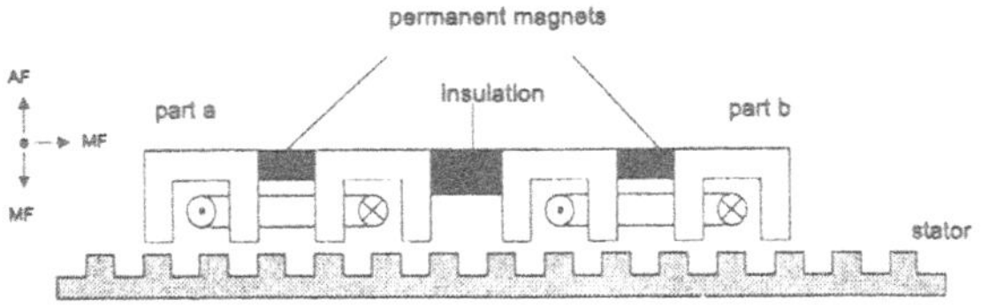

Figure 2: Structure of a driving unit of a hybrid stepper drive

The structure of one driving unit is shown in Figure 2. For periodical increasing and decreasing of the magnetic flux the current direction in the coils of part *a* and *b* must be changed periodical depending on the slide position. The simulation model of such a unit is shown in Figure 3. In various simulations the functional character of the model has been checked.

Figure 3: Simulation model of a driving unit of a hybrid stepper drive

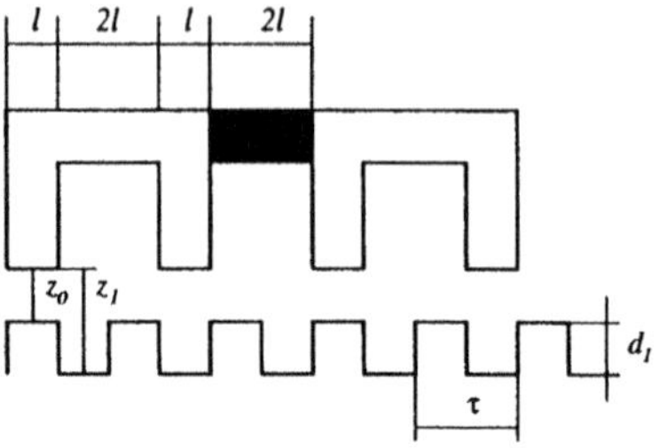

Figure 4: Geometrical parameters of the driving unit

For modelling of linear and planar drives two or more of such units are placed one after the other and side by side on a slide. A main part of the electromechanical model is the description of the interactions between mechanical and electrical substructure by means of so-called constitutive equations. These equations describe the behaviour of electrical components depending on mechanical coordinates. In this case these mechanical coordinates are the vertical and horizontal slide position over the stator. The interaction is described by the coil inductivities depending on the slide positions. The effects caused by permanent magnets are also described by coils with constant current. The constitutive equations and parameters are based on the electromechanical coenergy W'_m in the magnet gap between slide and stator. The magnet gap coenergy W'_m is a function of geometrical parameters of the driving system, shown in Figure 4 and the coil currents, respectively.

The magnetic coenergy can be expressed in the form

$$W'_m = \int_V w'_m dV = \sum_i w_i V_i = \frac{1}{2} \sum L_{ij} I^i I^j \tag{13}$$

where w_i is the energy density in the magnet gap volume V_i. The inductivities L_{ij} of the coils can be obtained like shown above.

Modelling of an Elastic Deformable Slide

For modelling of elastic deformable bodies by using the global Ritz-approach suitable shape functions are required. These shape-functions must satisfy at least the geometrical boundary conditions. The slides are regarded as free plates. The eigenfunctions of free plates (circular or rectangular) can be used as suitable shape-functions. To reflect the elastic behaviour appropriate eigenfunctions have to be selected. For circular plates with a free boundary, analytical given eigenfunctions can be used [6]:

$$\nabla^4 Z(r,\varphi) - \beta^4 Z(r,\varphi) = 0 \tag{14}$$

is the differential equation of plate theory in polar coordinates. In the case of symmetric boundary conditions the solutions can be represented using the following notations:

$$Z_m = R_m(r)\sin(m\varphi + \gamma_m) \tag{15}$$

with

$$R_m(r) = C_1 J_m(\beta r) + C_2 I_m(\beta r) \tag{16}$$

for circular plates without a hole. Using the boundary conditions for a completely free circular plate

$$M_r(r)|_{r=a} = 0 \quad \text{and} \quad V_r(r)|_{r=a} = 0 \tag{17}$$

the eigenfunctions can be expressed as

$$Z_{m,n}(r,\varphi) = \left(\frac{A+B}{C+D}\right)\cos(m\varphi) \tag{18}$$

with

$$\begin{aligned}
A &= I_m(\beta_{m,n})\,(\beta_{m,n}^2\,a^2 + m^2 + m(\nu-1)),\\
B &= I_{m+1}(\beta_{m,n}\,a)(\beta_{m,n}\,a(\nu-1))J_m(\beta_{m,n}\,r)),\\
C &= J_m(\beta_{m,n})\,(\beta_{m,n}^2\,a^2 + m^2 + m(\nu-1)),\\
D &= I_{m+1}(\beta_{m,n}\,a)(\beta_{m,n}\,a(\nu-1))I_m(\beta_{m,n}\,r)).
\end{aligned}$$

I_m, J_m are BESSEL-functions, a denotes the plate radius and the $\beta_{m,n}$ are solutions of the characteristic equation, respectively.

For rectangular plates various combinations of boundary conditions are possible. Several solutions for simple boundary conditions can be found in [7]. For completely free plates, functions for mode shapes of the type

$$Z(x,y) = X(x)Y(y) \tag{19}$$

with

$$\begin{aligned}
X_m(x) &= \frac{\cosh(k_m)\cos(\frac{k_m 2x}{a}) + \cos(k_m)\,\cosh(\frac{k_m 2x}{a})}{\sqrt{(\cosh(k_m)^2 + \cos(k_m)^2))}}\\
Y_n(y) &= \frac{\cosh(k_n)\cos(\frac{k_n 2y}{b}) + \cos(k_n)\,\cosh(\frac{k_n 2y}{b})}{\sqrt{(\cosh(k_n)^2 + \cos(k_n)^2))}}
\end{aligned}$$

if m, n even and

$$X_m(x) = \frac{\sinh(k_m)\sin(\frac{k_m 2x}{a}) + \sin(k_m)\,\sinh(\frac{k_m 2x}{a})}{\sqrt{(\sinh(k_m)^2 - \sin(k_m)^2)}}$$

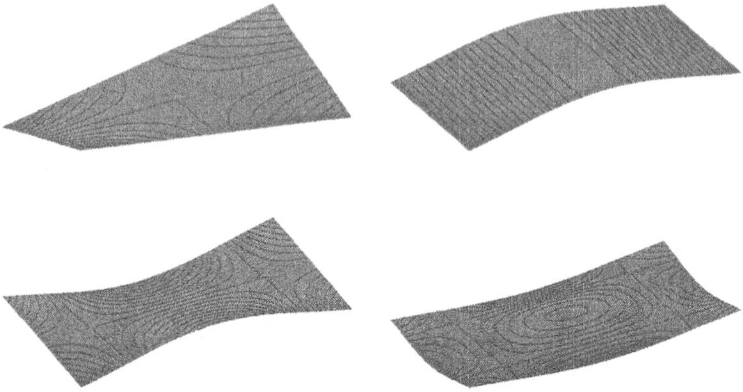

Figure 5: Some eigenmodes of a free rectangular plate

$$Y_n(y) \quad = \quad \frac{\sinh(k_n)\sin(\frac{k_n 2y}{b}) + \sin(k_n)\sinh(\frac{k_n 2y}{b})}{\sqrt{(\sinh(k_n)^2 - \sin(k_n)^2)}}$$

if m, n odd can be used. a is the plate dimension in x-direction, b in y-direction. The k_m can be obtained as the roots of the equations

$$\begin{aligned} \tan(k_m) + \tanh(k_m) &= 0 \quad \text{if } m \text{ even and} \\ \tan(k_m) - \tanh(k_m) &= 0 \quad \text{if } m \text{ odd,} \end{aligned}$$

k_n respectively. Some eigenmodes of a free rectangular plate are shown in Figure 5.

Results

Four of the driving units described above are located on a slide. Two driving units move along the x- and two along the y-direction. The slide is assumed to be a square plate. Figure 6 shows the positions of the driving units. On the bottom of the slide single forces acts vertically to describe the aerostatical bearings. The values of the forces depend on the size of the air gap between stator and slide. Using selected shapes-functions various shape-numbers have to be calculated if taking into account elastic bodies. All of this numbers are integrals of functions of the shapes-functions:

$$\Phi_{\nu_r \mu_s} = \rho \int_V \varphi_{\nu_r} \varphi_{\mu_s} \, dV \quad \text{and} \quad \Phi_{s\nu_r r\mu_s} = \rho \int_V \partial_s \varphi_{\nu_r} \partial_r \varphi_{\mu_s} \, dV \quad . \tag{20}$$

Such numerical integrations are extensively and should be done in a preprocessor. Therefore, a program in the programming language of MAPLEV was created, which computes the shape-functions for m, n-combinations and calculates the necessary

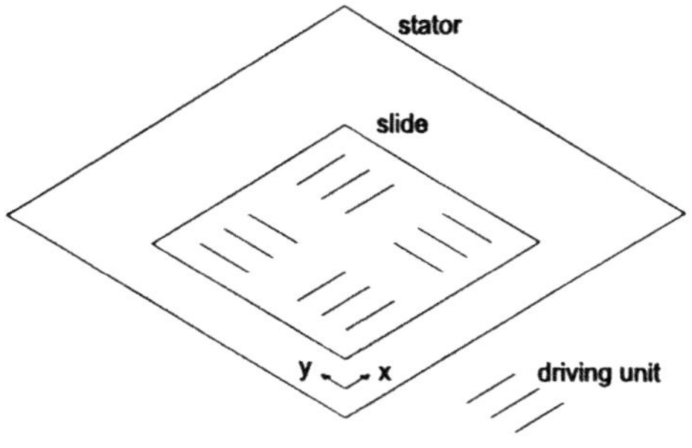

Figure 6: Driving unit positions on the slide

integrals. Also the required stiffness matrix of the elastic body depending on the selected shapes-functions will be calculated by a MAPLEV program. After the integration of the equations of motion of the HMBS with the generalized elastic coordinates x^{ν} and the selected shapes-functions the elastic displacement u of certain selected points on the elastic body can be displayed, and the elastic deformation of the complete body can be shown as an animated plot using a special MAPLEV program.

5. References

1 MAISSER, P.: Analytische Dynamik von Mehrkörpersystemen. ZAMM **68** (1988) 10, 463–481.

2 MAISSER, P.: Dynamik hybrider Mehrkörperdynamik aus kontinuumsmechanischer Sicht. ZAMM **76** (1996) 1, 15–33.

3 MAISSER, P., STEIGENBERGER, J.: Lagrange-Formalismus für diskrete elektromechanische Systeme. ZAMM **68** (1979) 59, 717-730.

4 MIU, D.K.: Mechatronics - Electromechanics and Controlmechanics. Springer-Verlag, New York/Berlin, 1993.

5 FREUDENBERG, H., MAISSER, P.: Modellierung, Simulation und Validierung integrierter Mehrkoordinatenantriebe. IfM-Report 3/1998, Institut für Mechatronik, Chemnitz.

6 HAGEDORN, P. : Technische Schwingungslehre, Bd. 2 Lineare Schwingungen kontinuierlicher mechanischer Systeme. Springer-Verlag, New York/Berlin, 1989.

7 LEISSA, A. W. : Vibrations of Plates. NASA SP-160, Washington, 1969.

CONTROL OF SHIPBOARD CRANES

M. S. Baptista[1,2], B. R. Hunt[3], C. Grebogi[1,2,3], E. Ott[2], and J. A. Yorke[1,3]
(1) Institute for Physical Sciences and Technology
(2) Institute for Plasma Research
(3) Department of Mathematics
University of Maryland, College Park, MD 20742

Abstract

We propose a new shipboard crane configuration for offloading cargo in open seas. We show that the "Maryland Rigging" crane of ship's configuration, used in association with a friction control mechanism, provides a very effective method for reducing load pendulation caused by the motion of the crane in the roll direction. The effectiveness of the control technique is obtained by comparing the performances of the "Maryland Rigging" configuration with the standard "rider block tagline system", commonly used in crane ships. In most of the cases studied for which different sea conditions are considered, the "Maryland Rigging" reduces the root mean square swing of the load by an order of magnitude as compared to the current rider block configuration.

1. Introduction

There are many situations in which cranes must be operated on a moving platform. One example is offloading ship's cargo in the open sea, using cranes that are mounted on another ship. When the sea is not calm, the crane ship rolls, pitches, heaves, etc., in response to the sea motion. Unlike a crane in a harbor, which is mounted on a fixed platform, the crane ship responds to the sea motion, imparting this motion to its load. It becomes imperative to develop motion control mechanisms to achieve safe, and effective offloading operations. The dynamics of the load is generally three dimensional and very complicated. However, most of the damage and problems are caused by load pendulation in the direction of the ship crane roll motion ([1], [2]). Current ship cranes have only crude pendulation control features that are effective in limited circumstances.

A rough model of the dynamics of a rocking crane system is given by the nonlinear pendulum equation,

$$\ddot{x} + \sin x = F(t). \tag{1}$$

Here $F(t)$ is a forcing signal due to the motion of the rocking platform, which is irregular but nonetheless has a strong periodic component. The fluctuations in

N. Van Dao and E.J. Kreuzer (eds.),
IUTAM Symposium on Recent Developments in Non-linear Oscillations of Mechanical Systems, 75–84.

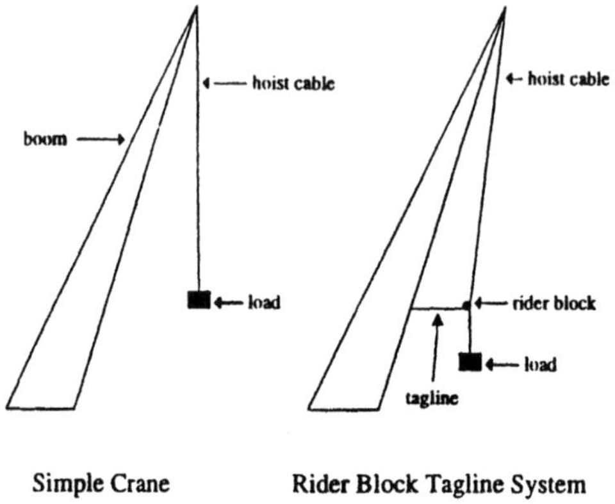

Figure 1: Current rigging configurations.

amplitude and frequency may be modeled as being random [3], [4], or chaotic. The pendulum experiences resonance when, for a certain period of time, the frequency of the forcing signal is closed to the natural frequency of the pendulum. Even with a small forcing amplitude, the amplitude of the pendulum oscillations grows, but it generally does not grow indefinitely (even with a periodic forcing signal) because of the nonlinearity. For the crane system, one expects large pendulations to occur when the roll frequency is close to the natural frequency of the system, which is of the order of 0.1 Hz for a pendulation length of around 30 meters. One can try to avoid large pendulations by dampening the system, or by adjusting the natural frequency of the system so that resonance does not occur.

Typical crane configuration attempt to reduce the pendulation in the cargo with a Rider Block Tagline System (RBTS), as illustrated in Figure 1. Instead of being connected directly to the cargo, the hoist cable is threaded through the rider block and then connected to the cargo. The rider block is adjusted by the tagline and the liftline (not shown) so that the effective pendulation length of the cargo, and consequently its natural frequency, is changed.

The RBTS can be effective if the roll motion of the ship is regular. However, wave motion, and the ship's response to it, may be broadband, hence resonance can occur for a variety of natural frequencies of the crane. In such cases, the pendulation of the cargo cannot be reduced simply by adjusting the position of the rider block. Another problem is that quick and precise implementation of a control strategy is not possible because it is difficult to adjust the rotational speed of a cable winch quickly and smoothly. Observations have shown that once the pendulation is built up in the heavy rigging gears, it is very difficult to dissipate energy under current rigging and control designs.

In this paper, we present an alternative strategy of pendulation control, which uses an adjustable frictional force to dissipate energy in the crane system. This strategy is realized by a new rigging configuration as shown in Figure 3.

This "Maryland Rigging" includes a pulley-brake assembly in addition to the usual components. We insist that any modification of the current cranes should meet the following requirements:

- minimum change from the current configuration;
- low cost;
- better control of load pendulation.

The pulley-brake assembly is intended to be a separate component, so that it can be added to the crane system easily. The rest of the rigging system is almost unchanged. The purpose of the pulley-brake assembly is to dampen the load pendulation. This alone is a major advantage over existing riggings, which do not offer any significant damping. Furthermore, the magnitude of the breaking can be adjusted easily. The pulley-brake assembly thus offers a practical mechanism for the implementation of a variety of control strategies, with the level of friction as a control parameter. The simplest strategy is to apply the brake uniformly, meaning constant friction. In this case the frictional force is assumed to act in the opposite direction of the rotation of the pulley with constant magnitude (independent of the speed of rotation).

A previous work [5] studied the performance of the "Maryland Rigging" configuration using, as input data for the roll oscillations of the ship crane, sinusoidal and chaotic signals. In that work, a constant friction force was applied in order to reduce the load pendulation. Given continuity to that work, we now consider a realistic roll input data measured in a real crane ship [1], [2] and the friction forcing, introduced to dampen load pendulation, depends not only on the direction of the rotation of the pulley (dry friction force) but also on the magnitude of the rotational speed (viscous friction force) of the pulley.

In Section 2, we show the measured series of data points that corresponds to the roll oscillation of a ship crane. In Section 3, we describe the models for the "Maryland Rigging" and the rider block configurations. In Section 4, we show how the "Maryland Rigging" with control, a combined dry-viscous friction force, proves to be very robust to different sea conditions, preventing large load pendulations. In Section 5, the performance of the "Maryland" configuration is compared to the standard rider block configuration, by analyzing the root mean square of the horizontal displacement of the load.

2. Roll Oscillation

The crane ships are considered to be under the action of sea oscillations which impart to the ship instabilities in the roll direction. These instabilities are measured in a real crane ship.

So, the measured data sets of the roll oscillation, $\beta(t)$, (in degrees) collected from a crane ship, can be seen in Figures 2(A-E). Over 8000 points are collected within a period of 4000 seconds. Each measurement is carried out at different

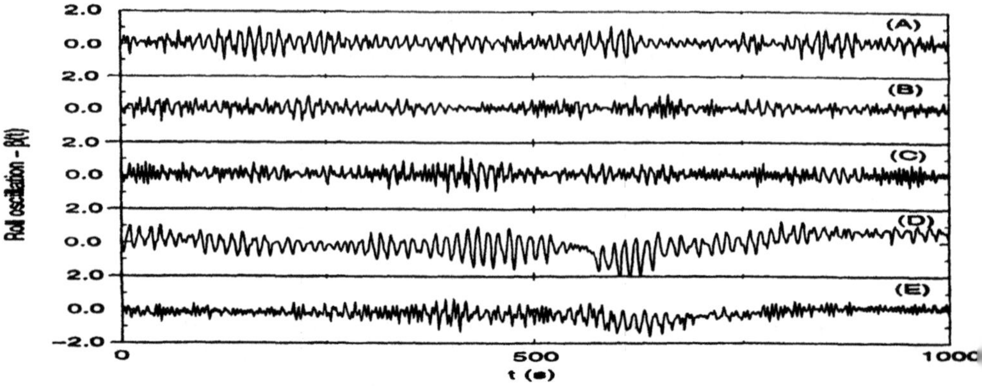

Figure 2: Experimental data $\beta(t)$ for the roll oscillation in degrees.

time but at the same place. For our simulations, we use intervals of time of 1000 seconds. Thus, for reasons that will be clear later, we pick 2000 consecutive points over these 8000 collected ones, such that the first point (that by definition represents the roll oscillation for the time t=0) in this arranged time series is close to zero. So, $\beta(t = 0) \approxeq 0$.

With these discrete data sets, we numerically obtain, by third-order interpolation, smooth functions for $\beta(t)$, $\dot{\beta}(t)$, $\ddot{\beta}(t)$ that are the evolution of the roll oscillation, the first derivative of the roll, and the second derivative of the roll, all quantities that are needed in our simulations.

The sea conditions, available for our simulations, are the ones showed in Figure 2. However, more general situations, where we would find a more aggressive, or calm sea than the ones measured [1], [2], would make our results even more tangible. So, we rescale the functions $\beta(t)$, $\dot{\beta}(t)$, $\ddot{\beta}(t)$, by a factor A, in order to change the amplitude of the oscillations but keeping the time scale invariance:

$$\beta'(t) \to A\beta(t), \quad \dot{\beta}'(t) \to A\dot{\beta}(t), \quad \ddot{\beta}'(t) \to A\ddot{\beta}(t), \tag{2}$$

where $0 < A \leq 2.5$.

3. Rider Block and "Maryland Rigging" Configurations

We introduce the standard rider block tagline system in Figure 3, where we see that the hoist cable is holded by a tagline at the rider block.

Considering the tagline cable and the hoist cable, from the upper part of the boom to the rider block, as a rigid body (what is very reasonable to assume once the considered sea conditions are not so drastic), the equation of motion for the crane of Figure 3 is

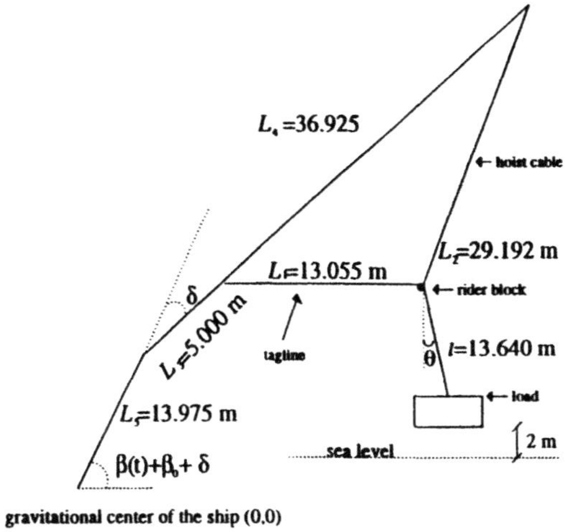

Figure 3: Standard rider block configuration.

$$
\begin{aligned}
\ddot{\theta} = \frac{1}{l}(&\ddot{\beta}(L_5 \sin(\beta + \delta + \theta) + L_3 \sin(\beta + \theta) + L_1 l \sin(\beta + \theta - \beta_0)) \\
&+ \dot{\beta}^2 (L_5 \cos(\beta + \delta + \theta) + L_3 \cos(\beta + \theta) + L_1 l \cos(\beta + \theta - \beta_0)) \\
&- g \sin \theta).
\end{aligned} \tag{3}
$$

In Equation (3), β [β represents $\beta(t)$] is the roll oscillation of the ship shown in Figure 2, β_0=45 deg, δ=15 deg, θ is the angle between the horizontal and the load cable l, and L_1, L_3, and L_5 are indicated in Figure 3.

The "Maryland Rigging" configuration is schematically shown in Figure 4. There is a pulley that moves along the cables L_1 and L_2 whose ends are attached to the points B and C of the boom. The load is sustained by a cable l connected to the pulley. The control strategy is to apply a break at the pulley in order to dampen load pendulation.

The equations of motion for the "Maryland Rigging" configuration are

$$
\begin{aligned}
m_1 \ddot{x}_p &= T_2 \cos \alpha_2 - T_1 \cos \alpha_1 + T_3 \sin \theta \\
m_1 \ddot{y}_p &= T_1 \sin \alpha_1 + T_2 \sin \alpha_2 - T_3 \cos \theta - m_1 g \\
T_1 - T_2 &= f(t) \\
l \ddot{\theta} &= -\ddot{x}_p \cos \theta - \ddot{y}_p \sin \theta - g \sin \theta \\
T_3 &= m_2(-\ddot{x}_p \sin \theta + \ddot{y}_p \cos \theta + g \cos \theta + l \dot{\theta}^2),
\end{aligned} \tag{4}
$$

where $f(t)$ is a combined dry-viscous friction force that models the action of a break over the pulley, given by

$$f(t) = \eta_1 sign(\dot{L}_1 - \dot{L}_2) + \eta_2(\dot{L}_1 - \dot{L}_2). \quad (5)$$

In Equation (4), T_1, T_2, T_3 are the tensions along cables L_1, L_2, and L_3, respectively. The mass of the pulley is m_1 and the mass of the load is m_2. They are assumed to be 0.01 and 1, respectively.

In Equation (5), η_1 and η_2 are the dry and viscous friction coefficient, respectively, and $\dot{L}_1$ and $\dot{L}_2$ are the cable velocities. The difference $\dot{L}_1 - \dot{L}_2$ is the rotational speed of the pulley with respect to the cable, and while the dry friction term depends only on the direction of the pulley velocity, the viscous friction term depends not only on the direction but also on the magnitude of the rotational speed of the pulley.

Our control strategy consists in finding the best range of values for η_1 and η_2 for which the load pendulation is minimum. A large variety of different friction laws from Equation (5) were tried, considering instead of $\dot{L}_1$, $\dot{L}_2$ different accessible parameters such as θ, $\dot{\theta}$, ξ, $\dot{\xi}$, β, $\dot{\beta}$. However, no better performance than Equation (5) was found.

Equations (4) have two degrees of freedom: one given by the pulley position x_p and y_p (see 6), and the other given by the angle θ from which is obtained the cargo position (x_q, y_q) by doing $(x_q,y_q)=(x_p - l\cos\theta,\ y_p - l\sin\theta)$.

The rider block as well as the "Maryland Rigging" parameters were chosen according to a prototype crane ship. So, the boom length, $L_3 + L_4$, is estimated to be the order of 42 m and the length L_5, from the gravitational center of the ship (the base for the crane) to the crane pedal, is to be the order of 14 m. Note that both configurations are supposed to have similar size and length scales.

For both configurations, we choose a very special value for L_1+L_2 in order to let the cargo position to be 2 meters above the sea level, the maximum cable length to be expected when offloading the cargo ship, and also the configuration that would give the largest possible load pendulation.

For t=0, both configurations are at the rest position, what means that at t=0, $\theta(t=0)$=0 and $\dot{\theta}$=0 (for both configurations), and $\dot{x}_p, \dot{y}_p$=0 (for the "Maryland" configuration). The position of the pulley for the rest position is calculated by assuming that the angle $\beta(t=0) + \beta_0 + \delta$=$\beta_0 + \delta$. Thus, $\beta(t=0)$ must be equal to zero. That is the reason why we have arranged the data for the roll oscillation to be such that $\beta(t=0) \approxeq 0$ (see Figure 2).

Our simulations are performed for an interval of time of 1000 seconds, for which the first 300 seconds are discarded in our analysis. Such transient time is considered because it takes some time to transfer momenta from the rolling ship to the load.

4. The "Maryland Rigging" Performance with Control

We obtain a series of profile curves relating the dependence of the root mean square (RMS) for the horizontal displacement of the cargo position (x_q) on

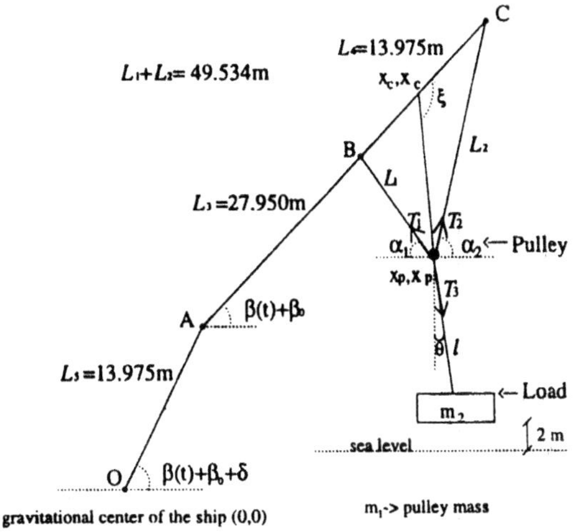

Figure 4: "Maryland Rigging" configuration.

the friction coefficients η_1 and η_2, for different rescaling factors A. We show in Figures 5(A-B) only three curves (for $\eta_1 = 0.00, 0.25, 0.50$) that show the RMS horizontal displacement in respect to η_2, for the rescaling factors A=1.0 and $A = 2.0$. Figure 5 represents well the conclusions derived from an intense analysis of a large number of profile curves.

From Figure 5, we see that the use of a single dry or viscous friction force works well for damped oscillations. If η_1=0.00 (viscous friction), we see that a small load pendulation is obtained for η_2=40.0. If η_2=0.0 (dry friction), good results are obtained for η_1=0.25. However, for a practical application of the kind of control by equations as given by (5), it is fundamental that good results are obtained not only for some values of the friction coefficient, but rather for a range of values giving always good results independent of the sea conditions. We seek a control strategy that is robust for every sea condition, even when a large variation on the friction coefficients is necessary.

5. Comparison Between "Maryland Rigging" and Rider Block Configuration

To compare the performance of the "Maryland Rigging" with the standard rider block configuration, we choose η_1=0.25 and η_2=10.0 (the black filled square in Figure 5A), a set of parameters for which we get nearly an optimal performance of the "Maryland Rigging" crane configuration.

As we change the amplitude factor A, we show in Figures 6(A-E) that the RMS for the horizontal displacement of the load position for the "Maryland" configuration is an order of magnitude lower than the rider block configuration. Each curve in Figures 6(A-E) is obtained for one roll data series. So, Figures 6(A-E) are obtained from the roll data shown in Figures 2(A-E), respectively.

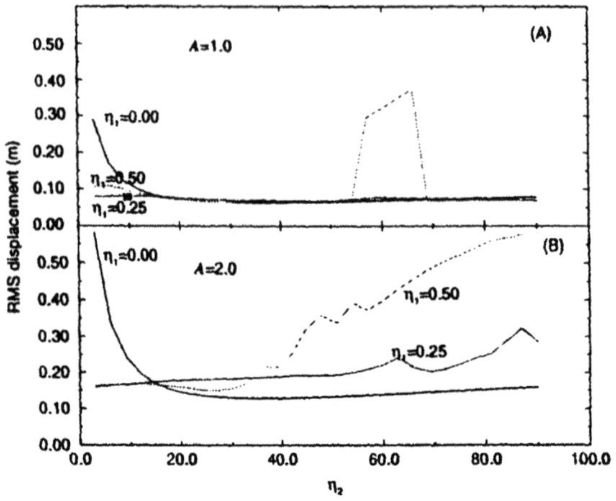

Figure 5: Dependence of the RMS for the horizontal displacement of the load position on η_1 and η_2 for A=1.0 (A) and A=2.0 (B).

Table 1: Ratio between the RMS displacement of rider block and the "Maryland Rigging" for the five considered roll data sets when A=1.0 and A=2.0.

data set	A=1.0	A=2.0
Figure 2A	5.38	5.02
Figure 2B	11.26	14.93
Figure 2C	10.84	8.99
Figure 2D	3.20	2.61
Figure 2E	8.28	10.67

The ratios between the RMS displacement of rider block and the "Maryland Rigging" for the five considered roll data sets are shown in Table , for two values of A, A=1.0 and A=2.0. We see that the data set of Figures 2A and 2D makes the ratio not too high. Those data sets give a characteristic frequency not so high as the others. Thus, the "Maryland" configuration with control is more effective in dampening the load pendulations caused by rapid ship roll oscillations.

6. Conclusions

We have presented the new "Maryland Rigging" crane ship configuration. This configuration is envisioned to be easily implementable into the current crane ships, and it also provides ways of applying a simple friction control technique, in order to reduce load pendulation.

The control of load pendulation is achieved by applying a combined dry-

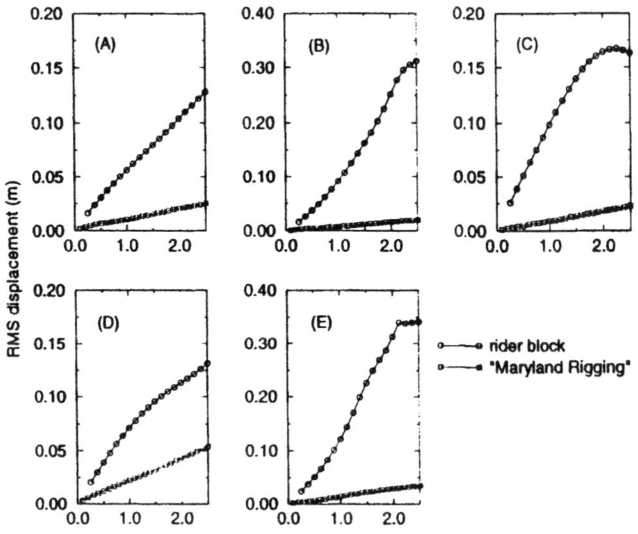

Figure 6: The "Maryland Rigging" and rider block RMS displacement with respect to the factor A. In Equation (5), η_1=0.25 and η_2=10.0.

viscous friction forcing, that provides a control technique which is robust for all sea conditions (considering only instabilities in the roll direction), even when there is a large variation of the friction coefficients.

The performance of the "Maryland Rigging" configuration to dampen load pendulation, measured by the root mean square of the horizontal displacement of the load, is an order of magnitude better than the rider block configuration.

In a recent work, in addition to the roll oscillation, the heave and the sway motions of the ship were introduced into the models presented in Sec. 3. These oscillations, also measured in a standard crane ships, change the position of the gravitational center of the ship in the horizontal direction (sway) and in the vertical direction (heave). The performance (RMS horizontal displacement) of rider block configuration with roll, sway and heave motion is nearly the same as the one obtained by considering only the roll oscillation. However, the performance of the "Maryland Rigging", when the roll, heave and sway oscillations are introduced, are worse than the performance obtained considering only the roll oscillation. Actually, with those three motions being considered, both configurations give nearly the same performance for the root mean square of the cargo horizontal displacement.

The reason for which the "Maryland Rigging" configuration, in presence of the three kinds of oscillations, presents a large load pendulation is because the load resonates with the sway oscillation. This resonance persists when the control is done by a law of the kind of Equation (5).

7. Acknowledges

M. S. Baptista is supported by the Fundação de Amparo a Pesquisa do Estado de São Paulo (FAPESP). This project is also supported by the Office of Naval Research and by an NSF/CNPq joint grant. We also thank A. Rausch and T. Vaughters for their help in obtaining the technical reports cited in the References.

8. References

1. JLOTSII: Analysis and Evaluation Report, Little Creek Amphibious Base, Norfolk, Virginia, August, 1985.
2. JLOTSIII: Display Determination 91 Test Report, Naval Surface Warfare Center, Bethesda, Maryland, November, 1992.
3. Meyers, J. J.: *Handbook of Ocean and Underwater Engineering*, McGraw-Hill, 1969.
4. Johnson, F. R.: Rating Lift Cranes Operating on Platforms In The Ocean Environment, Naval Civil Engineering Laboratory (NCEL), Port Hueneme, California, March, 1992.
5. Yuan, G. H., Hunt, B. R., Grebogi, C., Ott, E., Yorke, J. A., and Kostelich, E. J.: Proceedings of DETC'97, 1997 ASME Design Engineering Technical Conferences, Sacramento, California, September 14-17, 1997.
6. The point (x_p, y_p) lies on an ellipse determined by $L(t)=L_1+L_2$ and the foci on the points B and C of Figure 4. This ellipse can be parametrized by a number ξ as follows. First, let x' and y' be the coordinates of the pulley in the frame with origin (x_c, y_c) and x'-axis along the line $\underline{BC}$. We write x' and y' in terms of ξ as $x' = \frac{L(t)}{2}\cos\xi$, $y' = \frac{\sqrt{L(t)^2 - L_4^2}}{2}\sin\xi$. After changing the coordinates of the pulley into the rest frame, we have: $x_p = x'\cos\beta - y'\sin\beta + x_c$ and $y_p = x'\sin\beta + y'\cos\beta + y_c$.

A CONSISTENT AND EFFECTIVE METHOD FOR NONLINEAR RANDOM OSCILLATIONS OF MDOF SYSTEMS

GUO-KANG ER AND VAI PAN IU
Faculty of Science and Technology
University of Macao
P.O. Box 3001, Macao

Abstract. The difficulties in obtaining the probability solutions of nonlinear random multi-degree-of-freedom (MDOF) systems are underlined. For the MDOF systems which are excited by white noise, a new method proposed recently is applied and extended to obtain their approximate probability density function (PDF) solution. Numerical results are presented to validate the method for the highly nonlinear random MDOF systems.

1. Introduction

The random vibrations of nonlinear systems have attracted much attention in the past decades with no consistent and effective method proposed for highly nonlinear MDOF systems. Even though the investigation on the probabilistic solutions of nonlinear random single-degree-of-freedom (SDOF) systems has attracted much attention for half a century there has until recently been no effective method suitable for highly nonlinear systems.

The reason that random nonlinear MDOF systems troubled many researchers in various areas for almost half a century was that it was generally difficult with any available method to obtain desirable approximate PDF solutions of highly nonlinear random systems. For four- or higher-dimensional problems, there was even no method for reasonable approximate PDF solutions, except the *equivalent stochastic linearization method* which is suitable for weakly nonlinear systems without multiplicative excitation, or the *stochastic average method* which is suitable for weakly damped systems with weak excitations. Most practical problems are described by MDOF systems with four or more degree of freedom.

The literature on exact or approximate PDF solutions of nonlinear random systems is vast. Various methods have been proposed in past decades:

N. Van Dao and E.J. Kreuzer (eds.),
IUTAM Symposium on Recent Developments in Non-linear Oscillations of Mechanical Systems, 85–94.

equivalent stochastic linearization method [2], *stochastic average method* [11], *NonGaussian Hermite polynomial closure method* [1], *equivalent nonlinear system method* [8], *maximum entropy method* [10], and *multi-Gaussian closure method* [3], etc. The methods are limited by one or more of the following: (1) They are suitable only for weakly nonlinear system without multiplicative excitation, or weakly damped system with weak excitations, e.g., *equivalent stochastic linearization method, stochastic average method* and so on. (2) The PDF model does not satisfy the probability theory, i.e., the method may lead to negative PDF value, e.g., *Hermite polynomial closure method.* (3) They are suitable only for two-dimensional problems, or nonlinear random vibrations of SDOF systems, e.g., all methods except *equivalent stochastic linearization method, Hermite polynomial closure method* and *stochastic average method.* (4) Complicated nonlinear algebraic equations arise and it is difficult or practically impossible to formulate the algebraic equations because multi-dimensional integrals are needed, e.g., *maximum entropy method* for which multi-dimensional integrals are needed, and *multi-Gaussian closure method.*

A new method was reported recently for nonlinear random systems and applied to nonlinear random SDOF systems [4–6]. In this paper, the method is extended and applied to the PDF solution of nonlinear random vibrations of MDOF systems excited by white noise, or to the solution of the FPK equation in higher dimension without any limit on the degree of system nonlinearity. The nonlinear random vibration of a highly nonlinear two-degree-of-freedom systems is analyzed with the proposed method. Numerical results are given and compared with obtainable exact PDF solutions to show the effectiveness of the method for nonlinear random MDOF systems.

2. Statement of Problem

Consider the following nonlinear random MDOF system:

$$\ddot{Y}_i + H_{i0}(\mathbf{Y}, \dot{\mathbf{Y}}) = H_{ij}W_j(t) \tag{1}$$

where $\mathbf{Y} \in \Re^{n_y}, Y_i, (i = 1, 2, \cdots, n_y)$, is the component of the state vector $\mathbf{Y}$. $H_{i0} : \Re^{n_y} \times \Re^{n_y} \rightarrow \Re$; the function type of H_{i0} is polynomial; $H_{ij}(i = 1, 2, \cdots, n_y; j = 1, 2, \cdots, m)$ are constant; $W_j(t)$ is Gaussian white noise, $E[W_j(t)W_k(t + \tau)] = S_{ik}\delta(\tau)$, with $\delta(t)$ being Dirac's delta function and S_{ik} being the cross-spectral density of $W_i(t)$ and $W_k(t)$. It is noted that $W_j(t)$ may also be filtered white noise; in this case, more equations can be added to system (1). Without loss of generality, system (1) is analyzed in the following.

Denoting $Y_i = X_{2i-1}, \dot{Y}_i = X_{2i}, f_{2i-1} = X_{2i}, f_{2i} = -H_{i0}, g_{2i-1,j} =$

$0, g_{2i,j} = H_{ij}$ and $2n_y = n_x$, we can express Eq. (1) in Ito's sense as

$$\frac{d}{dt}X_i = f_i(\mathbf{X}) + g_{ij}W_j, \qquad i = 1, 2, \cdots, n_x; j = 1, 2, \cdots, m \tag{2}$$

where $\mathbf{X} \in \Re^{n_x}, X_i$ is a component of the vector process $\mathbf{X}$, $f_i : \Re^{n_x} \rightarrow \Re$. Generally, $f_i(\mathbf{X})$ is a nonlinear function of $\mathbf{X}$.

The system response $\mathbf{X}$ is a Markov vector and the probability density of the stationary Markov vector is governed by the following reduced FPK equation:

$$\frac{\partial}{\partial x_j}(f_j p) - \frac{1}{2}\frac{\partial^2}{\partial x_i \partial x_j}(G_{ij}p) = 0 \tag{3}$$

where $\mathbf{x}$ is the state vector, $\mathbf{x} \in \Re^{n_x}$, $p = p(\mathbf{x})$ and

$$G_{ij} = S_{ls} g_{il} g_{js} \tag{4}$$

The solutions to Eq. (3) are important for the statistical and reliability analysis of random systems. As the reliability analysis of random systems becomes more and more important and widely applied, much research work has been done on the approximate PDF solutions because exact solutions are very limited in practice.

Suppose that the PDF $p(\mathbf{x})$ of the stationary responses of the random system (2) satisfies the following relation:

$$\begin{cases} p(\mathbf{x}) \geq 0 & \mathbf{x} \in \Re^{n_x} \\ \lim_{x_i \to \infty} p(\mathbf{x}) = 0 & i = 1, 2, \cdots, n_x \\ \int_{\Re^{n_x}} p(\mathbf{x}) d\mathbf{x} = 1 \end{cases} \tag{5}$$

which are usually fulfilled by the PDF of the responses of system (2). If an approximate PDF $\tilde{p}\,(\mathbf{x}; \mathbf{a})$ is used, where $\mathbf{a} \in \Re^{N_p}$ and a_i, $(i = 1, 2, \ldots, N_p)$, are parameters to be determined, and N_p is the total number of the parameters, conditions (5) should also be fulfilled by the approximate PDF. Another requirement for the approximate PDF, as stated above, is that it must include many parameters so that high level of approximations can be reached.

There are four matters need to be addressed for the PDF solution of nonlinear random MDOF systems:

• Formulate a versatile and consistent PDF model so that condition (5) can be fulfilled.

• Arbitrary number of unknown parameters can be included in the PDF model and the precision of approximate PDF solution can be improved as the total number of unknown parameters increases.

• Formulate a solution procedure with the PDF model so that practical problems can be solved consistently.

• The PDF model and solution technique must be valid for MDOF systems.

Possible strategy attempting to address these four matters will be discussed in detail and used for the approximate PDF solution of MDOF systems in next sections.

3. Approximate PDF Solution of Random MDOF Systems

It is assumed that the approximate PDF of random MDOF system fulfills the following conditions:

$$\begin{cases} \tilde{p}\,(\mathbf{x};\mathbf{a}) > 0 & \mathbf{x}\times\mathbf{a}\in D_x^{n_x}\times\Re^{N_p} \\ \tilde{p}\,(\mathbf{x};\mathbf{a}) = 0 & \mathbf{x}\notin D_x^{n_x} \qquad i=1,2,\cdots,n_x \\ \int_{\Re^n}\tilde{p}\,(\mathbf{x};\mathbf{a})d\mathbf{x} = 1 & \end{cases} \tag{6}$$

where $\tilde{p}\,(\mathbf{x};\mathbf{a})$ denotes the approximate PDF of $\mathbf{x}$; $D_x^{n_x} = [m_1-c_1\sigma_1, m_1+d_1\sigma_1]\times[m_2-c_2\sigma_2, m_2+d_2\sigma_2]\times\cdots\times[m_i-c_i\sigma_i, m_i+d_i\sigma_i]\times\cdots\times[m_{n_x}-c_{n_x}\sigma_{n_x}, m_{n_x}+d_{n_x}\sigma_{n_x}]\subset\Re^{n_x}$ in which m_i and σ_i denote the mean value and standard deviation of X_i, respectively. $c_i>0$ and $d_i>0$ are defined such that $m_i-c_i\sigma_i$ and $m_i+d_i\sigma$ locate in the tails of the PDF of X_i and the derivatives of the PDF of X_i with respect to x_i at $m_i-c_i\sigma_i$ and $m_i+d_i\sigma$ are zero.

The approximate PDF solution $\tilde{p}\,(\mathbf{x};\mathbf{a})$ of Eq. (3) is assumed to be of the form

$$\tilde{p}\,(\mathbf{x};\mathbf{a}) = \begin{cases} c\exp^{Q_n(\mathbf{x};\mathbf{a})} & \mathbf{x}\times\mathbf{a}\in D_x^{n_x}\times\Re^{N_p} \\ 0 & \mathbf{x}\notin D_x^{n_x} \end{cases} \tag{7}$$

where c is normalization constant and $Q_n(\mathbf{x};\mathbf{a})$ is a polynomial in the state variables $x_1, x_2, \cdots, x_{n_x}$. The detailed form of the polynomial $Q_n(\mathbf{x};\mathbf{a})$ can be determined based on known information. Generally, the following form of $Q_n(\mathbf{x};\mathbf{a})$ may be used:

$$Q_n(\mathbf{x};\mathbf{a}) = \sum_{i=1}^{n_x} a_i x_i + a_{n_x+1}x_1^2 + a_{n_x+2}x_1x_2 + \cdots + a_{n_x(n_x+3)/2}x_{n_x}^2 + \cdots + a_{N_p}x_{n_x}^n \tag{8}$$

which is a n-degree polynomial in $x_1, x_2, \ldots, x_{n_x}$.

Eq. (3) can also be written as follows:

$$\frac{\partial f_j}{\partial x_j}p + f_j\frac{\partial p}{\partial x_j} - \frac{G_{ij}}{2}\frac{\partial^2 p}{\partial x_i \partial x_j} = 0 \tag{9}$$

Generally, the FPK equation can not be satisfied exactly with $\tilde{p}\ (\mathbf{x};\mathbf{a})$, because $\tilde{p}\ (\mathbf{x};\mathbf{a})$ is only an approximation to $p(\mathbf{x})$, and the total number N_p of unknown parameters is always limited in practice. Substituting $\tilde{p}\ (\mathbf{x};\mathbf{a})$ for $p(\mathbf{x})$ in Eq. (9) leads to the following residual error

$$\Delta(\mathbf{x};\mathbf{a}) = \frac{\partial f_j}{\partial x_j}\tilde{p} + f_j\frac{\partial \tilde{p}}{\partial x_j} - \frac{G_{ij}}{2}\frac{\partial^2 \tilde{p}}{\partial x_i \partial x_j} \tag{10}$$

Substituting Eq. (7) into Eq. (10) yields

$$\Delta(\mathbf{x};\mathbf{a}) = \delta(\mathbf{x};\mathbf{a})\,\tilde{p}\ (\mathbf{x};\mathbf{a}) \tag{11}$$

where

$$\delta(\mathbf{x};\mathbf{a}) = f_j\frac{\partial Q_n}{\partial x_j} - \frac{G_{ij}}{2}\left(\frac{\partial^2 Q_n}{\partial x_i \partial x_j} + \frac{\partial Q_n}{\partial x_i}\frac{\partial Q_n}{\partial x_j}\right) + \frac{\partial f_j}{\partial x_j} \tag{12}$$

which can be considered as local residual error of Eq. (9) if $p(\mathbf{x})$ is replaced by $\tilde{p}\ (\mathbf{x};\mathbf{a})$.

Because $\tilde{p}\ (\mathbf{x};\mathbf{a}) \neq 0$, therefore, the only possibility for $\tilde{p}\ (\mathbf{x};\mathbf{a})$ to satisfy Eq. (9) is $\delta(\mathbf{x};\mathbf{a}) = 0$. However, usually $\delta(\mathbf{x};\mathbf{a}) \neq 0$ because $\tilde{p}\ (\mathbf{x};\mathbf{a})$ is only an approximation to $p(\mathbf{x})$. In this case, another set of mutually independent functions $H_k(\mathbf{x})$ which span the space R^{N_p} can be introduced to make the projection of local residual error $\delta(\mathbf{x};\mathbf{a})$ on R^{N_p} vanish. Therefore, this method may be called a projection method. Suppose that $\delta(\mathbf{x};\mathbf{a})H_k(\mathbf{x})$, $k = 1, 2, \cdots, N_p$, is integrable in $\Re^{N_p}$, then, according to the above idea of projection method,

$$\int_{\Re^{n_x}} \delta(\mathbf{x};\mathbf{a})\,H_k(\mathbf{x})d\mathbf{x} = 0, \qquad k = 1, 2, \cdots, N_p \tag{13}$$

or

$$\int_{\Re^{n_x}} [f_j\frac{\partial Q_n}{\partial x_j} - \frac{G_{ij}}{2}(\frac{\partial^2 Q_n}{\partial x_i \partial x_j} + G_{ij}\frac{\partial Q_n}{\partial x_i}\frac{\partial Q_n}{\partial x_j}) + \frac{\partial f_j}{\partial x_j}]H_k(\mathbf{x})d\mathbf{x} = 0$$
$$k = 1, 2, \cdots, N_p \tag{14}$$

This means that Eq. (9) is satisfied with $\tilde{p}\ (\mathbf{x};\mathbf{a})$ in the average sense of integration if $\delta(\mathbf{x};\mathbf{a})\,H_k(\mathbf{x})$ is integrable in $\Re^{n_x}$.

By selecting $H_k(\mathbf{x})$ as $x_1^{k_1}x_2^{k_2}\cdots x_{n_x}^{k_{n_x}} f_N(\mathbf{x})$, $k = k_1 + k_2 + \cdots + k_{n_x}$, being $k_1, k_2, \cdots, k_{n_x} = 0, 1, 2, \cdots, N_p$ such that $\delta(\mathbf{x};\mathbf{a})\,H_k(\mathbf{x})$ is integrable in $\Re^{n_x}$, we give N_p quadratic nonlinear algebraic equations in terms of N_p undetermined parameters from Eq. (14). The algebraic equations can be solved with any available method to determine the parameters.

Numerical experience showed that a convenient and effective choice for function $f_N(\mathbf{x})$ is Gaussian PDF. Because of the particular choice of $f_N(\mathbf{x})$, the integration in Eq. (14) can be easily evaluated by using the relationships between higher and lower order moments of Gaussian stochastic processes.

4. Numerical Example

Example 1. Consider the following nonlinear random two-degree-of-freedom system:

$$\ddot{Y}_1 + \frac{1}{2}a_1(S_{11}\dot{Y}_1 + 2a_2S_{12}\dot{Y}_2) + 2a_3Y_1 + 4a_4Y_1^3 + 6a_5Y_1^5 = W_1(t) \quad (15)$$

$$\ddot{Y}_2 + \frac{1}{2}a_1[2(1-a_2)S_{12}\dot{Y}_1 + S_{22}\dot{Y}_2] + 2a_6Y_2 + 4a_7Y_2^3 + 6a_8Y_2^5 = W_2(t) \quad (16)$$

where $a_1, a_2, \cdots, a_8$ are some constants; $W_i(t), (i = 1, 2)$, is Gaussian white noise. Denoting $Y_1 = X_1, \dot{Y}_1 = X_2, Y_2 = X_3$ and $\dot{Y}_2 = X_4$, we can express the system by the following four-dimensional nonlinear random system:

$$\dot{X}_1 = X_2 \quad (17)$$

$$\dot{X}_2 = -\frac{1}{2}a_1(S_{11}X_2 + 2a_2S_{12}X_4) - 2a_3X_1 - 4a_4X_1^3 - 6a_5X_1^5 + W_1(t) \quad (18)$$

$$\dot{X}_3 = X_4 \quad (19)$$

$$\dot{X}_4 = -\frac{1}{2}a_1[2(1-a_2)S_{12}X_2 + S_{22}X_4] - 2a_6X_3 - 4a_7X_3^3 - 6a_8X_3^5 + W_2(t) \quad (20)$$

For this system, the exact stationary PDF solution [7, 9] is

$$p(x_1, x_2, x_3, x_4) = C\exp\{-a_1[\frac{1}{2}(x_2^2 + x_4^2) + a_3x_1^2 + a_4x_1^4 + a_5x_1^6 + a_6x_3^2 + a_7x_3^4 + a_8x_3^6]\} \quad (21)$$

where C is a normalization constant.

In the following analysis, the approximate PDF solutions obtained with the proposed projection method for different n values are compared with this exact solution. It is noted that the *equivalent stochastic linearization method* is a special case of the projection method if $f_N(\mathbf{x})$ is selected to be the PDF from an equivalent stochastic linearization and $n = 2$. In this example, the function $f_N(\mathbf{x})$ is selected to be the stationary PDF from equivalent stochastic linearization.

For $a_1 = a_3 = a_4 = a_6 = 1, a_5 = a_7 = a_8 = 0.5$ and arbitrary values of S_{11}, S_{12}, S_{22} and a_2, the system is highly nonlinear, and the approximate PDFs of X_1 and X_3 obtained with the presented method are compared with the exact logarithmic PDF solutions in Figures 1 and 2. It is apparent that the approximate solutions for $n = 4$ are very close to the exact solutions. For $n = 2$, the results coincide with those from equivalent stochastic linearization. The PDF solutions for $n = 4$ are a great improvement on those for $n = 2$.

In order to show the tail performance of the PDFs, the logarithmic PDFs are shown and compared in Figures 3 and 4. It is seen that the approximate PDFs for $n = 4$ are close to the exact PDF solution even in the tails. These results validate the method for nonlinear random MDOF systems.

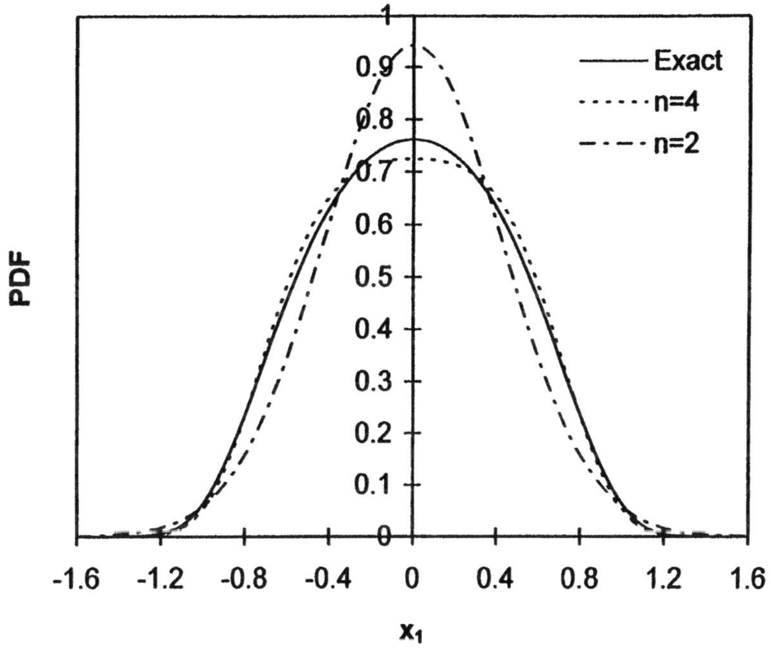

Figure 1 - Probability densities of X_1 in example 1

It is noted that the results from equivalent stochastic linearization much deviate from the exact solutions, specially in the tails of PDFs. Numerical experience showed that the results from equivalent stochastic linearization also much deviate from the exact solutions in the tails of PDFs even for weakly nonlinear systems. In other words, the tail behaviors of PDFs are sensitive to the system nonlinearity. Hence much attention must be paid when the *equivalent stochastic linearization method* is used in application, particularly when reliability analysis is concerned.

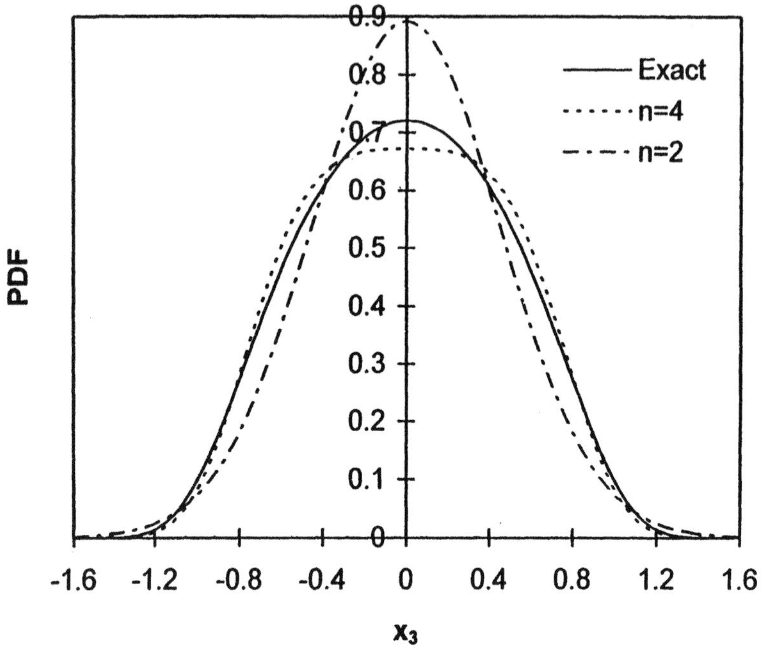

Figure 2 - Probability densities of X_3 in example 1

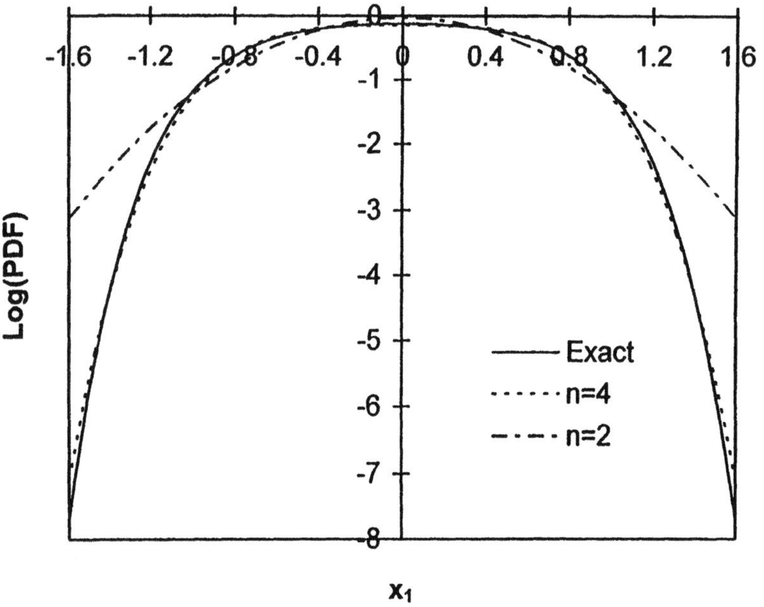

Figure 3 - Logarithmic probability densities of X_1 in example 1

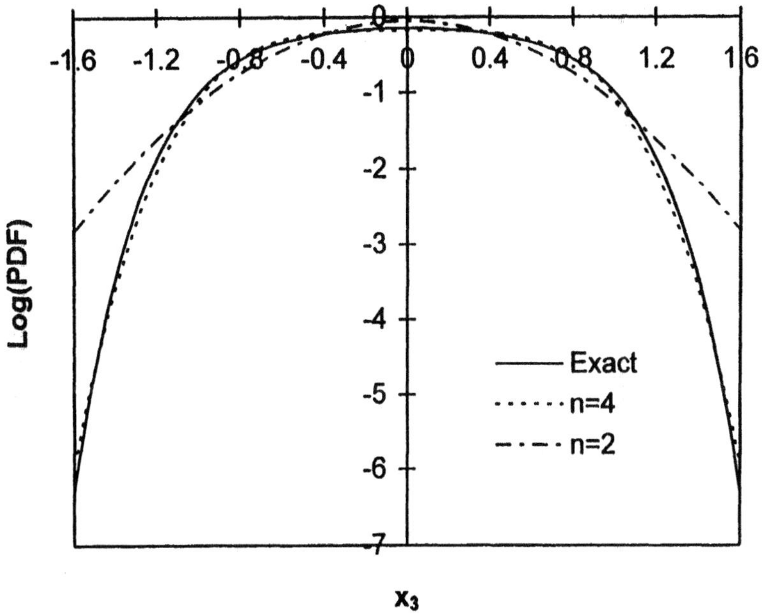

Figure 4 - Logarithmic probability densities of X_3 in example 1

From numerical experience we found that exact PDFs of X_1 and X_3 can be obtained if $n = 6$ and exact PDFs of X_2 and X_4 can also be obtained if $n \geq 2$ for this system. We further found from numerical experience that the projection method may provide a tool for the exact PDF solutions of the systems which exact PDF solutions are exponential functions of polynomial.

5. Conclusion

The recently proposed method [4, 5] for the approximate PDF solution of nonlinear random vibrations is extended to obtain approximate PDF solutions for random MDOF systems. The approximate PDF solution is taken as an exponential polynomial in the state variables of the system and a set of unknown parameters. Local residual error is determined by substitution of the approximate solution in the governing FPK equation. Then, a set of basis functions spanning a finite-dimensional real space is chosen, and the projection of the residual error is made to vanish on this space in order to formulate quadratic algebraic equations for the unknown parameters. The approximate PDF solution is thus obtained upon solving for the unknown parameters. The proposed solution procedure and numerical results show that (1) the method is not limited by the degree of nonlinearity of the system and thus suitable for highly nonlinear random systems; (2) the approximate PDF model meets the requirement of probability theory; (3) for systems excited by white noise, as encountered in many problems in science and engineering, the resulting algebraic equations are quadratic and easy to solve; (4) numerical results validated the method for MDOF systems; (5) the solution procedure is consistent, systematic and thus easy to implement on computers; and (6) the method may provide a tool for the exact PDF solutions of the systems which exact PDF solutions are exponential functions of polynomial.

6. Acknowledgements

The results presented in this paper were obtained in the course of research supported by the fundings of the Research Committee of the University of Macao.

References

1. S. A. Assaf and L. D. Zirkie, Approximate analysis of non-linear stochastic systems. *Int. J. Control* **23** (1976) 477–492.
2. R. C. Booton, Nonlinear control systems with random inputs. *IRE Transactions on Circuit Theory*, CT-1 **1** (1954) 9–19.
3. G. K. Er, Multi-Gaussian closure method for randomly excited non-linear systems. *Int. J. Non-Linear Mech.* **33** (1998) 201–214.
4. G. K. Er, A new nonGaussian closure method for the PDF solution of non-linear random vibrations. *Proceedings of 12th Engineering Mechanics Specialty Conference*, ASCE, San Diego (1998) 1403–1406.
5. G. K. Er, An improved closure method for analysis of nonlinear stochastic systems. *Nonlinear Dynamics* **17** (1998) 285–297.
6. G. K. Er and V. P. Iu, Probabilistic solutions to nonlinear random ship roll motion. *ASCE Journal of Engineering Mechanics* **125** (1999) 570–574.
7. Y. K. Lin and G. Q. Cai, *Probabilistic Structural Dynamics*. McGraw-Hill, New York (1995) 186–188.
8. L. D. Lutes, Approximate technique for treating random vibration of hysteretic systems. *J. Acoust. Soc. Am.* **48** (1970) 299–306.
9. A. Scheurkogel and I. Elishakoff, Non-linear random vibration of a two-degree-of-freedom system. In *Non-Linear Stochastic Engineering Systems*. eds. F. Ziegler and G. I. Schuëller, Springer-Verlag, Berlin (1988) 285–299.
10. K. Sobczyk and J. Trebicki, Maximum entropy principle in Stochastic Dynamics. *Prob. Eng. Mech.* **5** (1990) 102–110.
11. R. L. Stratonovich, *Topics in the Theory of Random Noise*. vol. 1, Gordon and Breach, New York (1963).

STABILITY AND HOPF BIFURCATION OF A FOUR-WHEEL-STEERING VEHICLE INVOLVING DRIVER'S DELAY

H. Y. HU and Z. Q. WU
Institute of Vibration Engineering Research,
Nanjing University of Aeronautics and Astronautics,
Nanjing 210016, China

1. Introduction

With increase of vehicle speed, there is an increasing demand for safety in driving. As the large yaw rate and sideslip angle of a vehicle body in high speed turning can not be well controlled by steering front wheels only, the four-wheel-steering (4WS) technique has been developed over the past decade. A great number of studies have been made on various control strategies for 4WS vehicles since the first 4WS system was reported. Yet, few archival publications dealt with the dynamics of 4WS vehicles with the time delay in driver's response and nonlinearity of the lateral tyre force taken into account. These effects on the vehicle dynamics, hence, are still open problems. Moreover, most control strategies designed for 4WS vehicles are based on the limit dynamics, i.e., the dynamics of a vehicle running at infinitely low or high speed, because the dynamic behavior of a vehicle at medium or high speed is very complicated and far from clear.

This paper presents a new mathematical model for 4WS vehicle-driver systems during turning, with the nonlinearity in lateral tyre force and the time delays in the driver's response and steering mechanism taken into consideration. The model is described by a set of 5 dimensional nonlinear differential equations with a time delay, which results in an infinite dimensional solution space. On the basis of this model, the steady state motion of the vehicle is determined. Then, the asymptotic stability condition of a typical steady state motion is given. The Hopf bifurcation of the steady state motion with the variation of vehicle speed, or preview distance and time delay of driver is discussed in two cases when the linear and bilinear control strategies of rear-wheel-steering are used, respectively.

N. Van Dao and E.J. Kreuzer (eds.),
IUTAM Symposium on Recent Developments in Non-linear Oscillations of Mechanical Systems, 95–104.

2. Model for the Vehicle-Driver System

2.1 LATERAL DYNAMIC EQUATIONS OF VEHICLE

Consider the vehicle model shown in Figure 1, where a symmetric rigid body of mass m with four wheels is moving at a constant speed U. Let G denote the center of mass, where the coordinate frame fixed on the vehicle body originates. For this model, the lateral velocity V and the yaw angular velocity r yield

$$\begin{cases} m(\dot{V}+rU)=2F_f\cos\delta_f+2F_r\cos\delta_r, \\ I_z\dot{r}=2aF_f\cos\delta_f-2bF_r\cos\delta_r, \end{cases} \tag{1}$$

where I_z is the inertia moment of rotation of the vehicle body with respect to the vertical axis z, a and b are the distances from G to the front and rear axles, δ_f and δ_r are the steering angles applied on the front and rear wheels, F_f and F_r are the lateral forces due to the contact between the tyre and the road surface at each front and rear wheel.

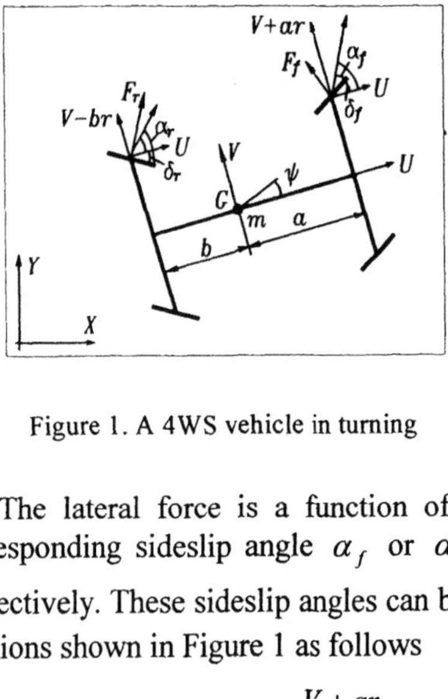

Figure 1. A 4WS vehicle in turning

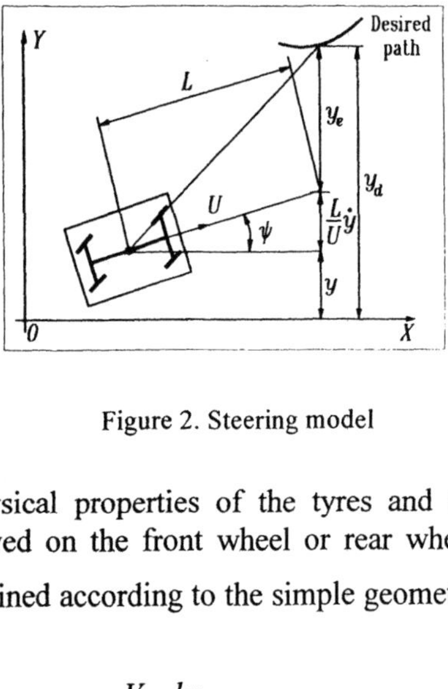

Figure 2. Steering model

The lateral force is a function of the physical properties of the tyres and the corresponding sideslip angle α_f or α_r observed on the front wheel or rear wheel, respectively. These sideslip angles can be determined according to the simple geometric relations shown in Figure 1 as follows

$$\alpha_f=\arctan(\frac{V+ar}{U})-\delta_f, \qquad \alpha_r=\arctan(\frac{V-br}{U})-\delta_r. \tag{2}$$

The most popular tyre model is the Magic Formula developed by Pacejka (1989). In this paper, the third order truncation of the formula will be used

$$F_f = -C_1\alpha_f + C_3\alpha_f^3, \qquad F_r = -D_1\alpha_r + D_3\alpha_r^3, \tag{3}$$

where C_1, C_3, D_1 and D_3 are positive parameters.

In addition, a fixed frame of coordinates (x, y, ψ) is defined, where (x, y) represents the location of G in driving and ψ the heading angle of the vehicle. Obviously, the following relations hold

$$\dot{y} = V\cos\psi + U\sin\psi, \qquad \dot{\psi} = r. \tag{4}$$

2.2 MODEL FOR DRIVERS

To investigate the effect of perceptual delay of a driver on the stability of vehicles, many mathematical models have been proposed for the driver, who senses the deviation from the desired path and steers the vehicle to reduce the deviation as shown in Figure 2. In this paper, the simplest driver model suggested by Nagai and Mitschke (1987) is used. The model includes an algebraic equation and a first order differential equation. They describe the deviation from the desired path and the retardation of driver and steering mechanism, respectively

$$\begin{cases} y_e(t+T_p) = y_d(t+T_p) - y(t) - \dfrac{L}{U}\dot{y}(t), & \text{(5a)} \\ \delta_f(t) + T_s\dot{\delta}_f(t) = K_m y_e(t - T_d), & \text{(5b)} \end{cases}$$

where L denotes the preview distance, $y(t)$ the lateral displacement of the vehicle, $y_d(t)$ the desired lateral displacement, $y_e(t)$ the error between desired and actual lateral displacement, T_p the preview time, T_s the time delay of the steering mechanism, T_d the time delay of the driver, K_m the steering gain, respectively.

Substituting Equation (5a) into Equation (5b) yields

$$T_s\dot{\delta}_f(t) + \delta_f(t) = K_m[y_d(t - T_d) - y(t - T) - \frac{L}{U}\dot{y}(t - T)], \tag{6}$$

where

$$T = T_p + T_d > 0 \tag{7}$$

represents the total time delay in the vehicle-driver system.

2.3 CONTROL STRATEGIES FOR REAR-WHEEL-STEERING

As shown in Equation (1), the lateral acceleration is composed of two components, the lateral velocity V and the yaw rate r. As the speed of vehicle increases, the lateral acceleration delays longer than the yaw response. This is mainly attributable to the fact that with an increase in vehicle speed U, the sideslip angles defined by Eq.(2) decline, and even become negative. Thus, a number of control strategies have been developed to make the steady state sideslip angles as close to zero as possible.

A popular control strategy is to steer the rear wheels on the basis of a pre-determined function as below

$$\delta_r = k_\delta \delta_f + k_r r . \tag{8}$$

There are two versions of this control strategy. One is the linear strategy, which requires

$$k_\delta = \frac{-b + \dfrac{ma}{C_1(a+b)}U^2}{a + \dfrac{mb}{D_1(a+b)}U^2}, \qquad k_r = 0. \tag{9}$$

This strategy features that $k_\delta \to -\dfrac{b}{a} < 0$ when $U \to 0$ and $k_\delta \to \dfrac{aD_1}{bC_1} > 0$ when $U \to +\infty$. The other version is the bilinear strategy with the coefficients given by

$$k_\delta = -\frac{C_1}{D_1} \neq 1, \qquad k_r = \frac{2(aC_1 - bD_1) + mU^2}{2D_1 U}. \tag{10}$$

2.4 COUPLED DYNAMIC EQUATIONS

In summary, the motion of the 4WS vehicle-driver system of concern yields a set of non-autonomous difference-differential equations of 5 state variables $(V, r, y, \psi, \delta_f)$ as following

$$\begin{cases} m\dot{V} = -mrU + 2F_f(V,r,\delta_f)\cos\delta_f + 2F_r(V,r,\delta_f)\cos(k_\delta\delta_f + k_r r), \\ I_z\dot{r} = 2aF_f(V,r,\delta_f)\cos\delta_f - 2bF_r(V,r,\delta_f)\cos(k_\delta\delta_f + k_r r), \\ \dot{y} = V\cos\psi + U\sin\psi, \\ \dot{\psi} = r, \\ \dot{\delta}_f = -\dfrac{\delta_f}{T_s} - \dfrac{K_m}{T_s}[y(t-T) + \dfrac{L}{U}\dot{y}(t-T)] + \dfrac{K_m}{T_s}y_d(t-T_d), \end{cases} \tag{11}$$

where $F_f(V,r,\delta_f)$ and (V,r,δ_f) can be determined from Eqs. (2), (3) and (8), while $y_d(t-T_d)$ can be considered as the external disturbance in the dynamic analysis.

3. Steady State Motions

Setting all derivatives in Eq.(11) as zero and making necessary manipulations, one obtains a set of algebraic equations governing the steady state motion of the vehicle

$$\begin{cases} r=0, \quad \delta_f = -K_m y, \quad V=-U\tan\psi, \\ [C_1(\psi+\delta_f)-C_3(\psi+\delta_f)^3]\cos(k_\delta\delta_f)=0, \\ [D_1(\psi+k_\delta\delta_f)-D_3(\psi+k_\delta\delta_f)^3]\cos(k_\delta\delta_f)=0. \end{cases} \tag{12}$$

As proved in Wu (1998), Equation (12) has a trivial solution $(0,0,0,0,0)$ corresponding to the vehicle motion along the straight line $y=0$, as well as 8 non-trivial solutions corresponding to the steady state motions due to the cubic nonlinearity of tyre forces. This fact is completely different from the case studied in current literature where the linear model of tyre forces was used.

4. Stability of Trivial Steady State Motion

By linearizing Equation (11) at $(0,0,0,0,0)$, one has the corresponding characteristic equation

$$\lambda^5 + a_4\lambda^4 + a_3\lambda^3 + a_2\lambda^2 + a_1\lambda + a_0 = 0, \tag{13}$$

where

$$a_i = a_{i0} + a_{i1}e^{-\lambda\tau}, \qquad i=1,2,3,4; \tag{14}$$

$$\begin{cases} a_{40} = \dfrac{1}{T_s} + 2\left(\dfrac{a^2C_1+b^2D_1}{I_zU} + \dfrac{bk_rD_1}{I_z} + \dfrac{C_1+D_1}{mU}\right), \qquad a_{41}=0, \\ a_{30} = 2\Big[\dfrac{bk_rD_1}{I_zT_s} + \dfrac{C_1+D_1}{mUT_s} + \dfrac{a^2C_1+b^2D_1}{I_zUT_s} + \dfrac{2k_rC_1D_1(a+b)}{mI_zU} \\ \qquad + \dfrac{bD_1-aC_1}{I_z} + \dfrac{2C_1D_1(a+b)^2}{mI_zU^2}\Big], \qquad a_{31} = \dfrac{2K_mL(C_1+k_\delta D_1)}{mUT_s}, \\ a_{20} = \dfrac{2}{I_zT_s}\Big[(bD_1-aC_1) + \dfrac{2C_1D_1(a+b)(a+b+k_rU)}{mU^2}\Big], \\ a_{21} = \dfrac{2K_m}{mT_s}\Big[\dfrac{2LC_1D_1(a+b)(ak_\delta+b+k_rU)}{I_zU^2} + (C_1+k_\delta D_1)\Big], \\ a_{10}=0, \qquad a_{11} = \dfrac{4K_mC_1D_1(a+b)}{mI_zT_s}\Big[\dfrac{(ak_\delta-k_\delta L+L+b)}{U} + k_r\Big], \\ a_{00}=0, \qquad a_{01} = -\dfrac{4K_mC_1D_1(a+b)(k_\delta-1)}{mI_zT_s}. \end{cases} \tag{15}$$

Because $a_{01} \neq 0$ when $k_\delta \neq 1$, none of the roots of Eq.(13) is zero. Thus, the trivial steady state motion becomes unstable only when Eq.(13) has a pair of imaginary roots $\lambda = \pm i\omega$. If this is the case, Eq.(13) is equivalent to the following conditions

$$\begin{cases} \mathrm{Re}(\omega,T) \equiv (a_{01} - a_{21}\omega^2)\cos\omega T + (a_{11}\omega - a_{31}\omega^3)\sin\omega T + a_{40}\omega^4 - a_{20}\omega^2 = 0, \\ \mathrm{Im}(\omega,T) \equiv (a_{11}\omega - a_{31}\omega^3)\cos\omega T - (a_{01} - a_{21}\omega^2)\sin\omega T + +\omega^5 - a_{30}\omega^3 = 0. \end{cases} \tag{16}$$

Solving Eq.(16) for $\cos\omega T$ and $\sin\omega T$, one has

$$\begin{cases} \cos\omega T = -\dfrac{(a_{01} - a_{21}\omega^2)(a_{40}\omega^4 - a_{20}\omega^2) + (a_{11}\omega - a_{31}\omega^3)(\omega^5 - a_{30}\omega^3)}{(a_{01} - a_{21}\omega^2)^2 + (a_{11}\omega - a_{31}\omega^3)^2}, \\ \sin\omega T = \dfrac{(a_{01} - a_{21}\omega^2)(\omega^5 - a_{30}\omega^3) - (a_{11}\omega - a_{31}\omega^3)(a_{40}\omega^4 - a_{20}\omega^2)}{(a_{01} - a_{21}\omega^2)^2 + (a_{11}\omega - a_{31}\omega^3)^2}. \end{cases} \tag{17}$$

Eliminating $\cos\omega T$ and $\sin\omega T$ in Eq.(17) results in the critical condition governed by the following algebraic equation of 10 orders in ω

$$\begin{aligned} &\omega^{10} + (a_{40}^2 - 2a_{30})\omega^8 + (a_{30}^2 - a_{31}^2 - 2a_{20}a_{40})\omega^6 \\ &+ (a_{20}^2 + 2a_{11}a_{31} - a_{21}^2)\omega^4 + (2a_{01}a_{21} - a_{11}^2)\omega^2 - a_{01}^2 = 0. \end{aligned} \tag{18}$$

If Eq.(18) has no positive real root, the system is asymptotically stable for arbitrary time delay $0 \le T < +\infty$. Otherwise, one can solve Eq.(18) for ω and substitute the positive solution into the second equation in Eq.(17), then one solves the equation for the minimal solution $T_{\min}$. Obviously,

$$T_c = \max(T_{\min}, 0) \tag{19}$$

gives the critical time delay when the vehicle undergoes instability.

5. Hopf Bifurcation

To make sure that the Hopf bifurcation occurs when the system undergoes instability with the variation of a system parameter, say, μ around the critical value μ_0 with $\lambda(\mu_0) = \pm i\omega$, one needs to check the transversality condition as following

$$\mathrm{Re}\left(\frac{\mathrm{d}\lambda}{\mathrm{d}\mu}\bigg|_{\mu=\mu_0}\right) \neq 0. \tag{20}$$

In what follows, let U, L and T be the control parameters and suppose them to be functions in μ. By differentiating Eq.(13) with respect to μ, one has

$$\frac{\mathrm{d}\lambda}{\mathrm{d}\mu}[\sum_{j=0}^{4} a_{j1}(-T\lambda^{j}+j\lambda^{j-1})e^{-\lambda T}+(5\lambda^{4}+\sum_{j=1}^{4} a_{j0}j\lambda^{j-1})]+\frac{\mathrm{d}U}{\mathrm{d}\mu}\cdots+\frac{\mathrm{d}L}{\mathrm{d}\mu}\cdots\frac{\mathrm{d}T}{\mathrm{d}\mu}\cdots=0. \quad (21)$$

By substituting $\lambda(\mu_0)=\pm i\omega$ into Eq.(21), one obtains

$$\left.\frac{\mathrm{d}\lambda}{\mathrm{d}\mu}\right|_{\mu=\mu_0}=\frac{1}{\sigma_{1R}+i\sigma_{1I}}[(\sigma_{2R}+i\sigma_{2I})\frac{\mathrm{d}U}{\mathrm{d}\mu}+(\sigma_{3R}+i\sigma_{3I})\frac{\mathrm{d}L}{\mathrm{d}\mu}+(\sigma_{3R}+i\sigma_{3I})\frac{\mathrm{d}T}{\mathrm{d}\mu}], \quad (22)$$

where

$$\begin{cases}\sigma_{1R}=[a_{41}\omega^4 T+(3a_{31}-a_{21}T)\omega^2+a_{01}T-a_{11}]\cos\omega T\\ \quad +[(4a_{41}-a_{31}T)\omega^3+(a_{11}T-2a_{21})\omega]\sin\omega T-5\omega^4+3a_{30}\omega^2-a_{10},\\ \sigma_{1I}=[(4a_{41}-a_{31}T)\omega^3+(a_{11}T-2a_{21})\omega]\cos\omega T\\ \quad -[a_{41}T\omega^4+(3a_{31}-a_{21}T)\omega^2+a_{01}T-a_{11}]\sin\omega T+4a_{40}\omega^3-2a_{20}\omega,\\ \sigma_{2R}=(a_{41,U}\omega^4-a_{21,U}\omega^2+a_{01,U})\cos\omega T-(a_{31,U}\omega^3-a_{11,U}\omega)\sin\omega T\\ \quad +a_{40,U}\omega^4-a_{20,U}\omega^2+a_{00,U},\\ \sigma_{2I}=-(a_{41,U}\omega^4-a_{21,U}\omega^2+a_{01,U})\sin\omega T-(a_{31,U}\omega^3-a_{11,U}\omega)\cos\omega T\\ \quad -a_{30,U}\omega^3+a_{10,U}\omega,\\ \sigma_{3R}=(a_{41,L}\omega^4-a_{21,L}\omega^2+a_{01,L})\cos\omega T-(a_{31,L}\omega^3-a_{11,L}\omega)\sin\omega T\\ \quad +a_{40,L}\omega^4-a_{20,L}\omega^2+a_{10,L},\\ \sigma_{3I}=-(a_{41,L}\omega^4-a_{21,L}\omega^2+a_{01,L})\sin\omega T-(a_{31,L}\omega^3-a_{11,L}\omega)\cos\omega T\\ \quad -a_{30,L}\omega^3+a_{10,L}\omega,\\ \sigma_{4R}=(a_{11}\omega^2-a_{31}\omega^4)\cos\omega T-(a_{41}\omega^5-a_{21}\omega^3+a_{01}\omega)\sin\omega T,\\ \sigma_{4I}=-(a_{41}\omega^5+a_{21}\omega^3-a_{01}\omega)\cos\omega T+(a_{31}\omega^4-a_{11}\omega^2)\sin\omega T.\end{cases} \quad (23)$$

$a_{ij,U}$ and $a_{ij,L}$ represent the partial derivatives of a_{ij} with respect to U and L, respectively. From Eq.(22), there follows the transversality condition

$$\sigma_{1R}(\sigma_{2R}\frac{\mathrm{d}U}{\mathrm{d}\mu}+\sigma_{3R}\frac{\mathrm{d}L}{\mathrm{d}\mu}+\sigma_{4R}\frac{\mathrm{d}T}{\mathrm{d}\mu})+\sigma_{1I}(\sigma_{2I}\frac{\mathrm{d}U}{\mathrm{d}\mu}+\sigma_{3I}\frac{\mathrm{d}L}{\mathrm{d}\mu}+\sigma_{4I}\frac{\mathrm{d}T}{\mathrm{d}\mu})\neq 0. \quad (24)$$

6. Numerical Simulations

To demonstrate the above analytic results, consider the case study of a 4WS vehicle-driver system in Nagai et al (1995), where the following system parameters were fixed as constants

$$\begin{cases} m = 1300\ \text{Kg}, & I_z = 3000\ \text{Kg}\cdot\text{m}^2, & a = 1\text{m}, & b = 1.6\text{m}, \\ C_1 = 44400\text{N}/\text{rad}, & D_1 = 43600\text{N}/\text{rad}, & K_m = 0.02, \\ C_3 = 44400\text{N}/\text{rad}^3 & D_3 = 44400\text{N}/\text{rad}^3, & T_s = 0.2\ \text{s}, \end{cases} \tag{25}$$

whereas

$$10\,\text{m}/\text{s} < U \le 80\,\text{m}/\text{s}, \qquad 10\,\text{m} < L \le 80\,\text{m} \tag{26}$$

were taken as the changeable parameters.

Given a pair of (U, L), one can determine a corresponding T_c numerically from Eqs.(17-19). In this way, one obtains a 2-dimensional surface $T_c(U, L)$ in the parameter space (U, L, T) as shown in Figure 3. The most flat part of surface $T_c(U, L)$ in Figure 3 coincides with the plane $T = 0$. i.e., the trivial steady state motion of the vehicle running at these combinations of (U, L) is not stable even though there is no time delay in the response of driver. It is obvious that the 4WS vehicles, especially the one with bilinear control strategy has the largest stable region in (U, L) plane, no matter whether the time delay is taken into account or not.

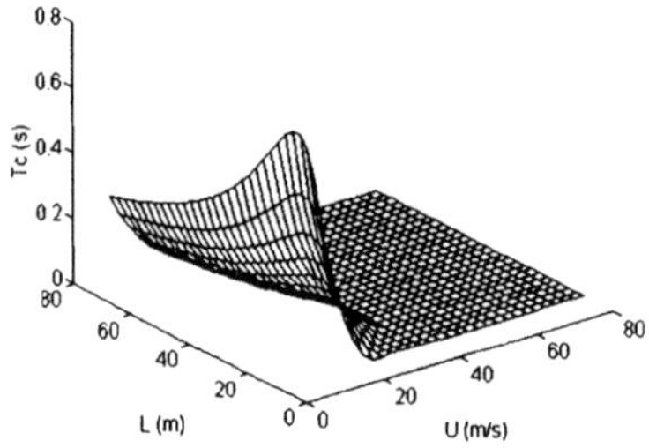

a. 2WS vehicle

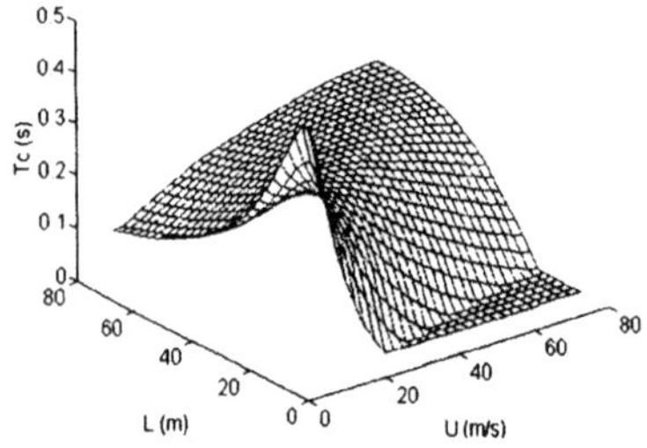

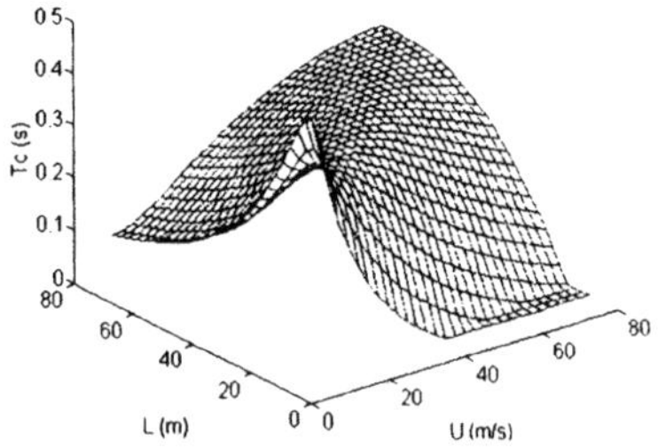

b. 4WS vehicle with linear control strategy

c. 4WS vehicle with bilinear control strategy

Figure 3. Critical time delays for various combinations of (U, L)

Furthermore, it can be numerical verified that the transversality condition holds in the surface $T_c(U, L)$ and hence the Hopf bifurcation exists in the trivial steady state motions of both 2WS and 4WS vehicle systems. To support this assertion, the Runge-Kutta approach with variable step was used to solve Eq.(11) in time domain. Figure 4 shows the Hopf bifurcation of the trivial steady state motion of a 4WS vehicle equipped with bilinear control strategy runs at a speed of $U = 30\text{m}/\text{s}$ when driver has the preview distance of $L = 40\text{m}$.

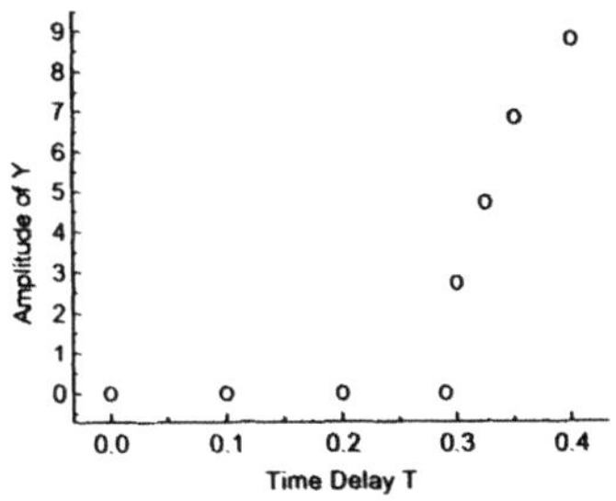

a. The Hopf bifurcation with variation of time delay

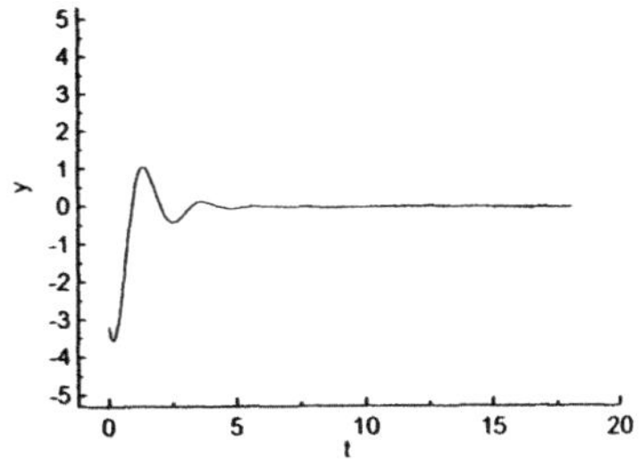

b. Time history of disturbed motion before the Hopf bifurcation ($T = 0.1\text{s}$)

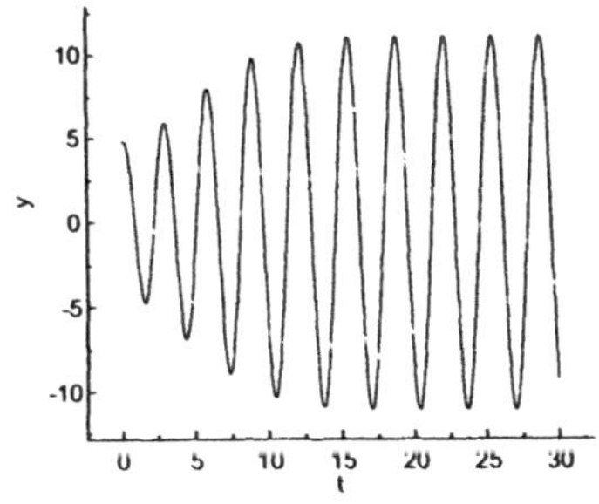

c. Time history of disturbed motion after the Hopf bifurcation ($T = 0.35\text{s}$)

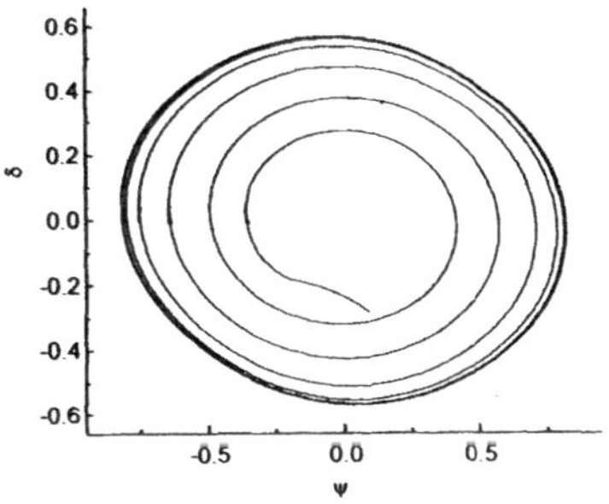

d. Phase portrait of disturbed motion after the Hopf bifurcation ($T = 0.35\text{s}$)

Figure 4. The Hopf bifurcation cased by the time delay

7. Conclusions

The 4WS vehicle-driver system of concern has a trivial steady state motion when it runs along a straight line, as well as 8 non-trivial steady state motions owing to the cubic nonlinearity in lateral tyre force.

With increase of driver's delay, the stable region for the trivial steady state motion shrinks in the plane spanned by the vehicle speed and the preview distance, and a supercritical Hopf bifurcation occurs. Thus, the lateral motion of vehicle will oscillate periodically if the driver's delay exceeds the critical value T_c given by Eq.(19).

Compared with 2WS vehicle, the 4WS vehicle equipped with both control strategies has better performance in the stability of the trivial steady state motion if there is no time delay in driver's response. Furthermore, the bilinear control strategy works better than the linear control strategy when the time delay is taken into consideration.

8. Acknowledgments

This work was supported in part by the National Natural Science Foundation of China under the Grant 59625511 and in part by the Funds of Nanjing Municipal Government for Returned Overseas Scientists.

9. References

1. Cho, Y. H. and Kim, J. (1995) Design of optimal four-wheel steering system, *Vehicle System Dynamics*, **24**, 661-682.
2. Liu Z., Payre G. and Bourassa P. (1996) Nonlinear Oscillations and Chaotic Motions in a Road Vehicle System with Driver Steering Control, *Nonlinear Dynamics*, **9**, 281-304.
3. Nagai, M. and Mitschke, M. (1987) An adaptive control model of a car-driver and computer simulation of the closed-loop System, *Proceedings of* 10-*th IAVSD Symposium*, Prague.
4. Nagai M., Ueda E. and Moran A. (1995) Nonlinear Design to Approach Four-Wheel-Steering System Using Neural Network, *Vehicle System Dynamics*, **24**, 329-342.
5. Pacejka, H. (1989) A new tyre model with an application in vehicle dynamics studies". *SAE Paper* No.89007.
6. Palkovics L. (1992) Effect of the controller parameters on the steerability of the four wheel steered car, *Vehicle System Dynamics*, **21**, 109-128.
7. Wu Z. Q. (1998) Research on Nonlinear Dynamics of Complicated Systems, *Postdoctoral Thesis*, No.9805, Nanjing University of Aeronautics and Astronautics.

MOORED CRANE VESSELS IN REGULAR WAVES

KATRIN ELLERMANN AND EDWIN KREUZER
Technical University Hamburg – Harburg
Ocean Engineering Section II – Mechanics
Eissendorfer Strasse 42
D-21073 Hamburg, Germany

1. Introduction

Floating cranes are used for a variety of tasks in marine technology. In coastal regions crane barges are the most commonly used vessels; in offshore engineering, larger crane ships or semisubmersibles can be found. All are used for lift operations, transportation, the construction of large offshore structures and for salvage operations.

The practical problems that arise during crane ship operations include the inability to position accurately the objects being handled, and collisions. The aim of this research project is to get a mathematical description for crane ship operations which enables predictions about the dynamical behavior of the vessel to be made. Then the results of the analysis can be used for safety considerations and to enlarge the operating range of crane ships.

The dynamical system consisting of the vessel, the crane and the load is frequently modeled as a system of coupled rigid bodies. Water waves provide the principal disturbing force. Waves, which drive the coupled system of vessel and load – usually represented by a pendulum – are the main source of such problems. The motion of the center of mass during lifting operations can also lead to unwanted dynamic behavior.

In recent years several publications have dealt with the dynamics of crane vessels. Rieckert, 1992, used a mathematical model with eight degrees of freedom and compared the results of the analysis of the linearized equations of motion with experiments. Numerical simulations and experimental investigations concerning lifting and lowering operations were made by Kreuzer and Mohr, 1997. The influence of nonlinearities arising from the mooring system and viscosity of the fluid are included in a model developed by Jiang and Schellin, 1990. The same model was used in a similar study by Kral, Kreuzer and Wilmers, 1996. They also showed that different phenomena, from period doubling to chaos, can be found in the dynamics of the crane vessels.

In this paper we consider the behavior of a mechanical model of a crane barge, based on the work of Jiang, 1991. We then show the results of simulations and bifurcation analyses, which are based on a software package developed by Baumgarten, 1998. The results are compared with experiments done at the Technical University Hamburg – Harburg and the Berlin University of Technology.

N. Van Dao and E.J. Kreuzer (eds.),
IUTAM Symposium on Recent Developments in Non-linear Oscillations of Mechanical Systems, 105–113.

2. Mathematical Model

To build our mathematical model, we have assumed that the crane vessel and its suspended load are confined to a plane.

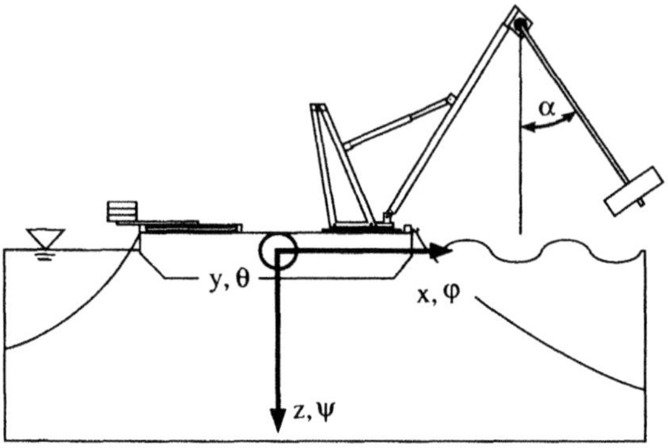

Figure 1. Crane vessel

The hull is modeled as a rigid body and the load is idealized as a mass point. This leads to the equations of motion of a multibody system:

$$\mathbf{M}(\mathbf{y})\ddot{\mathbf{y}} + \mathbf{k}(\mathbf{y}, \dot{\mathbf{y}}) = \mathbf{q}(\mathbf{y}, \dot{\mathbf{y}}) \tag{1}$$

with the mass matrix $\mathbf{M}$, the generalized gyroscopic forces $\mathbf{k}$, the generalized forces $\mathbf{q}$ and the generalized coordinates $\mathbf{y} = (x,\ \theta,\ z,\ \alpha)^T$, see Figure 1.

The generalized forces $\mathbf{q}$ include

- the hydrostatic forces

$$\mathbf{F_b}(\mathbf{y}) = (0, \quad -(m_p + m_l)gh_m\theta, \quad -\rho g A_w z, \quad 0)^T \tag{2}$$

  with the density of water ρ, gravitational constant g, cross section at the water surface A_w, heave motion z, the pitch angle θ, mass of the vessel and the load m_p, m_l and the metacentric height h_m,
- the forces resulting from the mooring system, approximated by a polynomial, see also Schellin and Mohr, 1998:

$$\mathbf{F_m}(\mathbf{y}) = (-c_1 x - c_2|x|x - c_3 x^3, \quad 0, \quad 0, \quad 0)^T, \tag{3}$$

  where c_i are the coefficients characteristic for the mooring system and x is the displacement in the surge direction,
- the forces due to viscous drag

$$\mathbf{F_d}(\dot{\mathbf{y}}) = (-\frac{1}{2}\rho c_D BT|\dot{x}|\dot{x}, \quad 0, \quad 0, \quad 0)^T, \tag{4}$$

  with the drag coefficient c_D, the width of the hull B and draught T,

- the exciting wave force due to regular waves of frequency ω, which is divided into a periodic part and a constant part representing the drag force

$$\mathbf{F_w}(t) = \begin{pmatrix} a p_{dyn\,x} \cos(\omega t) + a^2 p_{drag} \\ a p_{dyn\,\theta} \cos(\omega t) \\ a p_{dyn\,z} \cos(\omega t) \\ 0 \end{pmatrix} \tag{5}$$

  with the wave height a and the frequency dependent coefficients $p_{dyn\,i}$ and p_{drag},
- the linear hydrodynamic response forces due to the vessel's motion with the degrees of freedom of the crane barge collected in $\bar{\mathbf{y}} = (x,\ \theta,\ z)^T$

$$\mathbf{F_h}(\dot{\bar{\mathbf{y}}}) = -\mathbf{a}_\infty \ddot{\bar{\mathbf{y}}} + \mathbf{s_0}(\dot{\bar{\mathbf{y}}}), \tag{6}$$

  where in case of planar motion the frequency dependent part is described by the 3×1 vector $\mathbf{s_0}$, which is computed from the finite state space model

$$\dot{\mathbf{s}}_{\mathbf{3-k}} = \mathbf{s_{4-k}} - \mathbf{A_k s_0} - \mathbf{B_k} \dot{\bar{\mathbf{y}}} \quad k = 0, 1, 2, 3 \quad \mathbf{s_4}(t) = 0 \tag{7}$$

  with the 3×3 hydrodynamic coefficient matrices $\mathbf{A_i}$, $\mathbf{B_i}$ (Jiang, 1991).

This leads to a set of 20 differential equations summarized as follows:

$$\frac{d}{dt}\begin{pmatrix} \mathbf{y} \\ \dot{\mathbf{y}} \\ \mathbf{s_0} \\ \mathbf{s_1} \\ \mathbf{s_2} \\ \mathbf{s_3} \end{pmatrix} = \begin{pmatrix} \dot{\mathbf{y}} \\ (\mathbf{M} + \mathbf{a}_\infty)^{-1}(\mathbf{q_e} - \mathbf{k} + \mathbf{s_0} + \mathbf{f_w}) \\ \mathbf{s_1} - \mathbf{A_3 s_0} - \mathbf{B_3}\dot{\bar{\mathbf{y}}} \\ \mathbf{s_2} - \mathbf{A_2 s_0} - \mathbf{B_2}\dot{\bar{\mathbf{y}}} \\ \mathbf{s_3} - \mathbf{A_1 s_0} - \mathbf{B_1}\dot{\bar{\mathbf{y}}} \\ -\mathbf{A_0 s_0} - \mathbf{B_0}\dot{\bar{\mathbf{y}}} \end{pmatrix} . \tag{8}$$

3. Numerical Analysis

The numerical analysis is subdivided into two parts: (1) simulation of the system dynamics by integration of the equations of motion; and (2) bifurcation analysis.

3.1. SIMULATION

The simulation of a dynamical system allows for the investigation of the motion of a vessel, given the equations of motion and a set of initial conditions. Figure 2 gives an example for the motion of the crane barge. Starting from the equilibrium position of the unforced vessel with the length of the hoisting rope at 15m, the motion approaches a stable motion with period one (left).

In right part of Figure 2 the same system approaches a different motion due to a different set of initial conditions. Here the steady-state solution has the period two.

From these two examples it can be seen that mere simulation cannot reveal enough information to describe the system dynamics sufficiently. Therefore, bifurcation analysis has been used systematically to find different periodic solutions for the periodically forced system.

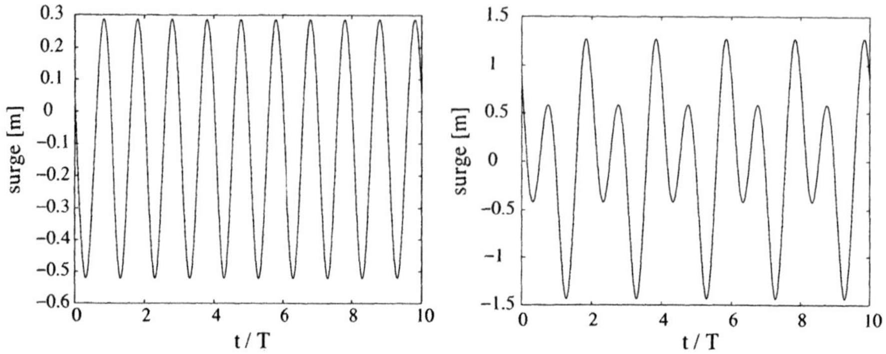

Figure 2. Time simulation of crane vessel

3.2. BIFURCATION ANALYSIS

Depending on the choice of the system's parameters, the dynamics of a crane vessel driven by regular waves could show periodic, quasi-periodic and chaotic behavior after transient motion has decayed. Here the analysis focuses on periodic solutions, which are characterized by a small number of points $\mathbf{x_i}$ in state space generated by a Poincaré map $\mathbf{P}$. For a periodic motion, these points $\mathbf{x_i}$ satisfy the algebraic equation

$$\mathbf{G}(\mathbf{x}) = \mathbf{P^l}(\mathbf{x}) - \mathbf{x} = \mathbf{0} \tag{9}$$

where l is the periodicity of the motion.

One means of characterizing the system behavior is a bifurcation diagram: A system state, that is characteristic for the solution, is plotted versus a control parameter. Other parameters of the system are kept constant. To follow a periodic solution, a control parameter λ is included in equation 9:

$$\mathbf{G}(\mathbf{x}, \lambda) = \mathbf{P^l}(\mathbf{x}, \lambda) - \mathbf{x} = \mathbf{0}. \tag{10}$$

Here the length of the hoisting rope was used as parameter, and the surge motion was used to characterize the specific solution.

With the program used for the bifurcation analysis, stable and unstable paths of periodic motion can be traced by means of path following algorithms, see Baumgarten, 1998. These algorithms give a parameter dependent solution of an underdetermined algebraic system of equations.

The solutions, which can be described as curves, are approximated by a set of points, see Figure 3. The first point of the solution is found by shooting methods. Then a predictor-corrector procedure is applied to calculate more points which correspond to periodic motion of the vessel.

Path following methods can be classified either by the type of the predictor, the corrector, or the parameterization. The parameterization determines the identification of points on the curve locally. Predictors give an estimation for a new point on the curve. Frequently used types are shown in Figure 4.

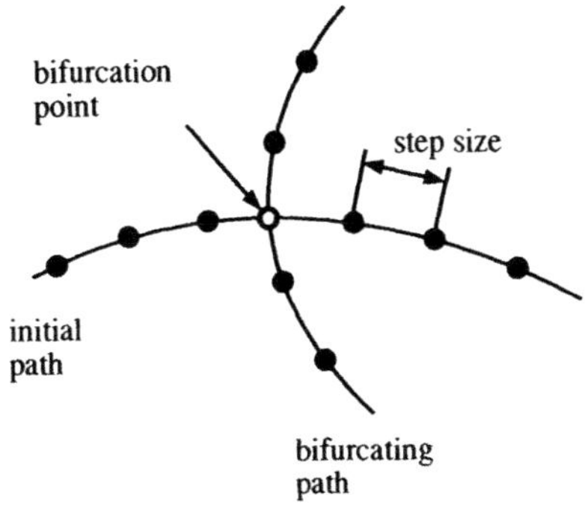

Figure 3. Path following

A corrector iteration is started from the estimated value. This requires solving the underdetermined equation

$$\mathbf{G}(\mathbf{x}, \lambda) = \mathbf{0}. \tag{11}$$

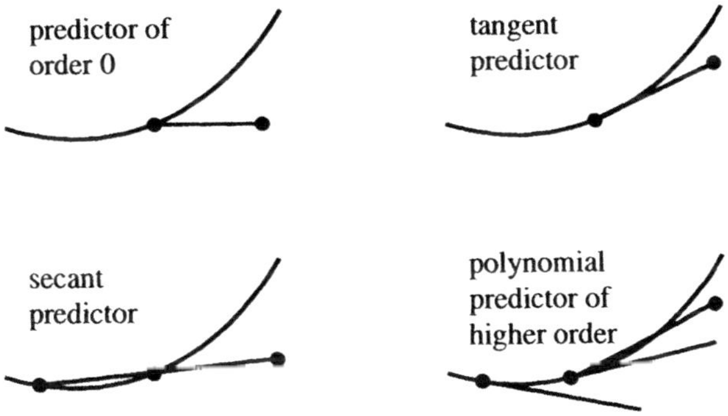

Figure 4. Predictor types

When following a periodic solution a change of stability indicates a bifurcation. This can be due to a new branch of a periodic motion or a chaotic attractor. By disturbing the state of the system, we found different types of periodic behavior near bifurcation points by a disturbance of the state of the system. Shooting methods were then applied to find a periodic solution near the initial value, and the curve continuation algorithm was restarted with the value found by the shooting methods. This method requires a high amount of manual modification of the values used in the analysis before applying the shooting techniques, but it is more likely to find a new solution near a critical point compared to other algorithms.

For the mathematical model of a crane barge, periodic motions with period one, two and three were found. The results of the bifurcation analysis are shown in Figure 5.

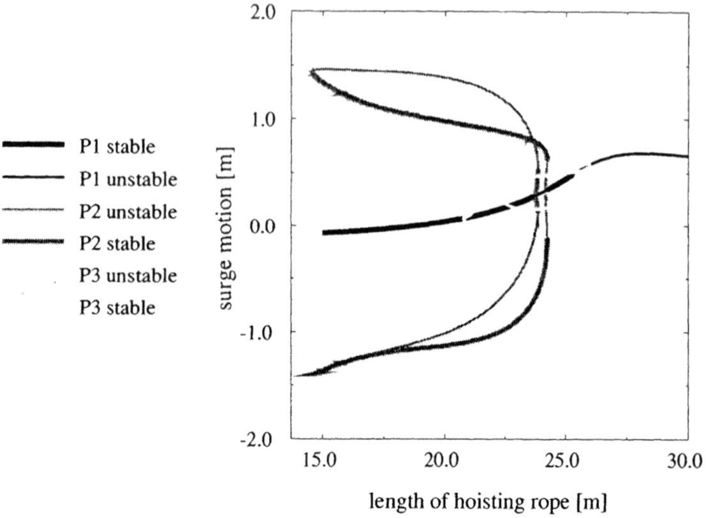

Figure 5. Bifurcation Diagram

4. Experiments

Experiments were made in the wave tank of the Technical University Hamburg–Harburg with a model of a crane barge scaled 1:100. The setup allows for variation of the main parameters also used in the mathematical model, and the motion of the vessel is measured during the experiment. Similar experiments with a model scaled 1:25 where carried out at the Berlin University of Technology.

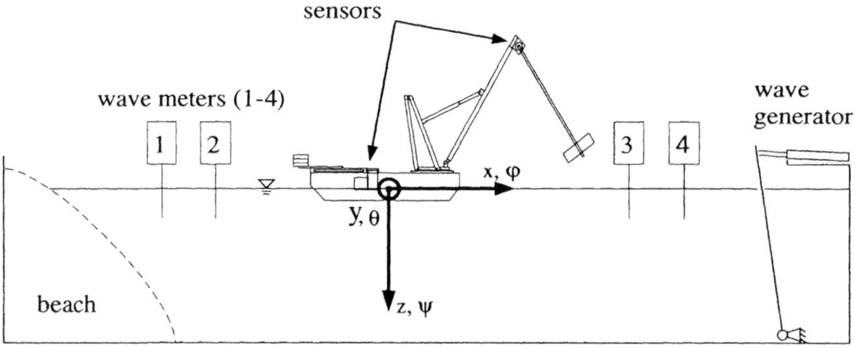

Figure 6. Experimental setup

The Sensors yield the signals for the position of the hull, its angular velocity and the angle between the hull and the hoisting rope.

4.1. EXPERIMENTAL SETUP

The model of the crane barge was positioned in the middle section of the wave tank. Sensors allowed for measurement of the position of the hull, its rotation and the position of the load relative to the hull. A wave generator, with a wave flap, was used to produce regular waves and to obtain a periodic forcing. The driving force was evaluated by measuring the height and frequency of the waves by means of wave meters, see Figure 6.

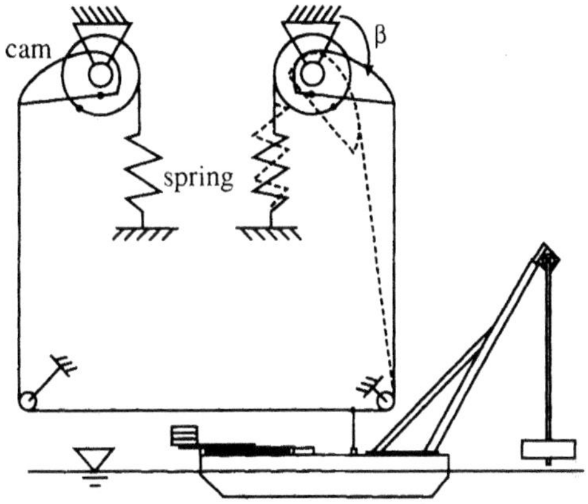

Figure 7. Mooring System

Several experiments with different mooring systems showed that the characteristics of the mooring line forces are crucial to the vessel's dynamics. It was found from calculations as well as experiments with mooring systems that an experimental setup with chains does not accurately represent real mooring systems. The spring-mechanism, which was developed in order to replace the chains, is shown in Figure 7; it consists of a spring and a combination of a cam and a roll. This yields a non-uniform ratio between the surge motion and the strain of the spring.

The measurements were made in regular waves after the transients had decayed.

4.2. NONLINEAR PHENOMENA

One concern in the experiments was finding coexisting attractors, i.e. a set of parameters with at least two different stable solutions. Model tests of the dynamical behavior of the crane barge revealed the existence of stable one- and two-periodic motions. Figure 8 shows phase portraits and Poincaré points for the motion of the model crane barge at a frequency of $0.9\,Hz$. Here the surge motion was used to visualize the type of motion of the entire system.

The period-one solution was obtained by starting the experiment from equilibrium, i.e. the vessel was not moving when the wave generator was started. The coexisting solution with period two was reached by disturbing the system significantly. In this case the initial position of the vessel was prescribed which led to high mooring forces at the beginning of the experiment.

Experimental bifurcation diagrams were obtained from a number of different measurements with a variation of one of the parameters. Here the length of the hoisting rope was

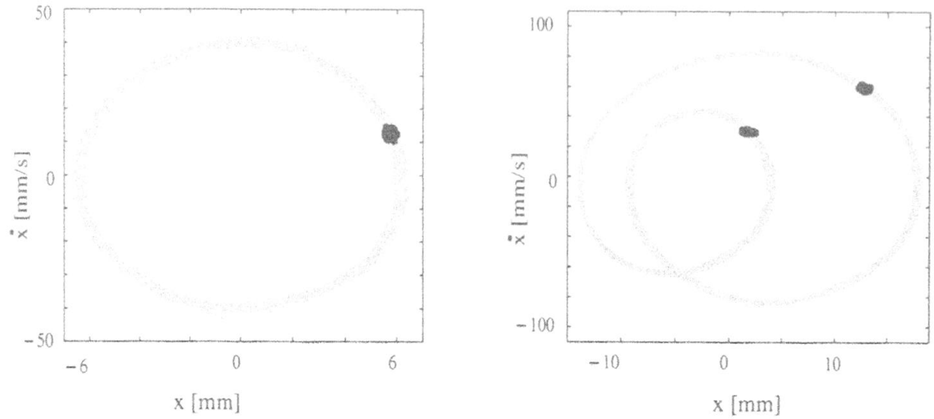

Figure 8. Phase diagrams of coexisting attractors

used as parameter for the experimental bifurcation diagrams. This led to the diagram shown in Figure 9.

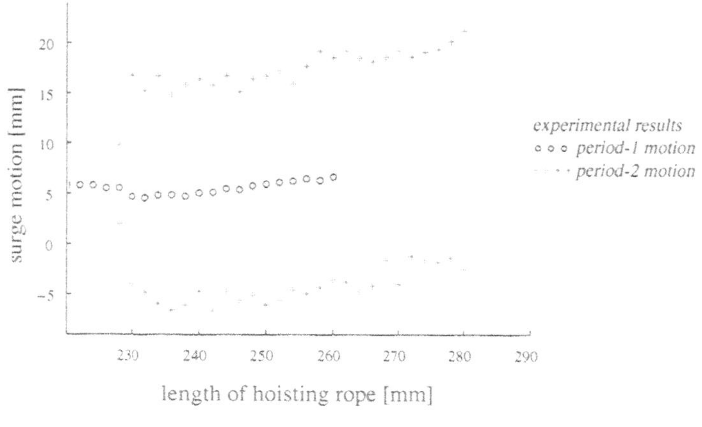

Figure 9. Experimentally determined bifurcation diagram

This experimental technique allows for determination of stable periodic motions with a sufficiently large bassin of attraction.

5. Conclusions

By using a planar model of a crane barge we have confirmed numerically and experimentally the existence of nonlinear phenomena in the dynamics of crane vessels – in the numerical investigation as well as in the experimental analysis. It was found that nonlinearities, especially in the mooring system, have significant influence on the vessel's dynamics so that linearized approximations of the equations of motion cannot describe the dynamics sufficiently. Tools from the theory of nonlinear dynamics can be applied successfully in order to trace different solutions or bifurcations.

6. Acknowledgments

This research is funded by the German Ministry of Education, Science, Research, and Technology. Furthermore, we would like to thank our partners in this research project Prof. G. Clauss and his research group at the Berlin University of Technology, and T.E. Schellin Ph. D., German Lloyds, for their assistance.

References

Baumgarten, R.: *Dynamisches Nachbeulverhalten parametrisch erregter, dünnwandiger Schalenfelder*. Dissertation, Technische Universität Hamburg - Harburg, 1998.

Clauss, G.F. and Vannahme, M.: An Experimental Study of the Nonlinear Dynamics of Floating Cranes, ISOPE-99 Brest, France, May 30 - June 4, 1999.

Ellermann, K.: *Experimentelle Verzweigungsanalyse der Dynamik eines Schwimmkranmodells*. Diplomarbeit, Technische Universität Hamburg - Harburg, Arbeitsbereich Meerestechnik II - Mechanik, 1998.

Jiang T.: *Untersuchung nichtlinearer Schiffsdynamik mit Auftreten von Instabilität und Chaos an Beispielen aus der Offshoretechnik*. Institut für Schiffbau der Universität Hamburg, Bericht Nr. 512, 1991.

Jiang T., Schellin, T. E.: *Analyse und Bewertung der Arbeitsbedingungen von schwimmenden Offshore-Kranen im Seegang*. Jahrbuch der Schiffbautechnischen Gesellschaft, Band 84, pp. 441-459, Springer-Verlag, Berlin, 1990.

Kral, R., Kreuzer, E. and Wilmers C.: Nonlinear oscillations of a crane ship. In: *Proceedings of the* 3rd *International Conference on Industrial and Applied Mathematics*, Hamburg, 1996.

Kral, R. and Kreuzer, E.: Dynamics of Crane Ships. In: *Proceedings of the International Conference on Applied Dynamics*, Science and Technics Publishing House, 1995.

Kral, R. and Kreuzer, E.: Multibody Systems and Fluid-Structure Interactions with Application to Floating Structures. In: *Multibody System Dynamics*, pp.1-19, 1999.

Kreuzer, E. and Mohr, A.: Nonlinear dynamics of a moored crane barge in regular waves. In: J. H. Vugts (ed.), *Proceedings of the* 8th *International Conference on the Behaviour of Offshore Structures (BOSS '97)*, Elsevier-Science, London, 1997.

Rieckert, T.: *Die Dynamik von Schwimmkranen mit hängender Last.*. PhD thesis, Technische Universität Berlin (D83), Verlag René F. Wilfer, Spardorf, 1992.

Schellin, T. E., Mohr, A.: *Rückstellkraftcharakteristik verankerter Schwimmkrane*. Technical Report FG 98.069, Germanischer Lloyd, 1998.

On the stability of systems
Of differential equations unsolved for derivatives

by

Lê Lương Tài

1. Introduction

Consider the system of differential equations

$$\frac{dy}{dt} = f(t, y, \frac{dy}{dt}), \tag{1}$$

or $\dot{y} = f(t, y, \dot{y})$, where $t \in R_+ = [0, +\infty)$; $y = (y_1, y_2, ..., y_n)$, $f = (f_1, f_2, ..., f_n)$ are vectors in the n - dimensional Euclidean space R_n. Let $f : R_+ \times \Omega \to R_n$; Ω is a domain defined as follows:

$$\Omega : \{(y, z) : y \in D \subset R_n ; z \in R_n : \|z\| < \infty\}.$$

Let $y(t; t_0, y_0) = \varphi(t)$ $t \in R_+$ be a solution of system (1). By putting

$$y = \varphi(t) + x, \tag{2}$$

we replace the system (1) by

$$\dot{x} = X(t, x, \dot{x}), \tag{3}$$

where $X(t, x, \dot{x}) = f(t, \varphi + x), \dot{\varphi} + \dot{x}) - f(t, \varphi, \dot{\varphi})$; $X(t, 0, 0) = 0$.

The system of equations (3) is called the system of equations of excited motion, while its non-zero solutions are called excited motions.

In this paper we suppose that the following two conditions hold:

i) $X(t, x, \dot{x})$ is a defined, continuous function satisfied all conditions of uniqueness of the solution in the domain

$$t \geq t_0 \geq 0; \ \|x\| \leq H, \ \|\dot{x}\| < +\infty, \tag{4}$$

where $\|.\|$ is the Euclidean norm;

ii) Solutions of the system (3) can be extendable; i.e. $x(t; t_0, x_0)$ is defined for all

$$t \geq t_0; \ \|x\| \leq H. \tag{5}$$

We will apply all known definitions on the stability of motion as well as the second Liapunov's method to study this problem. As in problems on the stability for systems of differential equations in standard form, in order to solve this problem we introduce functions $V(t, x)$ of real variables $(t, x) \in R_{n+1}$, defined in the domain $t \geq t_0$; $\|x\| \leq H$, and vanishing at the point $x = 0$

$$V(t, 0) = 0. \tag{6}$$

N. Van Dao and E.J. Kreuzer (eds.),
IUTAM Symposium on Recent Developments in Non-linear Oscillations of Mechanical Systems, 115–119.

Let $x = x(t; t_0, x_0)$ be a solution of the system of differential equations of excited motion (3). Then the total derivative of the function $V(t,x)$ is presented in the form

$$\dot{V}(t,x) = \left\{ \frac{d}{dt} V(\tau, x(\tau; t_0, x_0)) \right\}_{\tau = t} . \tag{7}$$

From here it follows that

$$V(t,x) = V(t_0, x_0) + \int_{t_0}^{t} \dot{V}(\tau, x(\tau; t_0, x_0)) d\tau . \tag{8}$$

In connection with the function $V(t,x)$, we will apply definitions on the sign-definiteness, the permission of indefinitely small extreme limit that are well - known from the theory of the stability of motion, for example in [1].

Together with the function $V(t,x)$ defined in the domain (5) we consider functions $V_1(t,x,z)$ of $2n+1$ variables (t,x,z) defined in the domain

$$t \geq t_0 \geq 0; \ \|x\| \leq H, \ \|z\| < +\infty . \tag{9}$$

In the following we will use definitions on the sign - definiteness of the function $V_1(t,x,z)$ in the recpect to the variable x (x - sign - definiteness) [2]

Definition [2]. The function $V_1(t,x,z)$ is called positive (negative) definite with respect to the variable x (x - positive definite) if there exists a positive definite function $W_1(x)$ dependent only on the variable x, such that the following inequality is satisfied

$$V_1(t,x,z) \geq W_1(x) \ \ (\text{or} \ \ V_1(t,x,z) \geq -W_1(x) \ ). \tag{10}$$

Notice that in [2] it is shown the function $V_1(t,x,z)$ is x - positive definite, if and only if there exists a continuous increasing function $c(r)$, $r \in [0,H]$, $c(0)=0$ such that in the domain (9) the following inequality is satisfied

$$V_1(t,x,z) \geq c(\|x\|). \tag{11}$$

2. Theorems on stability

Theorem 1. 1) *If* for the system of differential equations of excited motion (3) there exists a positive definite $V(t,x)$ such that its total derivative with respect to this system is either a negative semi - definite function or identically equal to zero, *then* the unexcited motion $x = 0$ is stable.

2) In addition, *if* the positive definite $V(t,x)$ admits an indefinitely small extreme limit i.e. there exists a continuous increasing function $b(r)$, $r \in [0,H]$, $b(0)=0$ such that, in the domain (5) the inequality

$$V(t,x) \leq b(\|x\|) \tag{12}$$

holds, *then* the unexcited motion $x = 0$ is uniformly stable.

Theorem 2. If for the system of differential equations of excited motion (3) there exists a positive definite $V(t,x)$ admitting an indefinitely small extreme limit, such that its total derivative with respect to this system is a x - negative definite function, i.e.

$$\dot{V}(t,x)=\frac{\partial V}{\partial t}+\frac{\partial V}{\partial x}X(t,x,\dot{x})=V_1(t,x,\dot{x})\leq -W_1(x),$$

where $W_1(x)$ is a positive - definite function dependent only on x, *then* the unexcited motion $x=0$ is asymptotically stable.

In the following we will use the definition of the domain $V>0$ in Tchetaev's sense and the positive definite function $V_1(t,x,z)$ in the domain $V>0$.

A set of points (t,x,z) in the domain (9) is called the domain $V>0$ if $V(t,x)>0$. A function $V_1(t,x,z)$ is called positive definite in the domain $V>0$, if for any given positive number $\varepsilon>0$ there exists a positive number $\delta(\varepsilon)$ such that for any points (t,x,z) in the domain (9), if $V(t,x)>\varepsilon$ then $V_1(t,x,z)\geq\delta$.

Theorem 3. If for the system of differential equations of excited motion (3) there exists a sign-indefinite $V(t,x)$ such that its total derivative with respect to this system is x - positive definite in the domain $V>0$ in Tchetaev's sense, *then* the unexcited motion $x=0$ is unstable.

Example. Consider the system of differential equations of the excited motion

$$\dot{x}=ay-ax\left(x^2+y^2\right)^{1/2}+x^2y^3f(t,x,y,\dot{x},\dot{y}),$$
$$\dot{y}=-ax-ay\left(x^2+y^2\right)^{1/2}+x^3y^2g(t,x,y,\dot{x},\dot{y}),$$

where f and g are bounded functions satisfying conditions for the existence and uniqueness of solutions of the system.

We take Liapunov's function in the form

$$V=\frac{1}{2}\left(x^2+y^2\right).$$

Then

$$\dot{V}--a\left(x^2+y^2\right)^{3/2}+x^3y^3(f+g),$$

where in a enough small neighbourhood of the origin $(0,0,0)$ the sign of $\dot{V}$ is completely determined by the sign of a. Thus, in this example, if $a>0$ the unexcited motion $x=0$ is asymptotically stable, and if $a<0$ the unexcited motion $x=0$ is unstable.

As is well known for problems on the stability of motion, sometimes we meet cases where we have to find conditions ensuring the asymptotic stability of the unexcited motion $x=0$ under any initial values. This stability is called asymptotic stability as a whole.

Theorem 4. If for the system of differential equations of excited motion (3), there exists a positive definite $V(t,x)$ admitting an indefinitely small extreme limit, such that its total derivative with respect to this system is a x - negative definite function, i.e.

$$\dot{V}(t,x)=\frac{\partial V}{\partial t}+\frac{\partial V}{\partial x}X(t,x,\dot{x})=V_1(t,x,\dot{x})\leq -W_1(x).$$

and *if* the function V satisfies the condition

$$\lim_{\|x\|\to\infty} V(t,x)=+\infty,$$

then the unexcited motion $x=0$ is asymptotically stable as a whole.

It is easily noticed that, if the right side of the system of differential equations (3) is not dependent on $\dot{x}$ we again get the well - known results published in handbooks on the theory of stability of motion.

Based on the preceding results, further research can be proceed in different ways, for example, by considering the problem on the stability of motion with respect to some of the variables, or applying these results to the study of complicated mechanical systems etc.

3.The problem of optimal stabilization.

Consider the system of differential equations of excited motion of a control system

$$\dot{x}=X(t,x,\dot{x};u),\quad u=(u_1,u_2,...,u_n), \tag{13}$$

where $X(t,x,z;u)$ is defined and continuous in the domain

$$t\geq t_0,\ \|x\|\leq H,\ \|z\|<\infty;\ \|u\|<\infty. \tag{14}$$

and u is a control action.

Suppose that there is a given control quality criterion in the form of the minimum of the following integral

$$I=\int_{t_0}^{+\infty}\omega(t,x[t];u[t])\,dt, \tag{15}$$

where $\omega(t,x;u)$ is a non-negative continuous function in the domain (14), $x[t]$ is the solution of the system of differential equations (13) responding to the control $u(t,x)$, and $u[t]=u(t,x[t])$.

The problem of optimal stabilization is as follows [3]:

To find the control action $u=u^0(t,x)$ assuring the asymptotic stability of the unexcited motion $x=0$ corresponding to the system (13). Furthermore, all other control actions $u=u^*(t,x)$ also guaranteeing the asymptotic stability of the unexcited motion $x=0$ must satisfy the following inequality

$$\int_{t_0}^{+\infty}\omega(t,x^0[t];u^0[t])\,dt\leq\int_{t_0}^{+\infty}\omega(t,x^*[t];u^*[t])\,dt. \tag{16}$$

The function $u=u^0(t,x)$ is called the *optimal control* of the problem (13), (15).

According to N.N. Krasovsky's method we introduce into the study the following expression

$$B[V;t,x,\dot{x};u]=\frac{\partial V}{\partial t}+\frac{\partial V}{\partial x}X(t,x,\dot{x};u)+\omega_1(t,x,\dot{x};u), \tag{17}$$

where

$$\omega_1(t,x,\dot{x};u)=\omega(t,x;u)+\dot{x}-X(t,x,\dot{x};u).$$

We have following theorem

Theorem 5. If for the system of differential equations of excited motion (13) there exist a positive definite $V^0(t,x)$ admiting an indefinitely small extreme limit and a function $u=u^0(t,x)$, satisfying following conditions:

i) The function $\omega_1(t,x,\dot{x};u^0(t,x))$ is x - positive definite with respect to x;

ii) The following equality holds:

$$B[V^0;t,x,\dot{x};u^0(t,x)]=0; \tag{18}$$

iii) For all values u, the inequality

$$B[V;t,x,\dot{x};u]\geq 0,$$

holds, *then* the function $u=u^0(t,x)$ is the solution of the problem of the optimal stabilization (13), (14), (15). Morever, for any control action $u=u^*(t,x)$ the following equality always holds:

$$\int_{t_0}^{+\infty}\omega(t,x^0[t];u^0[t])dt=\min\int_{t_0}^{+\infty}\omega(t,x^*[t];u^*[t])dt=V^0(t_0,x(t_0)).$$

REFERENCE

1. Malkin I.G. Theory of Stability of Motion. Nauka, Moscow, 1966. (in Russian)
2. Rumyantsev V.V., Oziraner A.S. Stability and Stabilization of Motion with Respect to a Part of Variables. Nauka, Moscow, 1987.
3. Krasovsky N.N., Some Problems of Stability of Motion. Fizmatgiz, Moscow, 1959.

Lê Lương Tài
Thainguyen University, Vietnam
Email: cprtnu@hn.vnn.vn

STABILITY OF CONTROLLED MOTION OF A GYMNAST IN HIGH-SPEED MID AIR MANEUVERS

P. MAISSER and U. JUNGNICKEL
Institute of Mechatronics
Chemnitz University of Technology
Reichenhainer Str. 88, 09126 Chemnitz, Germany

AMR (1989): 150H, 150K, 202C

Abstract

The twisting somersaults motion of a diver in free flight is simulated using a standard man model based on an anthropomorphic multibody system (MBS). The aim of the paper is to present a Lyapunov-stable dynamic control law of an MBS with n degrees of freedom moving along an m-dimensional submanifold ($0<m\leq n$). The prescribed motion defining that submanifold results from a kinematic analysis of video sequences. The inverse kinematics is solved by dynamic tracking. Consistent initial velocities for the free flight are evaluated by nonlinear optimization minimizing the deviation of the body fixed marker points from the measurement data. So, the total amount of the angular momentum of the MBS which has to be constant during the free flight is optimized. The conservation of the angular momentum of the MBS defines an intrinsic constraint manifold on which the diver is moving. This conservation law can also be used for nonholonomic motion planning. The approach essentially uses differential-geometric concepts and methods well-known from the Lagrangian Multibody Dynamics.

1. Introduction

Since the Olympic Games of Atlanta in 1996 the regulations in some disciplines have been changed in such a way that the degrees of difficulty are not limited as before. Coaches as well as athletes want to get support in creating new motions, they want to get information about the feasibility and stability of these new motions. This point of view is especially important in those disciplines where the athletes have to carry out a very short and high-speed mid-air maneuver; for example in diving off high boards or spring boards, figure skating and gymnastics. This paper deals with the motion of a diver from a high board or a spring board, and with its modelling, simulation and control by using multibody system dynamics.
The twisting somersault motion of a diver in free flight is simulated by using a standard

N. Van Dao and E.J. Kreuzer (eds.),
IUTAM Symposium on Recent Developments in Non-linear Oscillations of Mechanical Systems, 121–129.

man model based on an anthropomorphic multibody system (MBS). The main goal of the paper is to give a Lyapunov-stable dynamic control law of a tree-like MBS with n degrees of freedom and moving along an m-dimensional submanifold of the n-dimensional configuration space $(0 < m \leq n)$. The prescribed reference motion defining that submanifold is obtained by kinematic analysis of video sequences of real motions. The approach of defining a Lyapunov-stable dynamic control law is well-known in the nonlinear control theory of robotics. It is based on differential-geometric concepts and methods used in Lagrangian Multibody Dynamics.

2. A Standard Man Model in Multibody Dynamics

The following dynamical investigations are based on the Saziorski standard man model [1] completed and used by the Institute of Mechatronics, Chemnitz, in several research projects in biomechanics, especially in sports, rehabilitation, and accident mechanics [5], see fig. 1.
The position of the man model with respect to an inertial frame is described by 31 generalized coordinates q^a. Six of them describe the absolute position and orientation of the reference body pelvis, the others describe the relative positions of the bodies. Thus the man model has 31 degrees of freedom, 25 of which are intrinsic.

3. Definition of a Reference Motion

A special problem of motion control in sports is to define a reference motion. Which motion can be used as a reference motion? The answer could be a motion given high marks by the judges. A reference motion has to be defined by kinematic analysis of video sequences of real motions. The use of the ordinary inverse kinematics does not supply good results. The reason is that the measurement data of the time history of marker points fixed on the gymnast (in reality or a posteriori in the video record) generally leads to wrong driving torques calculated by inverse dynamics (even after smoothing data). Therefore, a special approach developed at the Institute of Mechatronics is used to get a sufficiently smoth motion and correct driving torques acting in the joints of the man model. That approach uses so-called dynamic tracking, that means the 3-dimensional man model is embedded in the set of marker points generated by video records, and between these video-generated marker points and the corresponding model-fixed marker points a visco-elastic force coupling is assumed . The numerical integration of Lagrange's equations of motion yields the desired smooth time history of the reference motion. The m-dimensional submanifold V^m, $m = n - r = 6$, of the n-dimensional configuration space V^n, $n = 31$, on which the man model has to move can be defined by rheonomic constraints

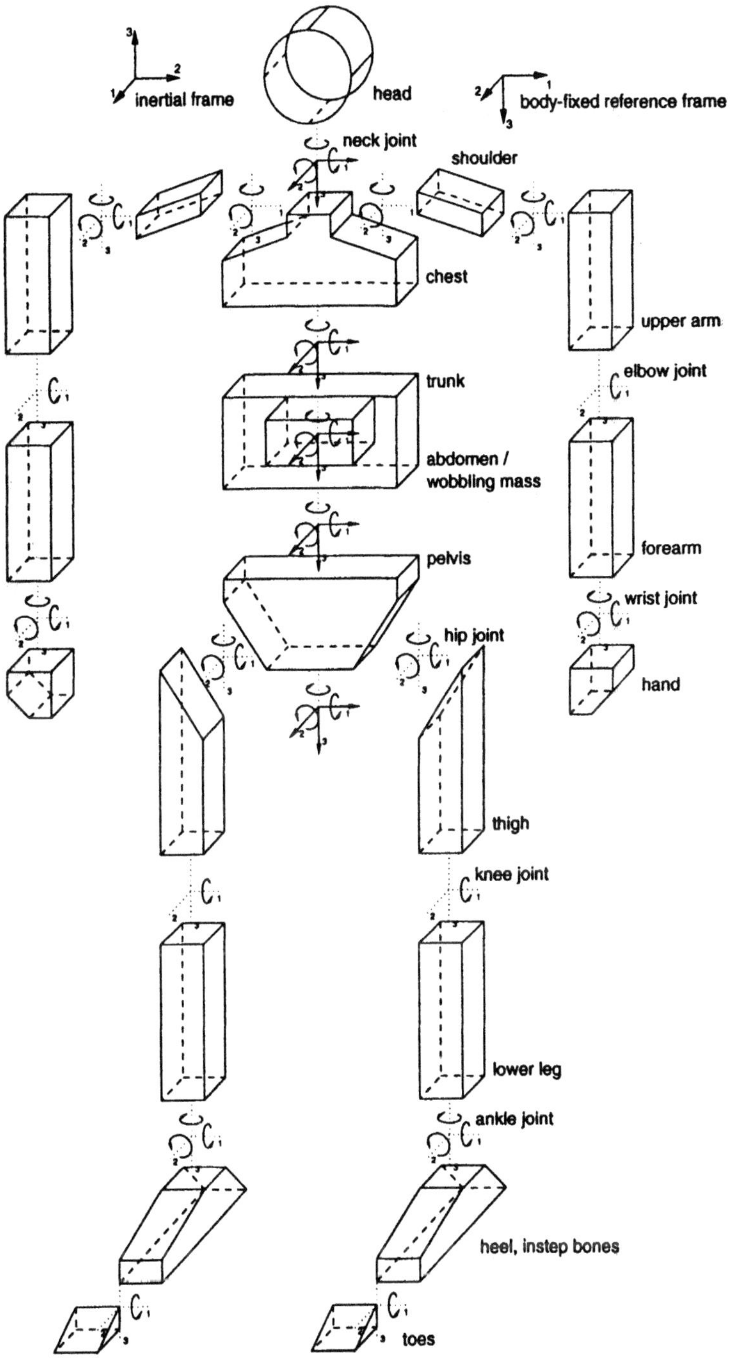

Figure 1. Kinematic scheme of the Saziorski standard man model

$$f^{a_1}(q,t) := q^{a_1} - q_0^{a_1}(t) = 0, \quad \left|\{a_1\}\right| = r := 25 \tag{3.1}$$

where q^{a_1} denote the intrinsic generalized coordinates, and $q_0^{a_1}(t)$ denotes the time history of the reference motion obtained by dynamic tracking , as described above.

4. A Lyapunov-stable Force Control Law

The main goal is to define a Lyapunov-stable motion of the man model moving along the submanifold V^m. The control strategy should be robust with respect to initial condition errors, sensor noise, and modelling errors.

In a first step, the calculated time history of the intrinsic coordinates $q_0^{a_1}(t)$ is used to simulate the free flying diver, i.e. the man model is partially kinematically controlled by prescribing the intrinsic coordinates q^{a_1} as functions of time: the system is a so-called underactuated or superarticulated mechanical system. Then, the remaining 6 external coordinates q^{a_2} characterize the free motion of the reference body pelvis. This motion is strongly influenced by the corresponding initial velocities: they define the total linear momentum as well as the total angular momentum of the man model. Therefore, for kinematic control of the intrinsic coordinates the initial velocities corresponding to the 6 external generalized coordinates q^{a_2} have to be defined in such a way that the motion of the diver prescribed in the 3-dimensional Euclidean space E^3 is approximated as well as possible. This can be done by using nonlinear optimization. The cost functional which has to be minimized is defined by the maximum of the squared distances between the marker points on the gymnast and the corresponding body-fixed marker points of the man model over a certain time interval.

A very important matter related to these investigations is the description of intrinsic nonholonomic constraints of the man model represented by classical conservation laws: the conservation of the total angular momentum with respect to the instantaneous center of mass of the man model; the total angular momentum with respect to the origin of the inertial frame as a square function of time; and the total linear momentum with respect to the inertial frame as a linear function of time. It is usual to use such nonholonomic intrinsic constraints for kinematic control design (nonholonomic motion planning). But, in reality, the human motion is controlled by forces/ torques. To get these corresponding generalized driving forces we can use the inverse dynamics. Under these calculated generalized forces, the motion of the man model will be unstable after a short time. These effects are well-known. Therefore, a dynamic feedback control law based on the Voronetz-equations (Lagrange's equations projected onto the submanifold of a constrained mechanical system) is used. The main idea - well-known from the nonlinear control theory in robotics [4] - is the nonlinear decoupling approach, i.e. an applied force acting on the MBS has to be designed which fulfils the following conditions: the force law consists of two components, the first is a so-called feedforward component which guarantees that the system is moving along its nominal submanifold of the n-dimensional configuration space V^n, and the second is a linear feedback component which provides (generalized) correction forces to reduce errors in the motion caused by different reasons. Our approach is more general than that found elsewhere because the nominal time history is prescribed only with respect to the intrinsic coordinates q^{a_1}, but not with respect to the external coordinates q^{a_2}.

We present a so-called augmented PD-control approach. The fundamental differential-geometric concepts and methods of the Lagrangian multibody dynamics and of the nonlinear control theory are assumed to be known, [2, 3, 4].

Consider an n dof anthropomorphic multibody system. The representing point of that MBS moves in the n-dimensional configuration space R^n which becomes a Riemannian space V^n by introducing a Riemannian metric g_{ab} and corresponding Christoffel symbols of the first kind Γ_{abc}. The motion equations are Lagrange's equations of the second kind in explicit form:

$$g_{ab}(q)\ddot{q}^b + \Gamma_{abc}(q)\dot{q}^b\dot{q}^c = Q_a(\dot{q},q,t). \tag{4.1}$$

Q_a denote the generalized forces. The nominal submanifold R^m, $m := n - r$, is defined by (3.1).

It can be shown that R^m is a Riemannian submanifold V^m also, but its metric depends on time, [3]. The task is to find a Lyapunov-stable position control law for tracking along V^m, $0 < m \leq n$, i.e. to define a control force R_a such that $q(t) \in V^m$ or $(q(t), \dot{q}(t)) \in T^m V^m$, the tangent bundle related to V^m. The motion equations (4.1) are divided into two parts with respect to the partitioning of the generalized coordinates $(q^a) = (q^{a_1}, q^{a_2})$; q^{a_1}, q^{a_2} denote the intrinsic and external coordinates, respectively:

$$g_{a_1b_1}(q)\ddot{q}^{b_1} + g_{a_1b_2}(q)\ddot{q}^{b_2} + \Gamma_{a_1bc}(q)\dot{q}^b\dot{q}^c = Q_{a_1} + R_{a_1}, \tag{4.2a}$$

$$g_{a_2b_1}(q)\ddot{q}^{b_1} + g_{a_2b_2}(q)\ddot{q}^{b_2} + \Gamma_{a_2bc}(q)\dot{q}^b\dot{q}^c = Q_{a_2} + R_{a_2}. \tag{4.2b}$$

We regard R_{a_1} as reaction forces due to the constraints (3.1); their structure yields a force control law which drives the system along the constraint manifold V^m. Then, with respect to the special type of constraints (3.1) the augmented PD control law is given by

$$R_{a_1} \equiv 0 \tag{4.3a}$$

$$R_{a_1} = g_{a_1 b_1}\ddot{q}_0^{b_1}(t) + g_{a_1 b_2} g^{b_2 a_2}\left[Q_{a_2} - g_{a_2 b_1}\ddot{q}_0^{b_1}(t) - \Gamma_{a_2 bc}\dot{q}^b \dot{q}^c\right] + \tag{4.3b}$$
$$+ \Gamma_{a_1 b c_1}\dot{q}^b \dot{q}_0^{c_1}(t) + \Gamma_{a_1 b c_2}\dot{q}^b \dot{q}^{c_2} - Q_{a_1} - K_{a_1 b_1}\dot{e}^{b_1} - C_{a_1 b_1} e^{b_1}.$$

$e^{b_1} := q^{b_1} - q_0^{b_1}(t)$ denotes the error, $K_{a_1 b_1}$ and $C_{a_1 b_1}$ denote symmetric and positive definite gain matrices characterizing the feedback component to reduce tracking errors. The remainder in (4.3b) describes the feedforward component of the control law to drive the system along V^m. Here, the Voronetz equations

$$\ddot{q}^{b_2} = g^{b_2 a_2}\left(Q_{a_2} - g_{a_2 b_1}\ddot{q}_o^{b_1}(t) - \Gamma_{a_2 bc}\dot{q}^b \dot{q}^c\right),$$

where $g^{b_2 a_2} g_{a_2 c_2} = \delta_{c_2}^{b_2}$ are implicitly used. The corresponding online control scheme is shown in fig. 2.

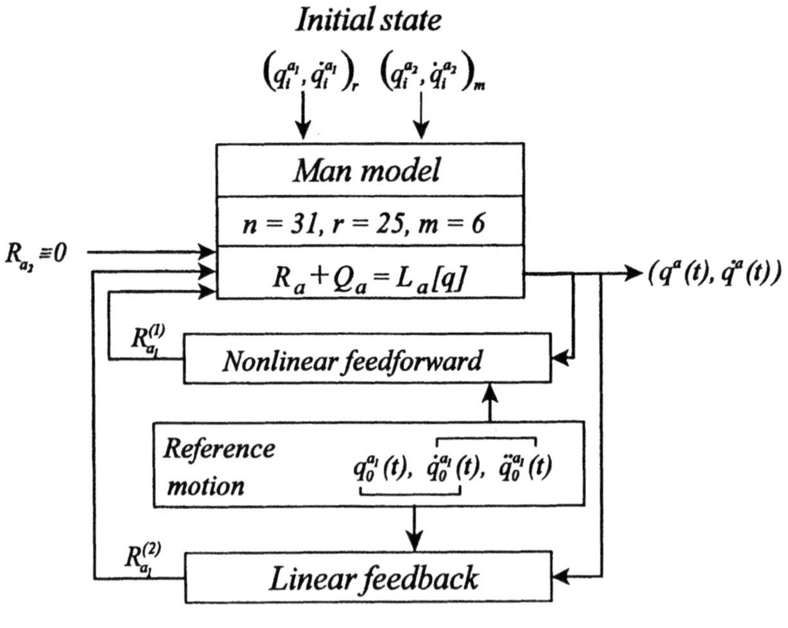

Figure 2. Closed loop control law of the diver model

Substituting R_{a_1} and (4.2b) into (4.2a) yields the error equation

$$\left[g_{a_1b_1}(q) - g_{a_1b_2}(q)\, g^{b_2a_2}(q)\, g_{a_2b_1}(q)\right]\ddot{e}^{b_1} + \\ + \left[\Gamma_{a_1bb_1}(q)\,\dot{q}^b + K_{a_1b_1}\right]\dot{e}^{b_1} + C_{a_1b_1}\, e^{b_1} = 0. \qquad (4.4a)$$

Using the matrix notations

$$g \equiv \begin{pmatrix} g_{a_1b_1} & g_{a_1b_2} \\ g_{a_2b_1} & g_{a_2b_2} \end{pmatrix} \equiv \begin{pmatrix} u & v \\ v^T & w \end{pmatrix},$$

$$\bar{g} := u - v \cdot w^{-1} \cdot v^T,\ \gamma := (\Gamma_{a_1bb_1}\dot{q}^b),$$

$$K := (K_{a_1b_1}),\quad C := (C_{a_1b_1}),\quad e := (e^{b_1}),$$

we find that (4.4a) reads

$$\bar{g}\,\ddot{e} + (\gamma + K)\,\dot{e} + Ce = 0. \qquad (4.4b)$$

Defining the function

$$V(\dot{e}, e, q) := \tfrac{1}{2}\dot{e}^T\bar{g}\dot{e} + \tfrac{1}{2}e^T Ce + \varepsilon e^T\bar{g}\dot{e} \qquad (4.5)$$

where ε denotes a small parameter we can show $V > 0$, $\dot{V} < 0$ along the trajectory given by (4.2) under the control law (4.3).

The first statement is clearly correct for sufficient small ε because C is assumed to be positive definite, and $\bar{g}$ denotes the inverse of the left upper submatrix of the block matrix

$$g^{-1} \equiv \begin{pmatrix} u & v \\ v^T & w \end{pmatrix}^{-1}$$ which is positive definite.

The second statement can be proved as follows: taking into account the skew-symmetry of $\dot{u} - 2\gamma$, i.e. $\dot{e}^T(\dot{u} - 2\gamma)\,\dot{e} \equiv 0,$ we find from (4.5)

$$\dot{V} = \dot{e}^T\bar{g}\ddot{e} + \tfrac{1}{2}\dot{e}^T\dot{\bar{g}}\,\dot{e} + \dot{e}^T Ce + \varepsilon\left[\dot{e}^T\bar{g}\dot{e} + e^T\left(\dot{\bar{g}}\,\dot{e} + \bar{g}\,\ddot{e}\right)\right] =$$

$$= -\dot{e}^T\left[(\gamma + K)\,\dot{e} + Ce\right] + \tfrac{1}{2}\dot{e}^T\dot{\bar{g}}\,\dot{e} + \dot{e}^T Ce + \varepsilon\left[\dot{e}^T\bar{g}\,\dot{e} + e^T\left(\dot{\bar{g}}\,\dot{e} + \bar{g}\,\ddot{e}\right)\right] =$$

$$= -\dot{e}^T\left[K + \tfrac{1}{2}(v \cdot w^{-1} \cdot v^T\dot{)} - \varepsilon\,\bar{g}\right]\dot{e} - \varepsilon\, e^T Ce + \varepsilon\, e^T\left(\dot{\bar{g}} - \gamma - K\right)\dot{e}.$$

That means that $\dot{V} < 0$ for sufficiently large K such that $K + \frac{1}{2}\dot{(v \cdot w^{-1} \cdot v^{T})}$ remains positive definite, and sufficiently small ε . Hence, V is a Lyapunov-function. In that case R_{a_1} is a Lyapunov-stable position control law.

The results of this procedure are shown in fig. 3. Indeed, the man model is moving along the manifold defined by the reference trajectory $q_0^{a_1}(t)$ thereby fulfiling the intrinsic constraints (conservation law of the angular momentum).

Figure 3. Free flight of a diver from a spring board, showing the conservation of the total angular momentum

5. Summary

The twisting somersault motion of a diver in free flight is simulated using a standard man model based on an anthropomorphic multibody system. The paper presents a Lyapunov-stable dynamic control law of a tree-like n dof multibody system moving along an m-dimensional submanifold $(0 < m \leq n)$ given by a reference motion. The approach uses differential-geometric concepts and methods well-known from Lagrangian Multibody Dynamics.

6. Acknowledgements

This research project is supported by the Federal Institute of Sport Science, Cologne, Germany under the Grant-No. VF 0408/06/03/98.
The authors would like to thank Prof. Dr. J. Steigenberger and Dr. H. Abeßer of the Institute of Mathematics at the Technical University of Ilmenau for their helpful discussions and advices.

7. References

1. Aurin, A.S.; Saziorski, W.M.: *Ergonomic Biomechanics* (in Russian), ISBN 5-217-00509-2, Mechanical Engineering, Moscow, 1889.

2. Enos, J.M.: *Dynamics and Control of Mechanical Systems. The Falling Cat and Related Problems*. American Mathematical Society. The Fields Institute for Research in Mathematical Sciences. Providence, Rhode Island, 1993.

3. Maisser, P.: Differential-Geometric Methods in Multibody Dynamics, *Nonlinear Analysis. Theory, Methods & Applications*, Vol. 30, No. 8, pp. 5127–5133, 1997.

4. Murray, R.M.; Li, X.; Sastry, S.S.: *A Mathematical Introduction to Robotic Manipulation*, CRC Press, Boca Raton, Ann Arbor, London, Tokyo, 1997.

5. Wolf, C.-D.; Parsche, U.; Bikowski, S.: Der Mensch als mechanisches Mehrkörpermodell in **alaska**, IfM-Report 7/98, Institute of Mechatronics at the Technical University Chemnitz, 1998.

ASYMPTOTIC METHODS FOR INVESTIGATING NONLINEAR WAVE PROCESSES

Yu.A. MITROPOLSKY
Institute of Mathematics
National Academy
of Sciences of Ukraine
252601, Kiev-4, Ukraine
3, Tereshchenkovskaya str.

As is known, the investigation of nonlinear equations describing the problem on oscillations of bounded objects is an important problem:

$$div[k(t,\vec{r})grad\, u] - q(t,\vec{r})u = \rho(t,\vec{r})\frac{\partial^2 u}{\partial t^2}. \tag{1}$$

In the case of variable coefficients k, q, ρ, depending on space coordinates $\vec{r}$ and time t such a system describes oscillation processes in an inhomogeneous medium. To that kind we can reduce problems on membrane oscillations, problems on electro-magnetic processes in non-conducting medium, problems on generation of electro-magnetic oscillations in closed hollow resonators and others.

In this paper we cover basic moments of application of asymptotic methods of nonlinear mechanics to the investigation of propagation of nonlinear wave in systems with weak nonhomogeneity of geometric and temporaly types under the action of small perturbing forces. Consider the following nonlinear equation with slowly varying coefficients for one-dimensional case:

$$\frac{\partial^2 u}{\partial t^2} - \alpha^2(\varepsilon x, \varepsilon t)\frac{\partial^2 u}{\partial x^2} + \beta^2(\varepsilon x, \varepsilon t,)u = \varepsilon f\left(\varepsilon x, \varepsilon t, \nu t, u, \frac{\partial u}{\partial t}, \frac{\partial u}{\partial x}\right), \tag{2}$$

where ε is a positive small parameter, $f\left(\varepsilon x, \varepsilon t, \nu t, u, \frac{\partial u}{\partial t}, \frac{\partial u}{\partial x}\right)$ is a function periodic in νt with a period 2π satisfying all the conditions necessary for the construction of asymptotic approximation.

Further, we denote $\varepsilon x = \varkappa$, $\varepsilon t = \tau$, $\frac{\partial u}{\partial t} = u_t$, $\frac{\partial u}{\partial x} = u_x$, $\frac{\partial^2 u}{\partial t^2} = u_{tt}$,

N. Van Dao and E.J. Kreuzer (eds.),
IUTAM Symposium on Recent Developments in Non-linear Oscillations of Mechanical Systems, 131–140.

$\dfrac{\partial^2 u}{\partial x^2} = u_{xx}$. Then for $\varepsilon = 0$ and k, τ as constant parameters the equation (2) is a well-known Klein – Gordon classical equation

$$u_{tt} - \alpha^2 u_{xx} + \beta^2 u = 0, \tag{3}$$

where $\alpha = \alpha(k, \tau)$, $\beta = \beta(k, \tau)$ are constant.

Equation (3) will be considered further as an unperturbed equation corresponding to the perturbed equation (2).

A solution of equation (3) has the following form

$$u(a, \psi) = acos\psi, \tag{4}$$

where $\psi = kx - \omega t + \varphi$, and a, φ are constant, $\psi_x = k$, $\psi_t = -\omega$,

$$\omega^2 = \alpha^2 k^2 + \beta^2 \tag{5}$$

is a dispersion relation.

For $\varepsilon \neq 0$ we seek asymptotic approximate solution of equation (2) by using the general scheme of the asymptotic method in the form of the series

$$u(\varkappa, \tau, a, \theta, \psi) = acos\psi + \varepsilon u_1(\varkappa, \tau, a, \theta, \psi) + \varepsilon^2 u_2(\varkappa, \tau, a, \theta, \psi) + \varepsilon^3 \ldots, \tag{6}$$

where $\dfrac{d\theta}{dt} = \nu$, the functions $u_i(\varkappa, \tau, a, \theta, \psi)$ to be found are periodic in θ and ψ with a period 2π, $\psi = kx - \nu t + \vartheta$, $\varepsilon \neq 0$, a and ϑ as functions of time t and the space coordinate x should be determined from the following system of equations:

$$\begin{aligned}
\frac{\partial a}{\partial t} &= \varepsilon A_1(\tau, a, \vartheta) + \varepsilon^2 A_2(\tau, a, \vartheta) + \ldots, \\
\frac{\partial a}{\partial x} &= \varepsilon B_1(\varkappa, a, \vartheta) + \varepsilon^2 B_2(\varkappa, a, \vartheta) + \ldots, \\
\frac{\partial \vartheta}{\partial t} &= -\omega + \nu + \varepsilon C_1(\tau, a, \vartheta) + \varepsilon^2 C_2(\tau, a, \vartheta) + \ldots, \\
\frac{\partial \vartheta}{\partial x} &= \varepsilon D_1(\varkappa, a, \vartheta) + \varepsilon^2 D_2(\varkappa, a, \vartheta) + \ldots,
\end{aligned} \tag{7}$$

where $A_i(\tau, a, \vartheta)$, $B_i(\varkappa, a, \vartheta)$, $C_i(\tau, a, \vartheta)$, $D_i(\varkappa,\ a,\ \vartheta)$ $(i = 1, 2, ...)$ as the functions of a and ϑ periodic in ϑ with a period 2π, which should be further determined. Evidently, for $\varepsilon = 0$ we can obtain $a = const$, $\vartheta = -\omega t + \nu t + \varphi$ $(\varphi = const)$ from the system (7) and the solution (6) which coincides with the solution (4) of the unperturbed equation (3) (Klein – Gordon equation).

Further, for determining terms of the series (6) as well as the right-hand sides of the equations (7) we use the well-known scheme of construction of asymptotic solutions. We differentiate with respect to t and x the right-hand side of the series (6), taking into account equations (7) and substituting results into the left-hand side of the original equation (2). The right-hand side of the equation (2) is expanded in powers of the parameter ε (after substituting u, u_t, u_x and expanding in Tailor series). Then equating the coefficients of the same powers of ε on the left-hand and the right-hand sides we find the following equation for determining $u_1(\varkappa, \tau, a, \theta, \psi)$:

$$\begin{aligned} u_{1\theta,\theta}\nu^2 - 2u_{1\theta\psi}\nu\omega + u_{1\psi\psi}\omega^2 - \alpha^2(\varkappa,\tau)k^2u_{1\psi\psi} + \beta^2(\varkappa,\tau)u_1 = \\ = f_0(\varkappa,\tau,a,\theta,\psi) - \left[\frac{\partial A_1}{\partial\vartheta}(-\omega+\nu) + 2a\omega C_1 + 2ak\alpha^2(\varkappa,\tau)D_1\right] \times \\ \times cos\psi - \left[\frac{\partial C_1}{\partial\vartheta}(-\omega+\nu)a + 2\omega A_1 + 2k\alpha^2(\varkappa,\tau)B_1\right] sin\psi, \end{aligned} \tag{8}$$

where

$$f_0(\varkappa,\tau,a,\theta,\psi) = f(\varkappa,\tau,acos\psi,a\omega sin\psi,-aksin\psi). \tag{9}$$

For determining the right-hand sides of the system (7) in the first approximation we obtain from the finiteness condition of functions $u_1(\varkappa,\tau,a,\theta,\psi)$ the following system of equations for determining $A_1(\tau,a,\vartheta)$, $B_1(\varkappa,a,\vartheta)$, $C_1(\tau,a,\vartheta)$, $D_1(\varkappa,a,\vartheta)$:

$$\begin{aligned} (-\omega+\nu)\frac{\partial A_1}{\partial\vartheta} + 2a[\omega C_1 + \alpha^2(\varkappa,\tau)kD_1] = f_{01}^{(1)}(\varkappa,\tau,a,\vartheta), \\ (-\omega+\nu)\frac{\partial C_1}{\partial\vartheta} + 2[\omega A_1 + \alpha^2(\varkappa,\tau)kB_1] = f_{01}^{(2)}(\varkappa,\tau,a,\vartheta), \end{aligned} \tag{10}$$

where

$$\begin{aligned} f_{10}^{(1)}(\varkappa,\tau,a,\vartheta) = \frac{1}{2\pi^2}\sum_\sigma e^{i\sigma\vartheta}\int_0^{2\pi}\int_0^{2\pi} f_0(\varkappa,\tau,a,\theta,\psi)e^{-i\sigma\psi_1}cos\psi d\theta d\psi, \\ f_{10}^{(2)}(\varkappa,\tau,a,\vartheta) = \frac{1}{2\pi^2}\sum_\sigma e^{i\sigma\vartheta}\int_0^{2\pi}\int_0^{2\pi} f_0(\varkappa,\tau,a,\theta,\psi)e^{-i\sigma\psi_1}sin\psi d\theta d\psi, \end{aligned} \tag{11}$$

$\psi_1 = \psi - \vartheta.$

Now we consider partial cases of equation (2). Assume that the right -hand sides of equation (2) do not depend on θ. In this way we simplify the determination of the right-hand sides of system (7). Some of them do not depend on ϑ. As a concrete example let us consider Klein-Gordon equation for a inhomogeneous medium

$$\frac{\partial^2 u}{\partial t^2} - \frac{\partial}{\partial x}\left\{\alpha^2(x,t)\frac{\partial u}{\partial x}\right\} + \beta^2(x,t)u = 0. \tag{12}$$

If x and t are normalized for a certain length of a wave and a period we assume that

$$\alpha = \tilde{\alpha}(\varepsilon x, \varepsilon t), \qquad \beta = \tilde{\beta}(\varepsilon x, \varepsilon t). \tag{13}$$

Then, instead of equation (12) we consider an equation with slowly varying coefficients

$$\frac{\partial^2 u}{\partial t^2} - \tilde{\alpha}^2(\varkappa,\tau)\frac{\partial^2 u}{\partial x^2} + \beta^2(\varkappa,\tau)u = 2\varepsilon\tilde{\alpha}(\varkappa,\tau)\tilde{\alpha}_{\varkappa}(\varkappa,\tau)\frac{\partial u}{\partial x}, \tag{14}$$

where $\tilde{\alpha}\varkappa = \dfrac{\partial\tilde{\alpha}}{\partial\varkappa}$.

For $\varepsilon = 0$, $\varkappa = const$, $\tau = const$, and hence $\tilde{\alpha}(\varkappa,\tau) = const$, $\tilde{\beta}(\varkappa,\tau) = const$ the equation turns into the well-known classical wave linear equation with constant coefficients, namely Klein – Gordon equation

$$\frac{\partial^2 u}{\partial t^2} - \tilde{\alpha}^2\frac{\partial^2 u}{\partial x^2} + \tilde{\beta}^2 u = 0, \tag{15}$$

which admits a solution of the form

$$u = acos(kx - \omega t + \varphi), \tag{16}$$

where a and φ are constant and a wave number k and frequency ω satisfy the dispersion relation

$$\omega^2 = \tilde{\alpha}^2 k^2 + \tilde{\beta}^2. \tag{17}$$

Denoting the phase function $\psi = kx - \omega t + \varphi$, we obtain $\psi_x = \dfrac{\partial\psi}{\partial x} = k, \psi_t = \dfrac{\partial\psi}{\partial t} = -\omega$. Moreover, by differentiating relation (17) with respect to k we obtain the so-called group velocity

$$\omega' = \frac{d\omega}{dk} = \frac{\tilde{\alpha}^2 k}{\sqrt{\tilde{\alpha}^2 k^2 + \tilde{\beta}^2}} = \frac{\tilde{\alpha}^2 k}{\omega}. \tag{18}$$

We seek an asymptotic solution of the Klein – Gordon equation for inhomogeneous medium in the first approximation in the form

$$u(a, \psi) = acos\psi, \quad \psi = kx - \omega t + \varphi, \tag{19}$$

where a and φ should be determined from the system of equations

$$2\frac{\partial a}{\partial t}\omega + 2\alpha^2 \frac{\partial a}{\partial x} k + (\omega_t + \alpha^2 k_x - 2\alpha\alpha_x k)a = 0, \tag{20}$$

$$\frac{\partial \varphi}{\partial t}\omega + \alpha^2 \frac{\partial \varphi}{\partial x} k = 0, \tag{21}$$

where (in the linear statement of the problem) the dispersion relation

$$\omega^2 - \alpha^2 k^2 - \beta^2 = 0, \tag{22}$$

and the compatibility condition

$$k_t + \omega_x = 0. \tag{23}$$

should also be satisfied.

Now we consider a dispersion relation in the case of inhomogeneous medium. According to our scheme of construction approximate solution, the changed wave number is $\bar{k} = k + \varphi_x$, and the changed frequency is $\bar{\omega} = \omega - \varphi_t$.

Thus, substituting these values into the dispersion relation (17) for the unperturbed equation (15) we obtain

$$(\omega - \varphi_t)^2 - \alpha^2 (k + \varphi_x)^2 - \beta^2 = 0, \tag{24}$$

or, taking into account equation (21), as well as the systems (7) in the first approximation we have:

$$\omega^2 - \alpha^2 k^2 - \beta^2 + \varepsilon^2 [C_1^2(\tau, a) - \alpha^2 D_1^2(\varkappa, a)] = 0, \tag{25}$$

which differs from the dispersion relation (17) by the presence of the term of the second order of smallness which introduces (25) some additional terms depending on $\tau = \varepsilon t$, $\varkappa = \varepsilon x$, and a into the relation (25).

In the first approximation, the condition of compatibility also changes by a value of the second order. Actually, we have $\bar{\psi}_x = k + \varphi_x$, $\bar{\psi}_t = -\omega + \varphi_t$ (wave number k and a frequency ω get components depending on $\varkappa$, τ and a), $\bar{\psi}_{xt} = (k + \varphi_x)_t$, $\bar{\psi}_{tx} = (-\omega + \varphi_t)_x$ and hence, the compatibility condition will be

$$(k + \varphi_x)_t + (\omega - \varphi_t)_x = 0,$$

or

$$k_t + \omega_x + \varphi_{xt} - \varphi_{tx} = 0. \tag{26}$$

Taking into account the system (7) we obtain

$$k_t + \omega_x + \frac{\partial}{\partial t}\left(\frac{\partial \varphi}{\partial x}\right) - \frac{\partial}{\partial x}\left(\frac{\partial \varphi}{\partial t}\right) =$$

$$= k_t + \omega_x + \varepsilon^2[D_{1a}(\varkappa, a)A_1(\tau, a) - C_{1a}(\tau, a)B_1(\varkappa, a)] + \varepsilon^3 \cdots = 0. \tag{27}$$

Thus, in the first approximation the asymptotic solution of the equation (14) can be determined by the following relations

$$u(a, \bar{\psi}) = acos\bar{\psi}, \quad \bar{\psi} = \bar{k}x - \bar{\omega}t, \tag{28}$$

$$2\left[\frac{\partial a}{\partial t}\omega + \alpha^2 k \frac{\partial a}{\partial x}\right] + (\omega_t + \alpha^2 k_x + 2\alpha\alpha_x k)a = 0 \tag{29}$$

$$\frac{\partial \varphi}{\partial t}\omega + \alpha^2 k \frac{\partial \varphi}{\partial x} = 0, \tag{30}$$

$$\omega^2 - \alpha^2 k^2 - \beta^2 + \varepsilon[C_1^2(\tau, a) - \alpha^2 D_1^2(\varkappa, a)] = 0, \tag{31}$$

$$k_t + \omega_x + \varepsilon^2[D_{1a}(\varkappa, a)A_1(\tau, a) - C_{1a}(\tau, a)B_1(\varkappa, a)] = 0. \tag{32}$$

The relations (28)– (32) completely coincide with the results obtained by G.G. Uizem, but our results allow us to consider nonlinear cases and nonstationary processes.

Now we consider an example characterizing the influence of nonlinear perturbation on a wave process. For simplicity, we assume that external perturbation does not depend on θ and there are no slowly varying parameters in the system.

Thus, we consider the first and the improved first approximation for the equation (for $\varepsilon = 0$, this is a well-known Bretherton equation):

$$\frac{\partial^2 u}{\partial t^2} + \frac{\partial^4 u}{\partial x^4} + \frac{\partial^2 u}{\partial x^2} + u = \varepsilon f(u, u_t, u_x), \tag{33}$$

where $f(u, u_t, u_x) = u^3$.

A solution of this equation will be written in the first approximation

$$u(t,x) = a(t,x) cos\psi(t,x), \tag{34}$$

where $\psi(t,x) = kx - \omega t + \varphi(t,x)$ and the slowly varying amplitude $a(t,x)$ and the phase $\varphi(t,x)$ should be determined from the system of the first approximation

$$\begin{aligned} \frac{\partial a}{\partial t} + \omega' \frac{\partial a}{\partial x} &= \frac{\varepsilon}{2\pi\omega} \int_0^{2\pi} f_0(a,\psi) sin\psi d\psi, \\ \frac{\partial \varphi}{\partial t} + \omega' \frac{\partial \varphi}{\partial x} &= \frac{\varepsilon}{2\pi\omega a} \int_0^{2\pi} f_0(a,\psi) cos\psi d\psi, \end{aligned} \tag{35}$$

where $\omega^2 = k^4 - k^2 + 1$ is a dispersion relation and $\omega' = \dfrac{2k^3 - k}{\omega}$ is a group velocity (both for the linear equation (33) for $\varepsilon = 0$), $f_0(a,\psi) = a^3 \cos^3 \psi$.

The improved first approximation, as usually in nonlinear mechanics, can be defined by the expression

$$u(t,x) = a(t,x) cos\psi(t,x) + \varepsilon u_1(a,\psi), \tag{36}$$

where $\varepsilon u_1(a,\psi)$ is calculated in accordance with the equations (8), and slowly varying amplitude $a(t,x)$ and a phase $\varphi(t,\varphi)$ are calculated in accordance with the equations of the first approximation.

Consider the first approximation in detail. For $u_1(a,\psi)$ not to contain secular terms, the functions $A_1(a)$, $B_1(a)$, $C_1(a)$ and $D_1(a)$ should satisfy the equations

$$A_1(a)\omega + (2k^3 - k)B_1(a) = \frac{h_1(a)}{2}, \tag{37}$$

$$C_1(a)\omega + (2k^3 - k)D_1(a) = \frac{g_1(a)}{2a}, \tag{38}$$

where $h_1(a)$ and $g_1(a)$ are determined by the formulas

$$g_1(a) = \frac{1}{\pi}\int_0^{2\pi} f_0(a,\psi)cos\psi d\psi,$$
$$h_1(a) = \frac{1}{\pi}\int_0^{2\pi} f_0(a,\psi)sin\psi d\psi. \tag{39}$$

We have

$$\frac{\partial^2\varphi}{\partial t\partial x} = \frac{\partial^2\varphi}{\partial x\partial t}, \quad \frac{\partial^2 a}{\partial t\partial x} = \frac{\partial^2 a}{\partial x\partial t},$$

and in accordance with the equations (7) we can write

$$A_1(a)\frac{dD_1(a)}{da} = B_1(a)\frac{dC_1(a)}{\partial a}, \quad A_1(a)\frac{dB_1(a)}{da} = B_1(a)\frac{dA_1(a)}{\partial a}. \tag{40}$$

The equations (37), (38) and (40) represent 4 relations for 4 unknown functions. As soon as these equations are solved and the right-hand sides of the equations (7) determining slowly varying $a(t,x)$ and $\varphi(t,x)$ are found, we can obtain $u_1(a,\psi)$.

But in the general case we cannot solve the system of the first approximation (37), (38) and (40) without taking into account certain details and restrictions of physical problem.

Having assumed that $u(t,x)$ has sine-shaped oscillations for all t, we can put $\varepsilon C_1(a) = \varphi_t(t,x) = 0$ and $\varepsilon A_1(a) = a(t,x) = 0$, then

$$D_1(a) = \frac{1}{2(2k^3-k)}g_1(a) \tag{41}$$

is a wave number displacement with respect to k ($\varepsilon D_1(a) = \varphi_x(t,x)$) and

$$B_1(a) = \frac{1}{2(2k^3-k)}h_1(a). \tag{42}$$

Otherwise, if we assume that for all x we have purely sine-shaped wave we can put $\varepsilon B_1(a) = a_x(t,\, x) = 0$, $\varepsilon D_1(a) = \varphi_t(t,\, x) = 0$ and solve equations (37) and (38) with respect to $\varepsilon A_1(a) = a_t(t,\, x)$ and $\varepsilon C_1(a) = \varphi_t(t,\, x)$ the corrections for the amplitude and frequency displacement.

Taking the Bretherton equation (the equation (33) with $\varepsilon = 0$) as an example we now consider the way how the external perturbation affects the dispersion relation established for a linear unperturbed equation. For the equation (33) (for $\varepsilon = 0$) the solution is of the form (34) and the dispersion relation

$$\omega^2 = k^4 - k^2 + 1. \tag{43}$$

is true. Assume, that $A_1(a)$ and $C_1(a)$ are equal zero.Then in the first approximation for $\varepsilon f_0(a\cos\psi) = \varepsilon u^3 = \varepsilon a^3 \cos^3\psi$ we obtain

$$\frac{\partial a}{\partial x} = 0, \qquad \frac{\partial \varphi}{\partial x} = \frac{3\varepsilon a^2}{8(2k^2 - k)}. \tag{44}$$

Let us calculate how the correction of the displacement of the wave number k affects the dispersion relation (43). To do that we must substitute $k + \varphi_x$ for the wave number k in the dispersion relation (43). We obtain

$$\begin{aligned} \omega^2 &= \left(k + \frac{3\varepsilon a^2}{8(2k^3 - k)}\right)^4 - \left(k + \frac{3\varepsilon a^2}{8(2k^3 - k)}\right)^2 + 1 = \\ &= k^4 - k^2 + \frac{3\varepsilon a^2}{4} + 1 + \varepsilon^3 \ldots . \end{aligned} \tag{45}$$

Thus, in the first approximation (to within values of order ε) the dispersion relation for the perturbed model Bretherton equation ($\varepsilon f(u) = \varepsilon u^3$) has the following form (under the assumption, that $A_1(a) = C_1(a) = 0$)

$$\omega^2 = k^4 - k^2 + 1 + \frac{3\varepsilon a^2}{4}. \tag{46}$$

In the first improved approximation, the perturbed Bretherton equation (under the assumptions mentioned above) has the form

$$u(t,x) = acos\psi + \frac{\varepsilon a^3}{32(9k^4 - 1)} cos3\psi, \tag{47}$$

where

$$\psi = \left(k + \frac{3\varepsilon a^2}{8(2k^3 - k)}\right) x - \omega t + \varphi. \tag{48}$$

After simple calculations, under the assumptions imposed on $A_1(a)$, $B_1(a)$ and $C_1(a)$, we obtain the following correction for the wave number k in the second approximation:

$$\frac{\partial \varphi}{\partial x} = \frac{3\varepsilon a^2}{8(2k^3 - k)} + \varepsilon^2 \frac{3a^4}{128(2k^3 - k)} \left[\frac{1}{2(9k^4 - 1)} + \frac{3(6k^2 - 1)}{(2k^3 - k)^2}\right]. \tag{49}$$

In conclusion, I would like to note that the method presented enables us to apply the asymptotic methods of nonlinear mechanics to the problem of studying Korteweg-de Vries equations with the further formulation of some new results of a complete investigation of wave processes. Stokes, in his studies that initiated the development of dispersive waves theory, obtained the following fundamental result : first, in nonlinear systems wave packets can exist ; and second, a dispersion relation contains the amplitude. The dependence of this relationship on the amplitude leads to important qualitative changes in the solution behaviour and introduces not only new phenomena but numerical corrections as well.

The purpose of my talk has been to focus scientists studying wave processes in various fields of natural sciences on the possibility of successful application of the asymptotic methods to the investigation of wave equations under small perturbations. In my lecture these methods were applied to the nonlinear Klein - Gordon equation was already considered by David Montgomery in connection with a solution of one wave problem in plasma by means of the asymptotic methods of nonlinear mechanics but our consideration is more general (perturbation depends both on τ and $\varkappa$). More precise solutions and corrections in dispersion relations were obtained. I also restricted myself to the formal approximate solution of already obtained equations, not going into the physical meaning of these equations.

References

1. *Bretherton F. P. , Garret C. J. R.* Wavetrain in inhomogeneous moving media. *Proc. Roy. Soc.* (London) - 1967. - A 302. - P.529, - 554.

2. *Bogoliubov N. N., Mitropolsky Yu. A.* Asymptotic methods in the theory of nonlinear oscillation. *M, Nauka,* 1974. - 503p.

3. *Mitropolsky Yu. A.* On construction of an asymptotic solution of perturbed Klein Gordon equation. *Ukr. math. journ.* - 1955. - 47, N^o3. - P. 1209 - 1216.

4. *Mitropolsky Yu. A.* On construction of an asymptotic solution of perturbed Bretherton equation. *Ukr. math. journ.* - 1998. -50. N° 1. - P. 58 - 73.

5. *Mitropolsky Yu. A., Limarchenko O. S.* On the problem of asymptotic approximation for slow wave processes in nonlinear dispersive media. *Ukr. math, journ,* - 1998, - 50, N° 3, - P.357 -371.

6. *Montgomery D., Tidman D. A.* Secular and nonsecular behavior for the cold plasma equation. *Phys. Fluids.* - 1964. - 7. - P. 242 - 249.

7. *Uizev J. B.* Linear and nonlinear waves. - M : Mir, 1977. - 320p.

SOME PROBLEMS ON NONLINEAR OSCILLATIONS

NGUYEN VAN DAO
Vietnam National University, Hanoi
19 Le Thanh Tong, Hanoi, Vietnam

1. Introduction

This paper presents our research on the interaction between nonlinear oscillations, Van-der-Pol's systems subjected to complicated excitations and quasilinear oscillations in systems with large static deflections [1-4]. It is well known that there is always an interaction of some kind between nonlinear oscillations, namely, between the forced, parametric and self-excited oscillators. Each of these oscillators demonstrates definite sustained oscillations, comprising one or a combination of several modes. The principal questions to be answered are: What will happen if these oscillators are coupled in some manner? Does a resultant nonlinear oscillation exist and is it stable? The stationary oscillations and their stability have been paid special attention.

2. Interaction between External and Parametric Excitations

In this section, we examine some quasilinear oscillating systems subjected to external and parametric excitations. We restrict ourselves to a class of quasilinear systems with two excitations. The following systems have been considered:

The system with external excitations in principal resonance and parametric excitation of the first degree in subharmonic resonance of the order one-half:

$$\ddot{x} + \omega^2 x = \varepsilon\{\Delta x - h\dot{x} - \gamma x^3 + 2px\cos 2\omega t + e\cos(\omega t + \sigma)\}. \tag{1}$$

The system with interaction between an external excitation and a parametric excitation of the second degree, both in the principal resonance :

$$\ddot{x} + \omega^2 x = \varepsilon\{\Delta x - h\dot{x} - \gamma x^3 + 2px^2\cos \omega t + e\cos(\omega t + \gamma)\}. \tag{2}$$

The system with interaction between an external excitation in principal resonance and a parametric excitation in subharmonic resonance:

$$\ddot{x} + \omega^2 x = \varepsilon\{\Delta x - h\dot{x} - \gamma x^3 + 2px^3\cos 2\omega t + e\cos(\omega t + \sigma)\}. \tag{3}$$

N. Van Dao and E.J. Kreuzer (eds.),
IUTAM Symposium on Recent Developments in Non-linear Oscillations of Mechanical Systems, 141–151.

The interaction between two parametric excitations of the first and third degree:

$$\ddot{x}+\omega^2 x=\varepsilon\{\Delta x-h\dot{x}-\gamma x^3+2px\cos 2\omega t+2qx^3\cos 2(\omega t+\sigma)\}. \tag{4}$$

The interaction between two parametric excitations of the first and second degree:

$$\ddot{x}+\omega^2 x=\varepsilon\{\Delta x-h\dot{x}-\gamma x^3+2px\cos 2\omega t+2qx^2\cos 2\omega t\}. \tag{5}$$

We now examine in more detail the equation (3) for $\sigma=0$. Its solution is found in the form [1] :

$$x=a\cos\psi,\qquad \dot{x}=-a\omega\sin\psi,\qquad \psi=\omega t+\theta. \tag{6}$$

The averaged equations are

$$\begin{cases} \dot{a}=\dfrac{-\varepsilon}{2\omega}f_0=\dfrac{-\varepsilon}{2\omega}\Big\{h\omega a+\dfrac{1}{2}pa^3\sin 2\theta+e\sin\theta\Big\},\\ a\dot{\theta}=\dfrac{-\varepsilon}{2\omega}g_0=\dfrac{-\varepsilon}{2\omega}\Big\{\Big(\Delta-\dfrac{3\gamma}{4}a^2\Big)a+pa^3\cos 2\theta+e\cos\theta\Big\}. \end{cases} \tag{7}$$

The amplitude and phase of stationary oscillations satisfy the equations:

$$f=ef_0-pa^3f_1=0\quad;\quad g=eg_0+2pa^3g_1=0; \tag{8}$$

where

$$\begin{cases} f_0=h\omega a+\dfrac{1}{2}pa^3\sin 2\theta+e\sin\theta=0,\\ g_0=\Big(\Delta-\dfrac{3\gamma}{4}a^2\Big)a+pa^3\cos 2\theta+e\cos\theta=0,\\ f_1=f_0\cos\theta-g_0\sin\theta,\quad g_1=f_0\sin\theta+g_0\cos\theta. \end{cases} \tag{9}$$

We can write

$$\begin{cases} f=A\sin\theta+B\cos\theta-E=0,\\ g=G\sin\theta+H\cos\theta-K=0, \end{cases} \tag{10}$$

where

$$\begin{aligned} &A=e^2-pa^4[pa^2-(\Delta-\frac{3\gamma}{4}a^2)]=T+pa^4X,\\ &X=\Delta-\frac{3\gamma}{4}a^2+3pa^2,\quad T=e^2-4p^2a^6,\\ &H=e^2+2pa^4[pa^2+(\Delta-\frac{3\gamma}{4}a^2)]=T+2pa^4X,\\ &B=-2ph\omega a^4,\quad E=-eh\omega a,\quad G=4ph\omega a^4,\quad K=-eaX. \end{aligned} \tag{11}$$

The transformation of the original equations (f_0,g_0) into the associated ones (f,g) has the matrix:

$$\{\tau\}=\begin{Bmatrix} e-2pa^3\cos\theta & pa^3\sin\theta\\ 4pa^3\sin\theta & e+2pa^3\cos\theta \end{Bmatrix}.$$

This matrix has an important characteristic; its determinant denoted by T depends only on a (and also Δ, in general)

$$T = |\{\tau\}| = e^2 - 4p^2a^6.$$

Hence, in the plane $R(\Delta, a^2)$ we can identify two regions :

- The equivalence region satisfying $T \neq 0$ *or* $a^6 \neq \frac{e^2}{4p^2}$.
- The non-equivalence region determined by $T = 0$ *or* $a^6 = \frac{e^2}{4p^2} = a_\gamma^6$.

In the equivalence region, the original equations (f_0, g_0) and the associated equations (f, g) are equivalent. Therefore, the corresponding parts of the original resonance curve C_0 and the associated resonance curve C coincide.

The non-equivalence region is only a line. For the system under consideration, it is a straight line which is parallel to the abscissa axis Δ with the ordinate a_γ^2. The non-equivalence line is a branch of the associated resonance curve C. It is not a branch of the original resonance curve. Almost all of this curve contains strange elements which belong to C, but do not belong to C_0. The method for determining the original resonance curve C_0 is to determine the associated resonance curve C then exclude the strange elements. Resonance curves for the systems with and without friction are presented in Figure 1 (for $h = 0$) and Figures 2, 3 (for $h = 0.003$) and ($h = 0.27$). The amplitude curves in Figures 1,2 are similar to those of the interaction between linearly parametric and forced oscillations [1].

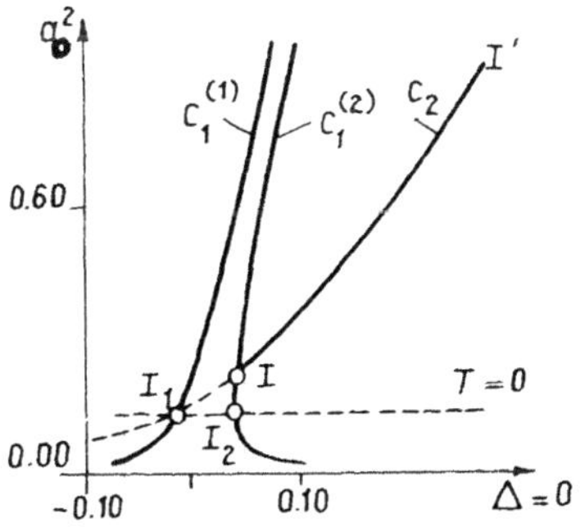

Figure 1. Resonance curves for the system without friction.

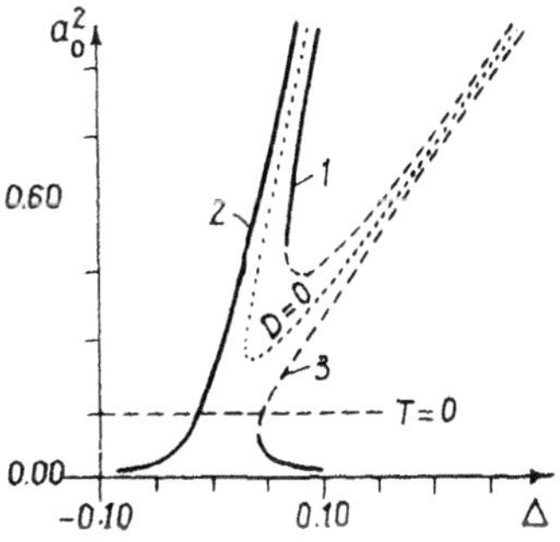

Figure 2. Resonance curves for the system with friction: h = 0.01.

The amplitude curves in Figure 3 characterize the nonlinear system under consideration. For small values of amplitude a_0, the forced component dominated the other components and the corresponding parts of resonance curves are similar to those of forced oscillation. For large values of a_0 the influence of the parametric component is clear and as a result of the interaction between two oscillations, the resonance curve has the form of a upward parabola.

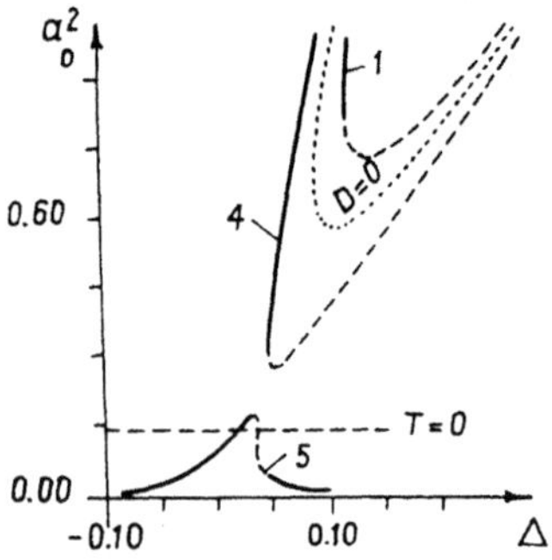

Figure 3. Resonance curves for the system with friction: $h = 0.027$.

Typical resonance curves for oscillating systems are given in Figure 4 (for equation (1)), Figure 5 (for equation (2)) Figure 6 (for equation (4)) and Figure 7 (for equation (5)).

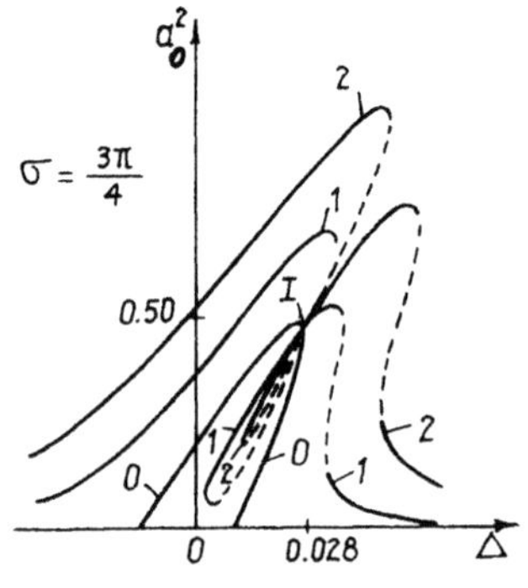

Figure 4. Resonance curves for equation (1).

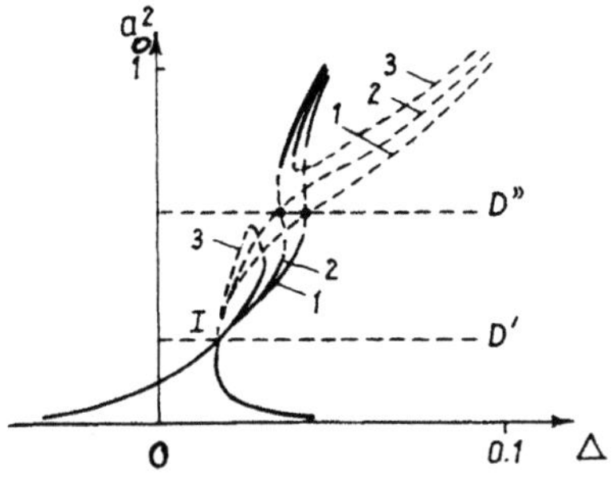

Figure 5. Resonance curves for equation(2).

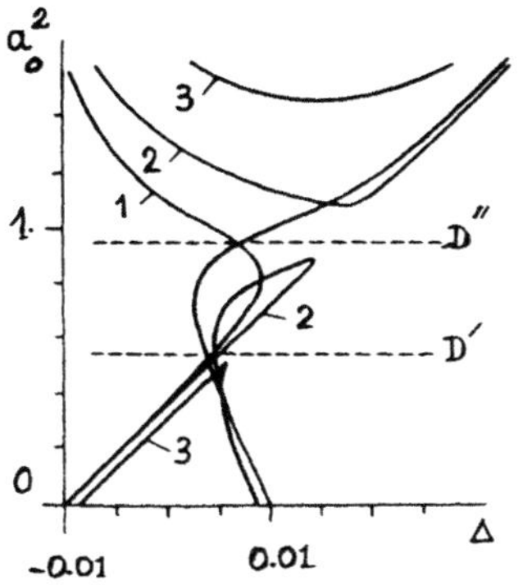

Figure 6. Resonance curves for equation(4).

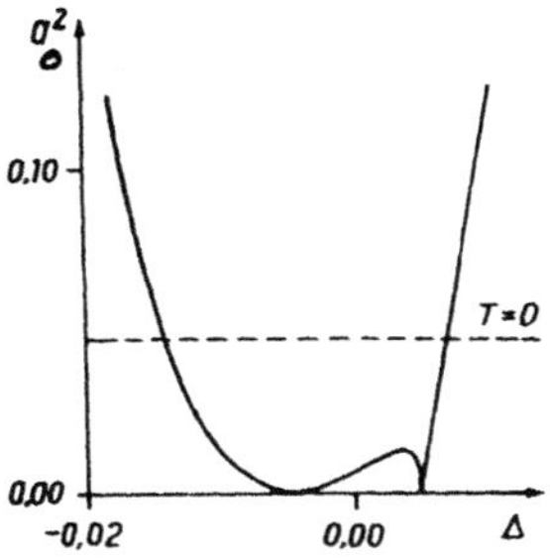

Figure 7. Resonance curves for equation(5).

3. Van-der-Pol's Systems Subjected to Complicated Excitations

Different kinds of resonance curves of stationary processes and the intermediate forms of the resonance curves have been examined with the aid of a computer. The systems under consideration are:

Van-der-Pol's system under the parametric excitation of the first degree and forced external excitation described by d.e.:

$$\ddot{x} + \omega^2 x = \varepsilon\{\Delta x + h(1 - kx^2)\dot{x} + 2px\cos 2\omega t + e\cos(\omega t + \sigma)\}. \tag{12}$$

Typical forms of the resonance curves are shown in Figures 8 and 9.

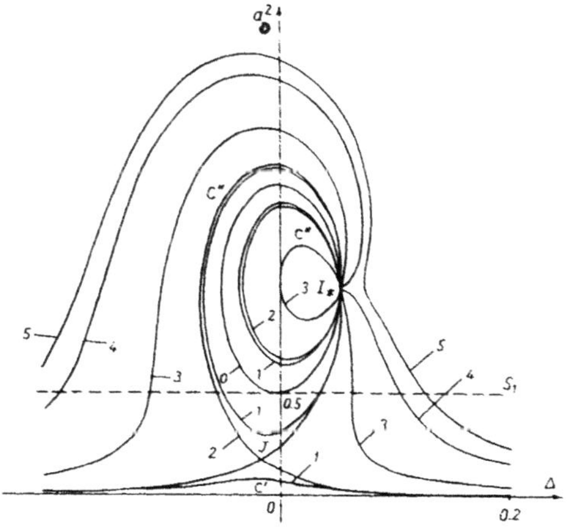

Figure 8. Resonance curves for $\sigma = 0$ and for $e = 0$ (curve 0), $e = 0.0150$ (curve 1), $e = 0.0177$ (curve 2), $e = 0.0500$ (curve 3), $e = 0.1000$ (curve 4), $e = 0.1200$ (curve 5).

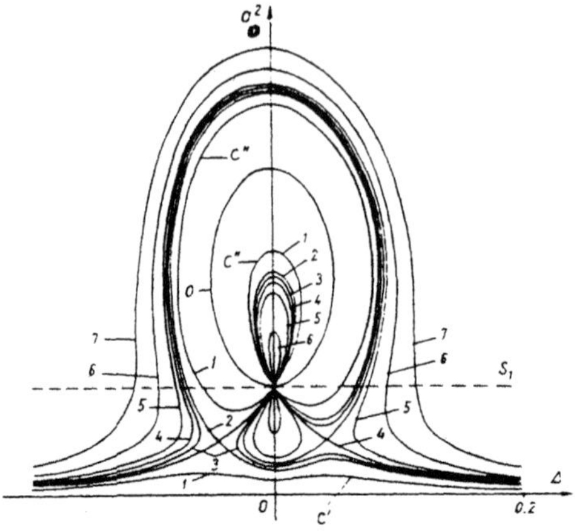

Figure 9. Resonance curves for $\sigma = \pi/4$ and for $e = 0$ (curve 0), $e = 0.0400$ (curve 1), $e = 0.0483$ (curve 2), $e = 0.0500$ (curve 3), $e = 0.0516$ (curve 4), $e = 0.0550$ (curve 5), $e = 0.0648$ (curve 6), $e = 0.0980$ (curve 7).

Van-der-Pol's system subjected to the parametric excitation of the second degree and external excitation:

$$\ddot{x} + \omega^2 x = \varepsilon\{\Delta x + h(1 - kx^2)\dot{x} + 2px^2 \cos\omega t + e\cos(\omega t + \sigma)\}. \qquad (13)$$

Typical resonance curves are given in Figures 10 and 11.

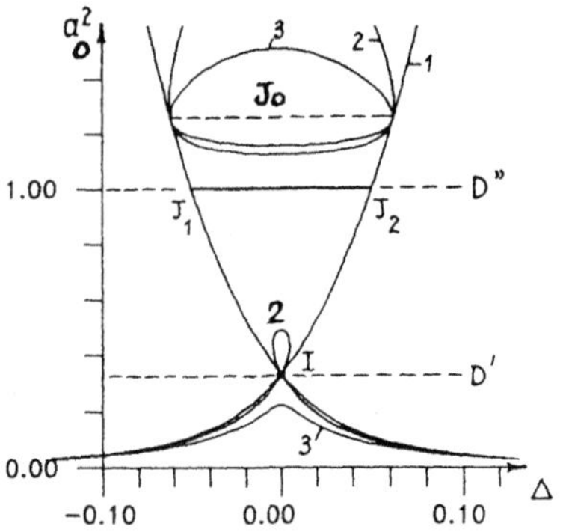

Figure 10. Resonance curves for $\sigma = \pi$ and $h = 0$ (curve 1), $h = 0.03$ (curve 2), $h = 0.05$ (curve 3).

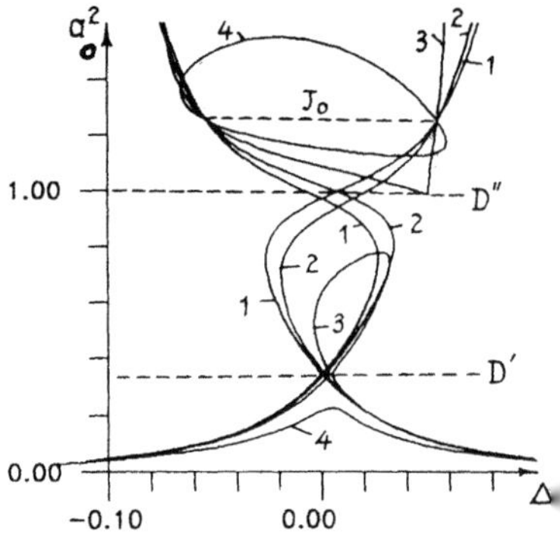

Figure 11. Resonance curves for $\pi \neq \sigma \in \left\{\frac{5\pi}{6}, \frac{7\pi}{6}\right\}$ and $h = 0$ (curve 1), $h = 0.006$ (curve 2), $h = 0.02$ (curve 3), $h = 0.05$(curve 4).

Van-der-Pol's system subjected to the parametric excitations of the first and third degrees :

$$\ddot{x} + \omega^2 x = \varepsilon\{\Delta x + h(1 - kx^2)\dot{x} + 2px\cos 2\omega t - 2qx^3 \cos 2\omega t\}. \qquad (14)$$

Some typical resonance curves are presented in Figures 12 and 13.

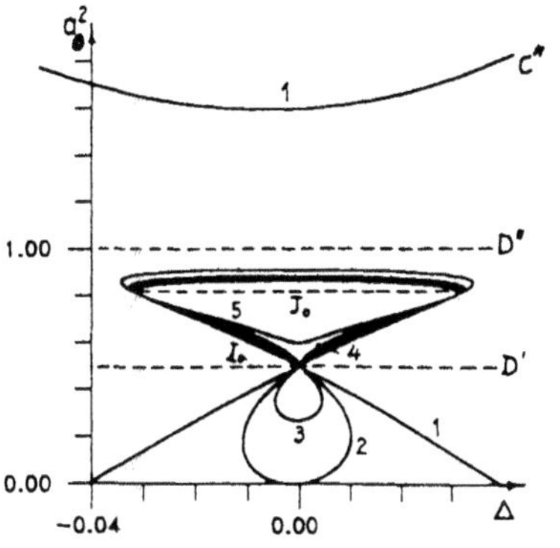

Figure 12. Resonance curves for p= 0.05 , q= 0.1 , 4< k <8 and h= 0.0300 (curve 1), h=0.0500 (curve 2), h=0.0550 (curve 3), h =0.0666 (curve 4), h = 0.0800 (curve 5).

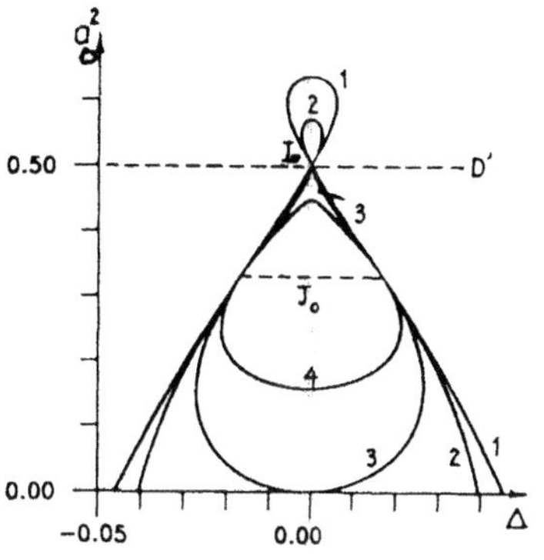

Figure 13. Resonance curves for k=8 and h=0.04 (curve 1), h= 0.05 (curve 2), h= 0.06 (curve 3).

Van-der-Pol's sytem with variable nonlinear friction described by d.e. :

$$\ddot{x} + \omega^2 x = \{\omega \Delta x + h[1 - k(x + q\cos\omega t)^2]\dot{x}\}. \quad (15)$$

Typical resonance curves are shown in Figure 14

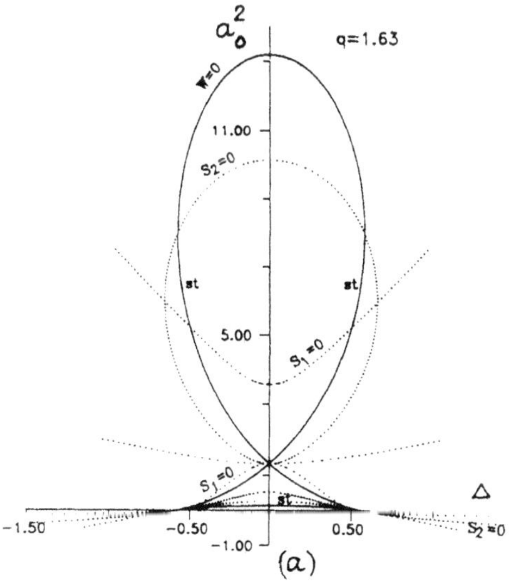

Figure 14. Typical resonance curves of equation (15), where "st" is stable branch.

4. Quasilinear Oscillations in Systems with large Static deflections

In mechanical systems the static deflection of the elastic elements is usually not apparent in the equations of motion. The reason is that either a linear model of the elastic elements or an assumption of too small static deflection was accepted. In the present section both a nonlinear model of elastic elements and their large static deflection are considered, so that the nonlinear terms in the equation of motion appear with different degrees of smallness. In this case the nonlinearity of the system depends not only on the the nonlinear characteristic of the elastic element but on its static deflection.

Let us consider the simplest oscillatory system which consists of a mass M and a spring as shown in Figure 15. The spring supporting the mass is assumed to be nonlinear with the characteristic

$$f(u) = c_0 u + \beta_0 u^3, \tag{16}$$

so that the spring force acting on the mass M is

$$c_0(\Delta - x) + \beta_0(\Delta - x)^3,$$

where Δ is the deformation of the spring in the static equilibrium position. When $x = 0$, the spring force $c_0\Delta + \beta_0\Delta^3$ is equal to the gravitational force Mg. That is:

$$c_0\Delta + \beta_0\Delta^3 = Mg.$$

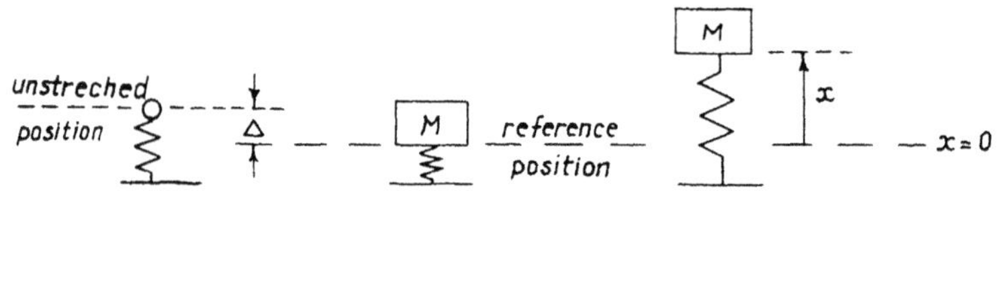

Figure 15. Oscilatory system with large static deflection.

We have the equation of the mass M in the form :

$$M\ddot{x} + c_0 x + 3\beta_0\Delta^2 x - 3\beta_0\Delta x^2 + \beta_0 x^3 = 0.$$

It is supposed that Δ is large and x small enough, so that in comparision with the linear term, $\beta_0 x^3$ is a small quantity of second degree and $\beta_0\Delta x^2$ is of the first degree of smallness :

$$\frac{x}{\Delta} = 0(\varepsilon), \quad \beta_0 x^3 = 0(\varepsilon^2), \quad \beta_0\Delta x^2 = 0(\varepsilon),$$

where ε is a small positive parameter. In this case $\beta_0\Delta x^2$ is finite. Taking into account the viscous damping force $h_0\dot{x}$ and the exciting force $P(t,x)$ which are both assumed to be small quantities of second degree and introducing the notation:

$$\omega^2 = \frac{c_0 + 3\beta_0\Delta^2}{M}, \quad \varepsilon\gamma = \frac{3\beta_0\Delta}{M}, \quad \varepsilon^2\beta = \frac{\beta_0}{M}, \tag{17}$$

$$\varepsilon^2 h = \frac{h_0}{M}, \quad \varepsilon^2 f(t,x) = \frac{1}{M}P(t,x),$$

we can write the equation of motion of the mass in the form :

$$\ddot{x} + \omega^2 = \varepsilon\gamma x^2 - \varepsilon^2(h\dot{x} + \beta x^3 - f(t,x)). \tag{18}$$

In comparision with the classical Duffing's equation, in equation (18) the small terms appear with different degrees; most of them are of second degree of smallness. From the structure of the equation (18) one can predict that the influence of the forces on the motion of the mass M can be found in the second approximation of the solution. A more general equation has also been investigated :

$$\ddot{x} + \omega^2 x = \varepsilon\gamma x^2 + \varepsilon^2 F(\tau, \varphi(\tau), x, \dot{x}). \tag{19}$$

The most interesting phenomenon in the systems under consideration is that their nonlinearity depends not only on the nonlinear characteristic of the spring as in the classical theory, but also on the static deflection Δ. Namely, if the spring has soft characteristic ($\beta < 0$) (curve 3, Figure 16), then the system under consideration also belongs to the soft type with more soft characteristic, because

$$\alpha = \frac{3}{4}\beta - \frac{5\gamma^2}{6\omega^2} < 0.$$

When the spring has hard characteristic ($\beta > 0$), the system under consideration belongs to the hard type if $\alpha > 0$ or $c_0 > 7\beta_0\Delta^2$ (curve 1, Figure 16) and to the soft type if $c_0 < 7\beta_0\Delta^2$ and to the neutral type if $c_0 = 7\beta_0\Delta^2$ (curve 2, Figure 16)

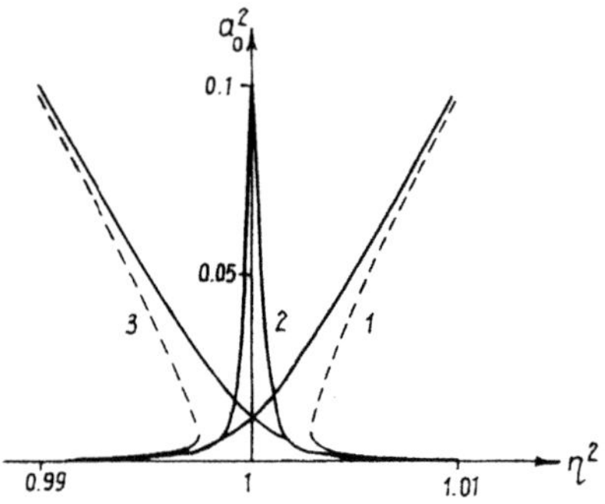

Figure 16. Stationary resonance curves of equation (18).

In addition to (19) the following problems have been considered :
The effect of ε^2-order due to the interaction between the excitation of ε-order in systems described by the equations :

$$\ddot{x} + \omega^2 x = \varepsilon\{h(1 - kx^2)\dot{x} + e\cos 3\omega t\} + \varepsilon^2(\Delta x - \gamma x^3), \tag{20}$$

$$\ddot{x} + \omega^2 x = \varepsilon[e\cos(2\omega t + \chi) + 2px\cos\omega t] + \varepsilon^2(\Delta x - 2h\dot{x} - \gamma x^3), \tag{21}$$

$$\ddot{x} + \omega^2 x = \varepsilon(e + 2px\cos\omega t) + \varepsilon^2(\Delta x - h\dot{x} - \gamma x^3). \tag{22}$$

The effect of ε^2-order caused by the interaction between the nonlinear restoring element and parametric excitation of ε-order in systems :

$$\ddot{x} + \omega^2 x = \varepsilon(\beta x^2 + 2px\cos\omega t) + \varepsilon^2(\Delta x - h\dot{x} - \gamma x^3), \tag{23}$$

$$\ddot{x} + \omega^2 x = \varepsilon(\beta x^2 + e\cos 2\omega t) + \varepsilon^2(\Delta x - h\dot{x} - \gamma x^3), \tag{24}$$

$$\ddot{x} + \omega^2 x = \varepsilon(\Delta x - \gamma x^3 + e\cos 3\omega t) + \varepsilon^2(-h\dot{x}). \tag{25}$$

The interaction of elements with two different orders in systems :

$$\ddot{x} + \omega^2 x = \varepsilon h(1 - kx^2)\dot{x} + \varepsilon^2[\Delta x - \gamma x^3 + f_i(x,t)], \tag{26}$$

$$f_1 = 2px\cos 2\omega t, \quad f_2 = e\cos\omega t;$$

$$\ddot{x} + \omega^2 x = 2\varepsilon px\cos\omega t + \varepsilon^2[\Delta x - \gamma x^3 + h(1 - kx^2)\dot{x}], \tag{27}$$

$$\ddot{x} + \omega^2 x = \varepsilon(\alpha x^2 + q\cos 2\omega t) + \varepsilon^2[\Delta x + D(1 - \delta x^2)\dot{x} - \beta x^3], \tag{28}$$

$$\ddot{x} + \omega^2 x = 2p\varepsilon x\cos\omega t + \varepsilon^2[\Delta x - h\dot{x} + 2qx\cos 2(\omega t + \sigma)], \tag{29}$$

$$\ddot{x} + \omega^2 x = \varepsilon(\alpha x^2 + q\cos 2\omega t) + \varepsilon^2[\Delta x - 2h\dot{x} - \beta x^3 + r\cos(\omega t - \eta)]. \tag{30}$$

$$\ddot{x} + \omega^2 x = \varepsilon pxcos\omega t + \varepsilon^2[\Delta x - 2h\dot{x} - \beta x^3 + rcos(\omega t - \eta)], \tag{31}$$

$$\ddot{x} + x = \varepsilon[\alpha x^2 + q\cos 2\varphi(t)] - \varepsilon^2(2h\dot{x} + \beta x^3). \tag{32}$$

5. Conclusion

In this report, the interaction between external and parametric excitations, Van-der-Pol's systems subjected to complicated excitations and quasilinear oscillations in systems with large static deflections have been studied. The asymptotic method in combination with the numerical method and a computer have been used to study the stationary oscillations and their stability. The amplitude frequency curves (resonance curves) of the systems under consideration are various, and the nonlinear characteristics are markedly changed in both quality and quantity in comparision with classical systems.

6. References

1. Mitropolskii Yu.A., Nguyen Van Dao. Applied asymptotic methods in nonlinear oscillations, *Kluwer Academic Publishers*, 1997.
2. Nguyen Van Dao, Nguyen Van Dinh, Tran Kim Chi. Interaction between nonlinear parametric and forced oscillations. *Vietnam Journal of Mechanics, NCNST of Vietnam*, T.XX, 1998, No.3.
3. Nguyen Van Dao. Quasilinear oscillations in systems with large static deflections. *Proceedings of the International Conference on Applied Dynamics, Hanoi*, 1995.
4. Nguyen Van Dao, Nguyen Van Dinh. Interaction between nonlinear oscillations. *Vietnam National University Publishing House, Hanoi*, 1999.

STATIONARY AND TRANSIENT PROCESSES IN OSCILLATING SYSTEMS WITH TWO DEGREES OF FREEDOM SUBJECTED TO PARAMETRIC AND FORCED EXCITATIONS

NGUYEN VAN DINH and TRAN KIM CHI
Institute of Mechanics, Hanoi
264 Doi Can Street, Ba Dinh, Hanoi, Vietnam

1. Introduction

In the present paper, the interaction between forced and parametric oscillations in a system of two degrees of freedom is considered. The main attention will be paid to some typical forms of the resonance curves [1], [3]. For stationary oscillation, the singular points have been used to identify the resonance curves (paragraph 3). The transient passages through the resonance are considered. These processes still have typical characteristics of nonlinear purely forced systems [2]. The presence of the indirect parametrically - excited components introduces special features of the resonance curves (paragraph 5).

2. The system under consideration

Consider a quasi-linear oscillating system with two degrees of freedom described by the differential equations

$$\ddot{x} + \lambda^2 x + \varepsilon\lambda^2(h_0\dot{x} + \alpha x^3 + cy^2 x) = Q\sin\nu t\,,$$
$$\ddot{y} + y + \varepsilon(h\dot{y} + \beta y^3 + bx^2 y) = \varepsilon p\cos(\nu t + \delta)\,, \qquad (1)$$

where $h_0 > 0$, $h > 0$, $b > 0$, $p > 0$, $\beta > 0, \alpha, c$, Q, δ are constants, and ε is a small parameter (the case $h = 0$ has been considered in [3]). Assuming that the frequency ν is

N. Van Dao and E.J. Kreuzer (eds.),
IUTAM Symposium on Recent Developments in Non-linear Oscillations of Mechanical Systems, 153–163.

far from the frequency λ of the mode x, but near that of the mode y, we introduce the detuning parameter :

$$\varepsilon\Delta = 1 - \nu^2 . \tag{2}$$

The oscillations are found in the form

$$\begin{aligned} x &= q\sin\nu t + a_1 \cos(\lambda t + \psi_1), \\ \dot{x} &= \nu q \cos\nu t - \lambda a_1 \sin(\lambda t + \psi_1), \\ y &= a\cos(\nu t + \psi), \\ \dot{y} &= -\nu a \sin(\nu t + \psi) \quad , \quad q = \frac{Q}{\lambda^2 - \nu^2} . \end{aligned} \tag{3}$$

The averaged differential equations are

$$\begin{aligned} \dot{a}_1 &= -\frac{\varepsilon\lambda}{2} h_0 a_1 , \\ \dot{a} &= -\frac{\varepsilon\nu}{2} f_0 = -\frac{\varepsilon\nu}{2}\{\nu h a + \frac{b}{4}q^2 a \sin 2\psi + p\sin(\psi - \delta)\}, \\ a\dot{\psi} &= \frac{\varepsilon\nu}{2} g_0 = \frac{\varepsilon\nu}{2}\{(\frac{3}{4}\beta a^2 + \frac{b}{2}q^2 + \Delta)a - \frac{b}{4}q^2 a\cos 2\psi - p\cos(\psi - \delta)\} . \end{aligned} \tag{4}$$

The oscillation in the system can be explained as follows: the non-resonance oscillation in x, through mixed cubic non-linearity, parametrically excites y and produces combined oscillation in y. The intensity of the indirect parametric excitation is characterized by the coefficient $\frac{b}{4}q^2$.

3. Stationary oscillations

In this paragraph, we briefly mention some typical cases of stationary oscillations, corresponding to the transient passages through resonance which will be investigated in the paragraph 5 (other cases were considered in detail in [1], [3]).

By setting the right hand side of (4) to zero, we obtain the equations for stationary oscillations :

- for the mode x : $a_1 = 0$,

- for the mode y : $f_0 = 0, \quad g_0 = 0$.

The last two equations can be transformed into

$$\begin{aligned} f &\equiv f_0 \sin\psi - g_0 \cos\psi = A\sin\psi + B\cos\psi - E = 0, \\ g &\equiv f_0 \cos\psi + g_0 \sin\psi = G\sin\psi + H\cos\psi - K = 0, \end{aligned} \tag{5}$$

where:

$$A = \nu ha,\ \ B = -(\frac{3}{4}\beta a^2 + \frac{b}{2}q^2 + \Delta - \frac{b}{4}q^2)\,a,\ E = -\,p\cos\delta\,,$$

$$G = (\frac{3}{4}\beta a^2 + \frac{b}{2}q^2 + \Delta + \frac{b}{4}q^2)\,a,\quad H = \nu ha,\ \ K = p\sin\delta\,.$$

The frequency - amplitude relationship is of the form

$$W = D_1^2 + D_2^2 - D_0^2\,, \tag{6}$$

where $$D_0 = \begin{vmatrix} A & B \\ G & H \end{vmatrix},\quad D_1 = \begin{vmatrix} E & B \\ K & H \end{vmatrix},\quad D_2 = \begin{vmatrix} A & E \\ G & K \end{vmatrix}. \tag{7}$$

From (6) the resonance curve can be identified. In general, it consists of two parts: the regular part C_1 and the critical part C_2.

The regular part C_1 lies in the regular region :

$$D_0 \neq 0\,. \tag{8}$$

The critical part C_2 lies in the critical region C_0:

$$D_0 = 0\,, \tag{9}$$

and satisfies two following conditions

- the compatibility condition (deduced from (6), (9))

$$D_1 = 0\,,\quad D_2 = 0\,, \tag{10}$$

- the solvability condition of the trigonometric system (5):

$$A^2 + B^2 \geq E^2\,,\quad G^2 + H^2 \geq K^2\,. \tag{11}$$

For $h > 0$ and for suitable δ, the critical part C_2 exists and consists of a single singular point $I_*(\nu_*^2, a_*)$ with coordinate (ν_*^2, a_*) in the plane (ν^2, a) determined from the equalities

$$\nu_* = -\frac{b}{4h}q^2\sin 2\delta\,,\quad \frac{3}{4}\beta a_*^2 = (\nu_*^2 - 1) - \frac{b}{2}q^2 - \frac{b}{4}q^2\cos 2\delta > 0\,, \tag{12}$$

and a_* must satisfy the condition

$$a_*^2 \geq a_c^2 = \frac{4p^2}{b^2q^4}\,. \tag{13}$$

It is easy to see that if $a_*^2 > a_c^2$, I_* is a nodal point; if $a_*^2 = a_c^2$, I_* is either a returning point or a degenerated nodal point; if $a_*^2 < a_c^2$, I_* is an isolated point (I_* does not belong to the resonance curve).

Depending on the change of the parameters, different forms of the resonance curves can be obtained [3].

If the indirectly - excited parametric oscillation does not exist (large h, small $\frac{b}{4}q^2$), the critical region C_0 lies under the axis ν^2 (i.e. there is no critical part C_2) and the resonance curve is similar to that of the purely forced oscillation, except, it may exist as a "crevasse".

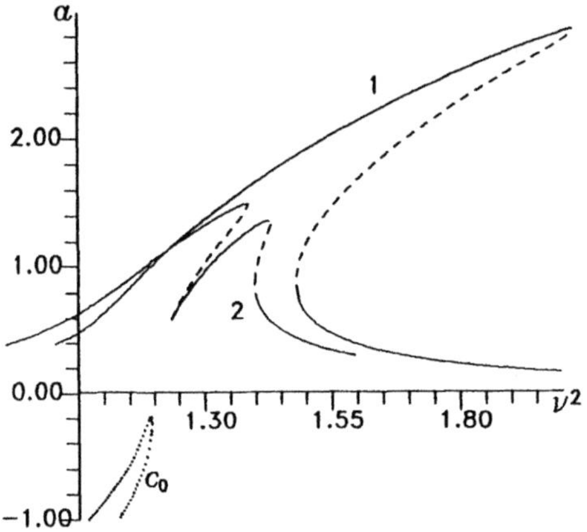

Figure 1: $h = 0.0915$, $p = 0.11759$, $\frac{b}{4}q^2 = 0.1$, $\frac{3}{4}\beta = 0.1$.

The resonance curves 1, 2 correspond to $\delta = \frac{\pi}{2}, \frac{3\pi}{4}$, respectively.

When the indirectly parametric excitation is intensive enough, then for small p, the resonance curve consists of two branches: the upper C' and the lower C", lying above and under C_0 respectively. If $a_*^2 > a_c^2$, C' and C" are joined at point I_* and the resonance curve has a loop (Figure 2a). If $a_*^2 < a_c^2$, C' and C" are separated by C_0, and C" has the form of an oval (Figures 3a, 3b).

As p increases, the upper branch moves up, while the lower branch becomes narrower. When p reaches the value $\frac{1}{2}bqa^2$, the loop disappears, and the nodal point I_* changes into a returning point (Figure 2b). At a certain value p, the oval disappears too.

As p increases further, the resonance curve becomes similar to that for no indirectly excited parametric oscillation (Figures 2c, 3c).

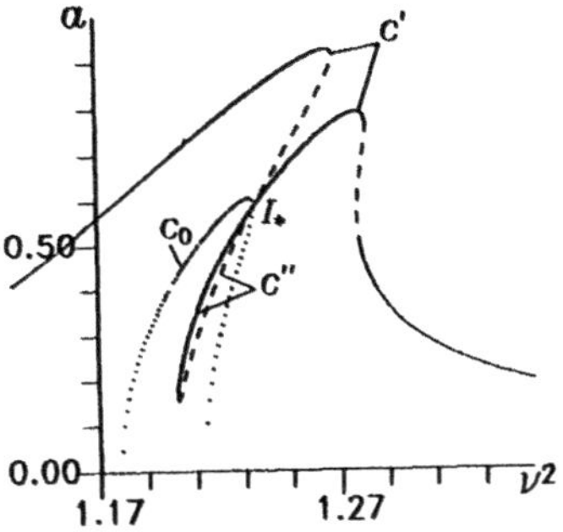

Figure 2a: p=0.03

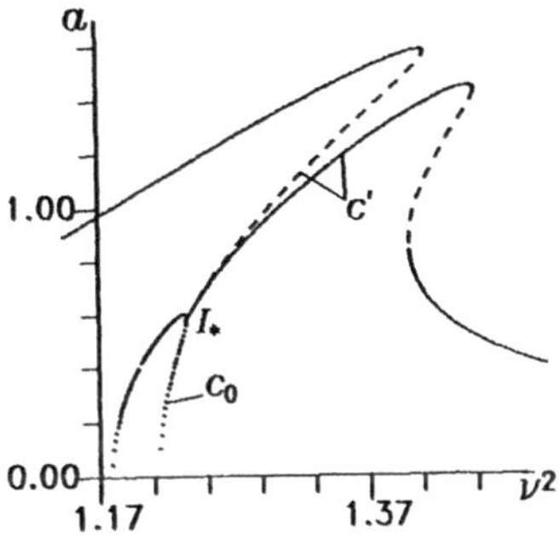

Figure 2b: p=0.11759

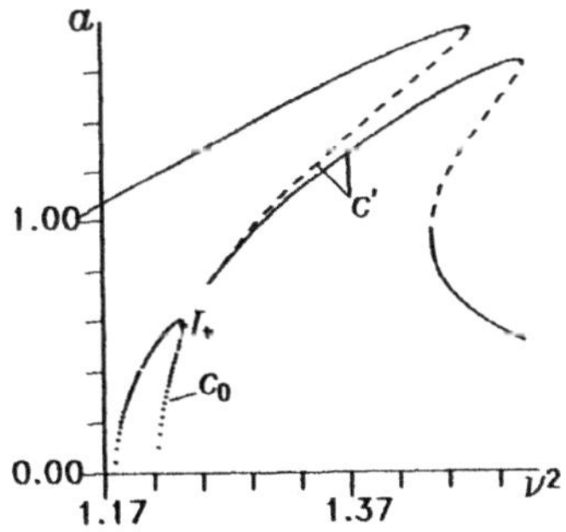

Figure 2c: p=0.15

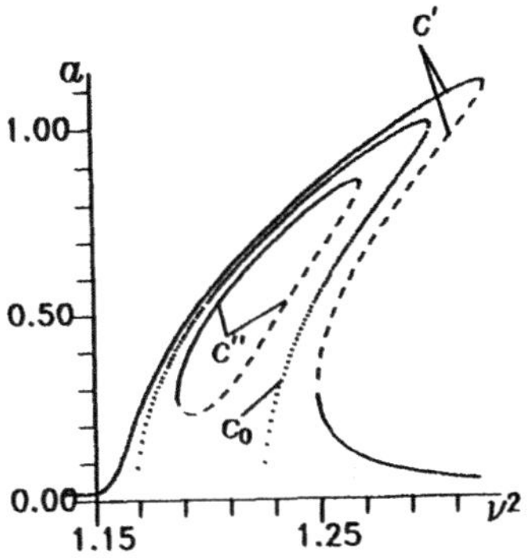

Figure 3a: p=0.004

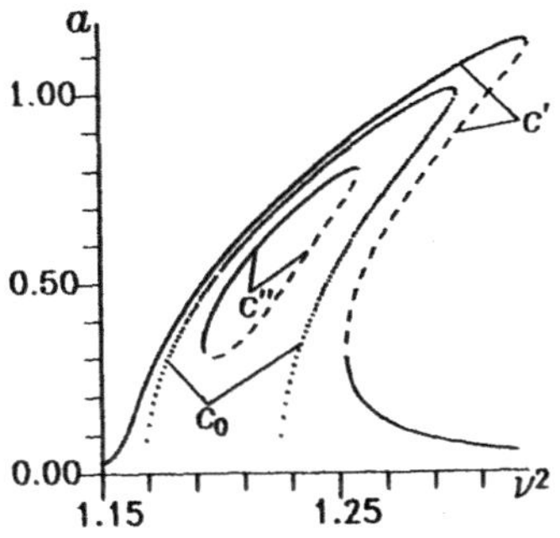

Figure 3b: p=0.005

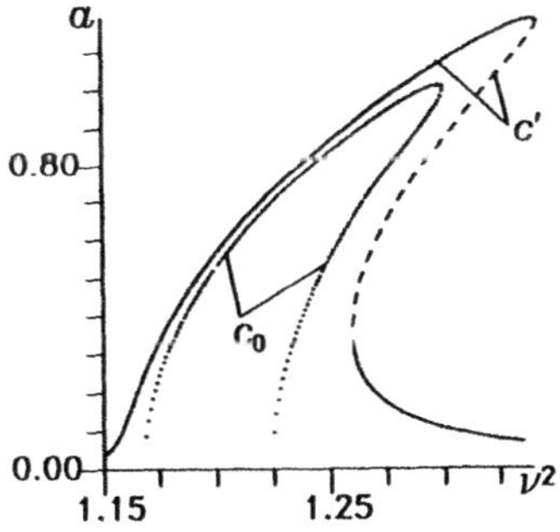

Figure 3c: p=0.007

Figures 2a, 2b, 2c correspond to the parameters: $\delta = \frac{3\pi}{4}$, $h = 0.009$, $\frac{b}{4}q^2 = 0.1$, $\frac{3}{4}\beta = 0.1$.

Figures 3a, 3b, 3c correspond to the parameters: $\delta = \frac{2\pi}{3}$, $h = 0.0877$, $\frac{b}{4}q^2 = 0.1$, $\frac{3}{4}\beta = 0.1$.

1. Stability condition

Since $h_0 > 0$, then $a_1 \to 0$, we have to study only the stability character of the mode y. Sufficient conditions for asymptotic stability of the oscillation under consideration are as follows :

$$S_1 = a\frac{\partial f_0}{\partial a} + \frac{\partial g_0}{\partial \psi} = 2a\nu h > 0, \tag{14}$$

$$S_2 = \frac{\partial f_0}{\partial a}\frac{\partial g_0}{\partial \psi} - \frac{\partial f_0}{\partial \psi}\frac{\partial g_0}{\partial a} > 0. \tag{15}$$

With $h > 0$, $\nu > 0$ the first condition is automatically satisfied. For the regular part C_1, the condition (15) can be transformed into

$$\frac{1}{D_0}\frac{\partial W}{\partial a} > 0, \quad W = D_1^2 + D_2^2 - D_0^2, \tag{16}$$

from which, ordinary stable portions bounded by vertical tangents can be determined. The stability character of the critical nodal point I_* is directly deduced from that of the ordinary portion, considered as containing it.

In the figures presented above, heavy (broken) lines correspond to stable (unstable) stationary oscillations.

5. The transient passages through the resonance

To study the transient passages through resonance, we consider the last two equations of the system (4), that is, the following equations $(a \neq 0)$:

$$\dot{a} = -\frac{\varepsilon\nu}{2} f_0 = -\frac{\varepsilon\nu}{2}\{\nu h a + \frac{b}{4}q^2 a \sin 2\psi + p\sin(\psi - \delta)\},$$

$$a\dot{\psi} = \frac{\varepsilon\nu}{2} g_0 = \frac{\varepsilon\nu}{2}\{(\frac{3}{4}\beta a^2 + \frac{b}{2}q^2 + \Delta)a - \frac{b}{4}q^2 a\cos 2\psi - p\cos(\psi - \delta)\} \tag{17}$$

We suppose that the frequency ν of the external force is a function of time t and that the derivative of ν with respect to time t $(\frac{d\nu}{dt})$ is proportional to ε. For simplicity, we take

$$\nu = \nu_0 + \alpha t, \quad \alpha = 0(\varepsilon). \tag{18}$$

To construct the curves describing the amplitude changes with different speeds of passing through resonance, we substitute ν by its expression (18) into equations (17), then we integrate these equations by using the Runge-Kutta method.

Figures 4a, 4b, 5a, 5b, 5c, 6a, 6b show the results. From these figures we draw the following conclusions: The passage through resonance of the non-linear system subjected simultaneously to forced and parametric excitations still has typical characteristics of a nonlinear forced system [2]. Namely, by increasing the speed (α) of passing through resonance, the maximum amplitude decreases and the sharpness of the first extremum of the resonance curve is smaller than that of the corresponding stationary resonance curve. If the passage through resonance is faster, then this sharpness decreases.

The presence of the indirect parametrically - excited component introduces special features not only for the stationary resonance curves, but also for the passage through resonance. The common feature of this process is that the resonance curves decrease strongly after reaching maximum. Near small enough amplitudes, the resonance curves are attracted to and asymptotically reach the stationary resonance curve. This phenomenon takes place for both processes of increasing $(\alpha > 0)$ and decreasing $(\alpha < 0)$ the frequency of the external force.

If the indirect parametric excitation does not exist (when h is large, $\frac{b}{4}q^2$ is small), the resonance curves (also in the case of stationary oscillations (Figure 1)) have forms which are similar to purely forced oscillations. That is, after reaching the first maximum, the resonance curve has some extrema with smaller amplitudes. The oscillation has a damped characteristic. The parametrically excited component makes more sudden changes of the resonance curves and, finally, these curves reach asymptotically to the stationary resonance curve. In Figures 4a, 4b, curves 1, 2, 3 correspond to the process of increasing the frequency of external force $(\alpha > 0)$, and curves 4, 5, 6 correspond to the inverse process $(\alpha < 0)$. The parameters for Figures 4a, 4b are the same as for Figure 1.

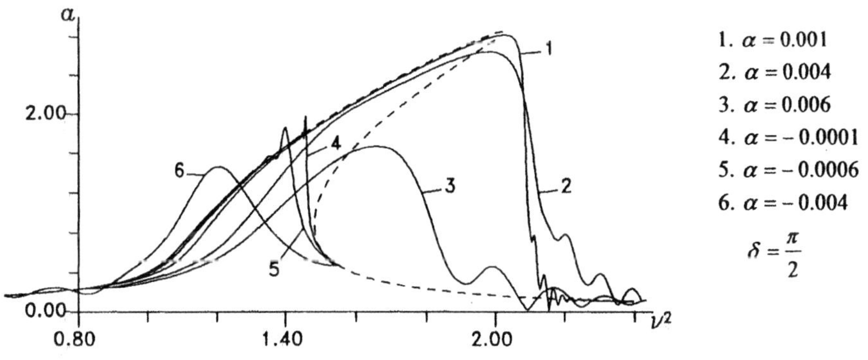

Figure 4a

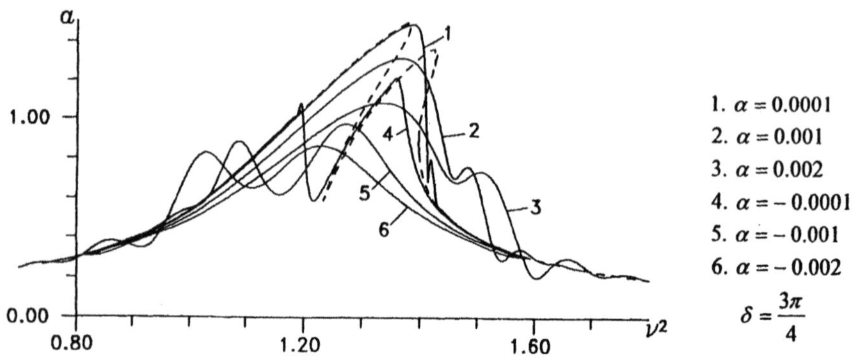

Figure 4b

With the parametric excitation large enough, with p small enough and $a_*^2 > a_c^2$ (the stationary resonance curve has singular point I_* as a node), the resonance curves decrease strongly after reaching the first maximum. Then, these curves have another maximum with very small amplitude and asymptotically reach the stationary resonance curves (curves 1, 2, 3 in Figure 5a correspond to the parameters of Figure 2a). When p has reached the value $\frac{1}{2}bq^2a_*$, node I_* becomes a returning point, and the change of resonance curves is smoother. The resonance curves decrease slowly after reaching the first maximum. Some extrema appear with smaller amplitudes (see curves 1, 2, 3, Figure 5b). Increasing p further until the forced excitation dominates the parametric excitation, the resonance curves have the form of those for purely forced oscillation. However, the parametric excitation still has influence. This excitation makes the resonance curves asymptotically reach the stationary resonance curve (see curves 1, 2, 3 in Figure 5c).

In the inverse process ($\alpha < 0$) the first two maxima of the resonance curve are very clearly expressed and the first maximum is connected to the existence of the loop or the crevasse. With parametric excitation large enough and with small p, the vertices of the extrema are nearly based on the left branches of the stationary resonance curve. After reaching two maxima, the resonance curve decreases monotonically (see curves 4, 5, 6 in Figure 5a). I ncreasing p, the vertices of the extrema move to the left and, finally, asymptotically reach the stationary resonance curve (see curves 4, 5, 6 in Figures 5b, 5c).

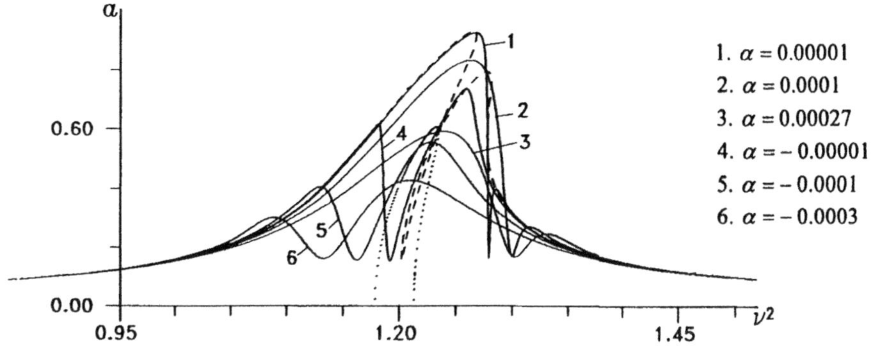

Figure 5a

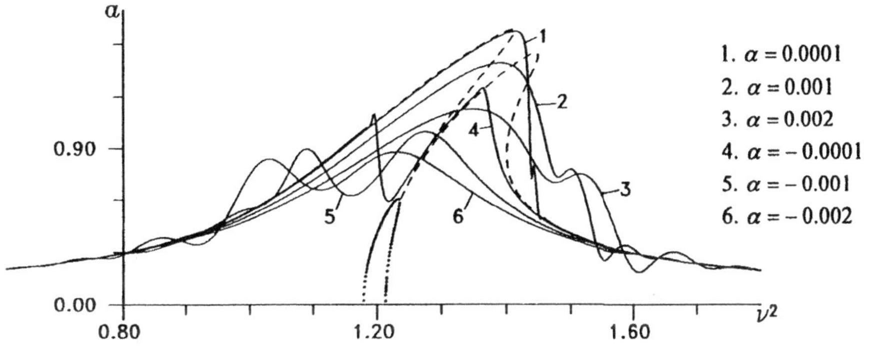

Figure 5b

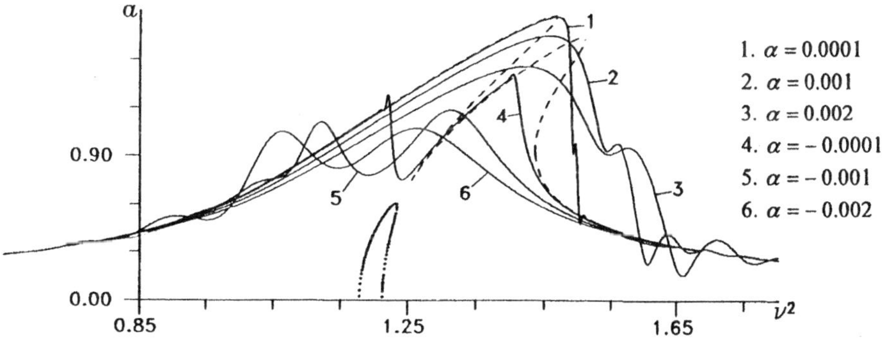

Figure 5c

With the parameters corresponding to Figures 3b, 3c (when $a_*^2 < a_c^2$ and I_* does not belong to the stationary resonance curve) we can see that the resonance curves decrease quickly and asymptotically reach the stationary resonance curve, independently of the relationship between the intensities of parametric and forced excitations. The vertices of the extrema are nearly based on the stationary resonance curve. With increasing (decreasing) the frequency of external force, these vertices are based on the right (left) branch of the stationary resonance curve (see curves 1, 2, 3 in Figures 6a, 6b for increasing ν, and curves 4, 5, 6 for decreasing ν).

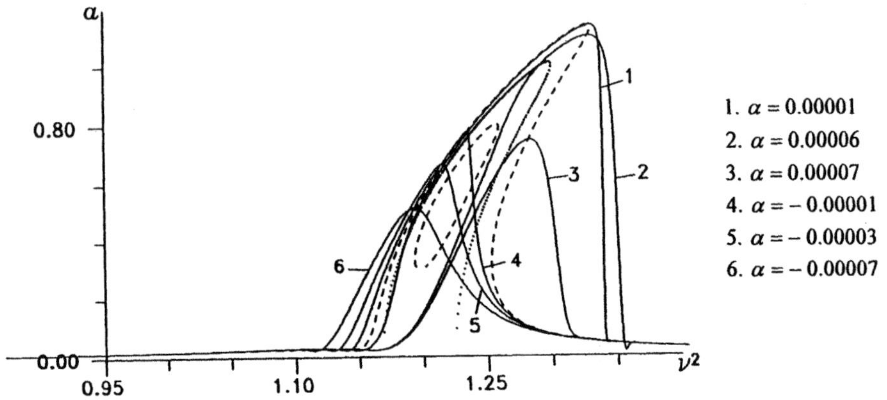

Figure 6a

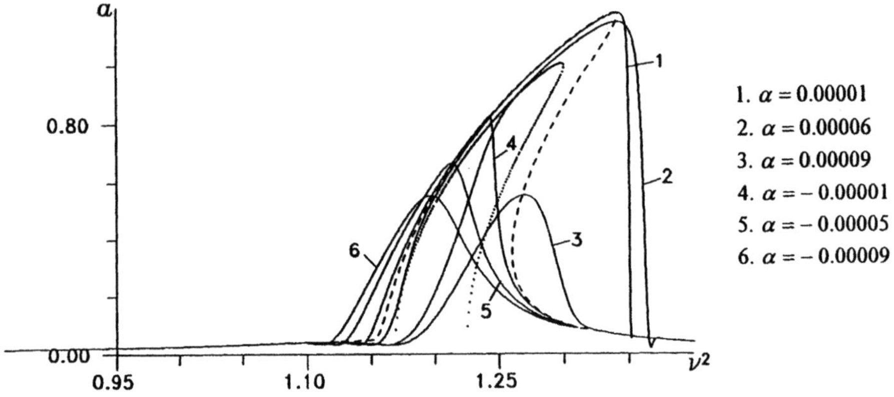

Figure 6b

In Figures 4a, 4b, 5a, 5b, 5c, 6a, 6b, 6c the continuous lines correspond to the transient passages through resonance and the broken lines correspond to the stationary resonance curves. The dotted lines are curves C_0.

6. Conclusion

The interaction between forced and parametric oscillations in a system with two degrees of freedom has been investigated for both stationary and non-stationary processes. The appearance of the parametrically excited component introduced special features in the resonance curves. For stationary oscillation, the singular points have been used to identify the resonance curves. In the passage through resonance, the resonance curves have fundamental characteristics of purely forced oscillations. The parametric excitation introduces a distinguishing character of the resonance curves: after reaching the maximum, the amplitude of oscillation decreases very quickly and finally tends asymptotically to the stationary resonance curve.

References

1. Mitropolskii, Yu.A., Nguyen Van Dao. *Applied Asymptotic Methods in Nonlinear Oscillations*. Kluwer Academic Publishers, Dordrecht, Netherlands, 1997.
2. Mitropolskii, Yu.A. *Problems of Asymptotic Methods of Non-Stationary Oscillations*. (in Russian), Nauka, Moscow,1964.
3. Tran Kim Chi, Nguyen Van Dinh. On the interaction between forced and parametric oscillations in a system with two degrees of freedom. *Journal of Mechanics, NCNST of Vietnam,* 1, XIX (1997), 12-21.

ON THE APPLICATION OF TWO PERTURBATION METHODS TO NONLINEAR SYSTEMS

H. NGUYEN♥ and G.L. OSTIGUY♣
♥*School of Engineering, Vietnam National University-Ho Chi Minh City, 268 Ly Thuong Kiet St., Dist. 10, Ho Chi Minh City, Vietnam.*
♣*Department of Mechanical Engineering, Ecole Polytechnique, Montreal, Quebec, Canada H3C 3A7.*

1. Introduction

It is well known that the response of a mechanical system depends on the type of excitation, or energy source, as well as on the natural frequencies of the system, the order of nonlinearity, and the type of damping mechanism.

Concerning the excitation, two types are often considered: (1) the excitation appears as an inhomogeneous term in the equations governing the system motion, and (2) the excitation appears as a time-varying coefficient in the governing equations of motion. The first type, which is called an external excitation, is the cause of forced oscillations. In this case, a small excitation cannot produce a large response unless the frequency of the excitation is close to one of the natural frequencies of the system, which is called a primary resonance. The second type, which is called a parametric excitation, is related to the dynamic instability and parametric oscillations of a system. In contrast to the preceding case, a small parametric excitation can produce a large response when the excitation frequency is close to twice one of the natural frequencies of the system, which is called a principal parametric resonance.

Considering the nonlinearity of a system, it is known that this property can occur for both discrete and continuous systems. For the former, exact solutions are quite limited; most of the analyses deal with weakly nonlinear systems which are amenable to perturbation methods. For strongly nonlinear systems, recourse is often made to geometrical methods to obtain a qualitative description of the behavior of the system, to perturbation techniques for which a basic nonlinear solution exists, or to numerical analysis. For the latter, since exact solutions are generally not available, recourse is made to approximate analyses by using analytical techniques, numerical methods, and numerical-perturbation techniques.

It is also known that governing equations with quadratic and/or cubic nonlinearities are associated with many physical systems. Flat plates, for which stretching is significant, possess cubic nonlinear terms [1-4]. The presence of nonlinear terms has an important influence upon the behavior of the system, especially under a condition of internal resonance. As is shown in [3], when a parametric resonance is excited in the

N. Van Dao and E.J. Kreuzer (eds.),
IUTAM Symposium on Recent Developments in Non-linear Oscillations of Mechanical Systems, 165–174.

presence of an internal resonance, the coincidence of the two types of resonances will give rise to simultaneous resonances. These kinds of resonances are characterized by the fact that several modes may exist in the response, even though only one mode is directly excited by the excitation. Internal resonance is responsible for this phenomenon and, as a consequence, for a significant transfer of energy from the directly excited mode to other modes of vibration.

This work deals with parametrically excited systems possessing cubic nonlinearities. The nonlinear temporal equations of motion are analyzed using the method of asymptotic series expansion developed by Mitropolskii [5], and the method of multiple scales popularized by Nayfeh [6, 7]. Attention is focused on both principal parametric resonance and simultaneous resonances. Numerical evaluation of the solutions was performed in order to get more insight into the differences between the two techniques.

2. Nonlinear Temporal Equations of Motion

Consider a parametrically excited system having a single degree of freedom, a cubic nonlinearity and linear viscous damping; the temporal equation of motion is governed by

$$\ddot{x} + 2c\dot{x} + \omega^2(1 - 2\mu\cos\theta)x + \gamma x^3 = 0 \tag{1}$$

where c is the coefficient of linear viscous damping, ω the natural frequency of the system, μ the excitation parameter, $\theta(t)$ total phase angle of the harmonic excitation, i.e., $\dot{\theta}(t) = \lambda$ the instantaneous excitation frequency, and γ the coefficient of the nonlinear (cubic) term.

In contrast with the above system, which has only a single natural frequency and a single mode of vibration, an n-degree-of-freedom system has n natural frequencies and n corresponding modes. For simplicity, the set of differential equations of motion is represented by a two-degree-of-freedom system as follows:

$$\ddot{x}_1 + \omega_1^2 x_1 = -2c_1\dot{x}_1 + 2\mu_1\omega_1^2\cos\theta\, x_1 - \left(\gamma_{11}x_1^3 + \gamma_{12}x_1^2x_2 + \gamma_{13}x_1x_2^2 + \gamma_{14}x_2^3\right) \tag{2a}$$

$$\ddot{x}_2 + \omega_2^2 x_2 = -2c_2\dot{x}_2 + 2\mu_2\omega_2^2\cos\theta\, x_2 - \left(\cdots\gamma_2\cdots\right) \tag{2b}$$

The second-order nonlinear differential equations with periodic coefficients (1) and (2) may be considered as extensions of the standard Mathieu-Hill equation. Equation (1) is used for analyzing the principal parametric resonance associated with any single spatial form, and equations (2) for studying simultaneous resonances involving two modes of vibration.

3. Solutions of the Temporal Equations of Motion

3.1. FIRST-ORDER ASYMPTOTIC APPROXIMATION

If the present system is weakly nonlinear, i.e., the excitation, the damping and the nonlinearity are small, and if the instantaneous frequency of excitation and the load parameter vary slowly with time, equation (1) can be rewritten as

$$\ddot{x}+\omega^2 x=\varepsilon\left[2\mu\omega^2\cos\theta\, x-2c\dot{x}-\gamma x^3\right] \tag{3}$$

where ε is a small parameter, for which $\tau=\varepsilon t$ represents the *slow* time.

Confining ourselves to the first order of approximation in ε, we seek a solution of equation (3) in the form

$$x=a(\tau)\cos\psi(\tau) \tag{4}$$

where a and ψ are functions of time defined by the system of differential equations

$$\dot{a}=\varepsilon A_1(\tau,\theta,a,\psi), \tag{5a}$$

$$\dot{\psi}=\omega(\tau)+\varepsilon B_1(\tau,\theta,a,\psi). \tag{5b}$$

Functions $A_1(\tau,\theta,a,\psi)$ and $B_1(\tau,\theta,a,\psi)$ are selected in such a way that equation (4) would, after replacing a and ψ by the functions defined in equations (5), represent a solution of equation (3).

Following the general scheme of constructing asymptotic solutions for vibrating systems, Nguyen [8] finally arrived at a system of equations which completely describes the nonstationary vibrational process. The stationary response associated with the principal parametric resonance $\lambda\approx 2\omega$ may be calculated as a special case of the non-stationary motions, and is given by

$$a=\left\{\left(8\omega^2/3\gamma\right)\left(s\pm\tfrac{1}{2}\left[\left(\mu/(s+1)\right)^2-\left(\Delta/\pi\right)^2\right]^{1/2}\right)\right\}^{1/2} \tag{6}$$

where $\Delta=2\pi c/\omega$ is the decrement of viscous damping and $s=(\lambda/2\omega)-1$ the detuning parameter. It is evident from (6) that the "±" sign upon the inner radical indicates the possibility of two nontrivial solutions.

The base width of the stationary response is the only region in which vibrations may normally initiate. By setting $a=0$ in equation (6), one obtains

$$s^4+2s^3+\left[1+\left(\Delta/2\pi\right)^2\right]s^2+\left[\left(\Delta/\pi\right)^2/2\right]s+\left[\left(\Delta/\pi\right)^2-\mu^2\right]/4=0. \tag{7}$$

This expression makes it possible to locate in the (μ, s) parameter space the boundaries of principal parametric instability region. Moreover, combining the expression

$$\mu=(s+1)\,\Delta/\pi, \tag{8}$$

calculated from the limited condition of the inner radical of (6), with the cut-off value of the excitation parameter

$$\mu_{co}=\Delta/\pi, \tag{9}$$

calculated from (8) when $s=0$ (that is, $\lambda\equiv 2\omega$), we may divide the remaining parameter space into two regions of stable trivial solutions [2].

The solution of equations (2) indicates, besides the possibility of principal parametric resonances, the presence of internal resonances. As previously mentioned, the combination of an internal resonance with a principal parametric resonance will give rise to simultaneous resonances. In this work, we will consider an internal resonance of the type $3\omega_1 \approx \omega_2$. Consequently, the two following cases of simultaneous resonances are possible: (1) $\lambda \approx 2\omega_1$ and $3\omega_1 \approx \omega_2$; and (2) $\lambda \approx 2\omega_2$ and $\omega_2 \approx 3\omega_1$.

For the sake of simplification, only the first case is presented here. In this case, it is supposed that the principal parametric resonance $\lambda \approx 2\omega_1$ and the internal resonance $3\omega_1 \approx \omega_2$ occur simultaneously. Then, performing numerous transformations and manipulations of the asymptotic solution, Nguyen [8] finally arrived at a system of equations describing the stationary response of the discretized system as follows:

$$-c_1 a_1 + \frac{1}{\lambda}\mu_1 \omega_1^2 a_1 \sin\psi + \frac{\gamma_{12}}{4(\omega_2 - \omega_1)} a_1^2 a_2 \sin\psi' = 0, \tag{10a}$$

$$\lambda - 2\omega_1 - \frac{3\gamma_{11}}{4\omega_1} a_1^2 - \frac{\gamma_{13}}{2\omega_1} a_2^2 + \frac{2}{\lambda}\mu_1 \omega_1^2 \cos\psi - \frac{\gamma_{12}}{2(\omega_2 - \omega_1)} a_1 a_2 \cos\psi' = 0, \tag{10b}$$

$$-c_2 a_2 - \frac{\gamma_{21}}{4(3\omega_1 + \omega_2)} a_1^3 \sin\psi' = 0, \tag{10c}$$

$$3\omega_1 - \omega_2 + \left[\frac{9\gamma_{11}}{8\omega_1} - \frac{\gamma_{22}}{4\omega_2}\right] a_1^2 + \left[\frac{3\gamma_{13}}{4\omega_1} - \frac{3\gamma_{24}}{8\omega_2}\right] a_2^2 - \frac{3}{\lambda}\mu_1 \omega_1^2 \cos\psi$$
$$+ \left[\frac{3\gamma_{12}}{4(\omega_2 - \omega_1)} a_1 a_2 - \frac{\gamma_{21}}{4(3\omega_1 + \omega_2)} \frac{a_1^3}{a_2}\right] \cos\psi' = 0 \tag{10d}$$

where $\psi = \theta - 2\psi_1$ is the phase angle associated with the principal parametric resonance involving the first spatial mode, and $\psi' = 3\psi_1 - \psi_2$ represents the phase angle corresponding to the specified internal resonance. The steady-state amplitudes, a_1 and a_2, and the phase angles, ψ and ψ', can be obtained by solving expressions (10) by a numerical technique.

It appears from (10) that there are two possibilities for a nontrivial solution: either a_1 is different from zero and a_2 is zero, or both are nonzero. The first possibility indicates that the specified internal resonance has no effect on the system response and only the principal parametric resonance involving the first mode may occur. For the latter possibility, as the first mode is the only one excited by the parametric excitation, the presence of the second mode in the response is possible only by the transfer of energy from the excited first mode to the second mode through an internal mechanism.

3.2. FIRST-ORDER MULTIPLE SCALES APPROXIMATION

The basic idea of the method of multiple time scales is to consider the expansion representing the response to be a function of multiple independent variables, or scales, instead of a single variable. Those multiple time scales are defined as

$$T_n = \varepsilon^n t \text{ for } n = 0,1,2,\cdots \tag{11}$$

It follows that the derivatives with respect to t become expansions in terms of the partial derivatives with respect to T_n according to

$$\frac{d}{dt}=\frac{dT_0}{dt}\frac{\partial}{\partial T_0}+\frac{dT_1}{dt}\frac{\partial}{\partial T_1}+\cdots=D_0+\varepsilon D_1+\cdots \tag{12a}$$

$$\frac{d^2}{dt^2}=D_0^2+2\varepsilon D_0D_1+\cdots \tag{12b}$$

Assume that the solution of equation (3) can be represented by an expansion having the form

$$x(t;\varepsilon)=x_0(T_0,T_1)+\varepsilon x_1(T_0,T_1)+\cdots \tag{13}$$

where $T_0=t$ and $T_1=\varepsilon t$. Knowing that $\theta=\lambda t$, substituting (12) and (13) into (3) and equating the coefficients of ε^0 and ε on both sides, one obtains

$$D_0^2x_0+\omega^2x_0=0, \tag{14}$$

$$D_0^2x_1+\omega^2x_1=-2D_0D_1x_0-2cD_0x_0-\gamma x_0^3+2\mu\omega^2\cos\lambda t\,x_0. \tag{15}$$

The solution of (14) is sought in the form

$$x_0=A(T_1)e^{i\omega T_0}+\bar{A}(T_1)e^{-i\omega T_0} \tag{16}$$

where A, which will be determined by eliminating the secular terms from x_1, is an unknown complex function, and $\bar{A}$ is the complex conjugate of A. Substituting (16) into (15) and expressing $\cos\lambda t$ in complex form, we obtain

$$\begin{aligned}D_0^2x_1+\omega^2x_1=&-\left[2i\omega\left(A'+cA\right)+3\gamma A^2\bar{A}\right]e^{i\omega T_0}-\gamma A^3e^{3i\omega T_0}\\&+\mu\omega^2Ae^{i(\lambda+\omega)T_0}+\mu\omega^2\bar{A}e^{i(\lambda-\omega)T_0}+cc\end{aligned} \tag{17}$$

where cc stands for the complex conjugate of the preceding terms and the prime denotes the derivative with respect to the slow time scale $T_1=\varepsilon t$.

To express the nearness of λ to 2ω in (17), we let

$$\lambda=2\omega+\varepsilon\rho \tag{18}$$

where ρ is the detuning parameter. Then it follows that secular terms can be eliminated by writing A in the form

$$A=\tfrac{1}{2}ae^{i\psi} \tag{19}$$

where a is the amplitude of vibration and ψ the phase lag. Separating the result into its real and imaginary parts, one obtains

$$a'=-ca+\tfrac{1}{2}\mu\omega a\sin\varphi, \tag{20a}$$

$$\varphi'=\rho-\frac{3\gamma}{4\omega}a^2+\mu\omega\cos\varphi \tag{20b}$$

where $\varphi=\rho T_1-2\psi$.

Steady-state motions occur when $a'=\varphi'=0$, and redefining the detuning parameter according to $s=\rho/2\omega$, we finally obtain

$$a = \left\{ \left(8\omega^2 / 3\gamma \right) \left(s \pm \tfrac{1}{2} \left[\mu^2 - \left(\Delta / \pi \right)^2 \right]^{1/2} \right) \right\}^{1/2}. \tag{21}$$

This expression represents the stationary response associated with the principal parametric resonance. For a steady-state solution to exist, a must be positive. The inner radical indicates that the excitation parameter must be greater than the damping coefficient to produce a sustained motion. The condition

$$\mu_{co} = \Delta / \pi \tag{22}$$

is the cut-off value of the excitation parameter. As before, the "±" sign upon the inner radical indicates the possibility of two nontrivial solutions.

Letting $a = 0$ in (21), one obtains

$$s = \mp \tfrac{1}{2} \left[\mu^2 - \left(\Delta / \pi \right)^2 \right]^{1/2} \tag{23}$$

corresponding to the base width of the principal parametric instability region in the (μ, s) parameter space. The (μ, s) plane is divided into three regions by the three curves defined by equations (22) and (23). As in the preceding method, it is noted that the boundaries of these regions are independent of γ, the coefficient of the nonlinear term.

As previously stated, two cases of simultaneous resonances are considered in this work. For simplicity, the second case of simultaneous resonances, i.e., $\lambda \approx 2\omega_2$ and $\omega_2 \approx 3\omega_1$, is presented here. Following the general scheme of constructing the method of multiple scales for equations (2), and introducing a detuning parameter for the principal parametric resonance according to

$$\lambda = 2\omega_2 + \varepsilon\rho \tag{24}$$

and another one for the internal resonance according to

$$3\omega_1 = \omega_2 + \varepsilon\sigma, \tag{25}$$

one can finally arrive at a system of equations describing the stationary motions:

$$-c_2 a_2 + \tfrac{1}{2}\mu_2\omega_2 a_2 \sin\varphi_1 - \tfrac{1}{8}\frac{\gamma_{21}}{\omega_2} a_1^3 \sin\varphi_2 = 0, \tag{26a}$$

$$\rho - \tfrac{1}{2}\frac{\gamma_{22}}{\omega_2} a_1^2 - \tfrac{3}{4}\frac{\gamma_{24}}{\omega_2} a_2^2 + \mu_2\omega_2 \cos\varphi_1 - \tfrac{1}{4}\frac{\gamma_{21}}{\omega_2}\frac{a_1^3}{a_2}\cos\varphi_2 = 0, \tag{26b}$$

$$-c_1 a_1 + \tfrac{1}{8}\frac{\gamma_{12}}{\omega_1} a_1^2 a_2 \sin\varphi_2 = 0, \tag{26c}$$

$$\sigma + \frac{1}{4\omega_2}\left\{ \left[\tfrac{27}{2}\gamma_{11} - \gamma_{22} \right] a_1^2 + \left[9\gamma_{13} - \tfrac{3}{2}\gamma_{24} \right] a_2^2 \right\} + \frac{1}{8\omega_2}\left[9\gamma_{12} a_1 a_2 - \gamma_{21}\frac{a_1^3}{a_2} \right]\cos\varphi_2 + \tfrac{1}{2}\mu_2\omega_2\cos\varphi_1 = 0 \tag{26d}$$

where $\varphi_1 = \rho T_1 - 2\psi_2$ and $\varphi_2 = \sigma T_1 + 3\psi_1 - \psi_2$.

From equations (26), it appears that there are two possibilities: either a_1 is zero and a_2 is nonzero, or neither is zero. The first possibility means that only the principal parametric resonance involving the second mode may exist; the second indicates the presence of two spatial forms in the system response resulting from a transfer of energy from the second excited mode to the first mode through internal resonance.

4. Results and Discussion

In order to get more insight into the differences between the two perturbation methods used in this work, numerical evaluation of the solutions was performed for a variety of cases, and the results presented in Figures 1 to 3 are typical of those obtained. In the figures, solid lines are associated with the first-order asymptotic approximation while phantom lines correspond to the method of multiple time scales.

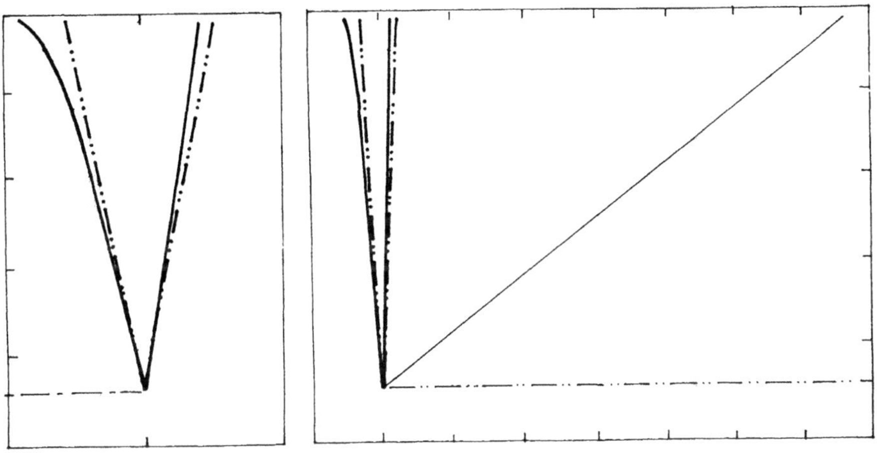

Figure 1 A. and B.

The principal region of incipient instability associated with the principal parametric resonance of any mode shape is shown in Figure 1(a), and the trivial solution in this instability zone is known to be unstable. As can be seen, if the excitation parameter is quite weak, there is no difference between the two perturbation methods. However, the difference is more pronounced when the excitation parameter μ is large. As stated before, the remaining parameter space is divided into two regions of stable trivial solutions. The three regions of two different trivial solutions are illustrated in Figure 1(b). Region III obtained by the two methods is a clear indicator of the difference between them. The results show that Region III evaluated by the asymptotic approximation is a function of the detuning parameter, and hence is limited by this parameter. Meanwhile, the one obtained by the multiple scales method is independent of this parameter, and therefore can extend indefinitely.

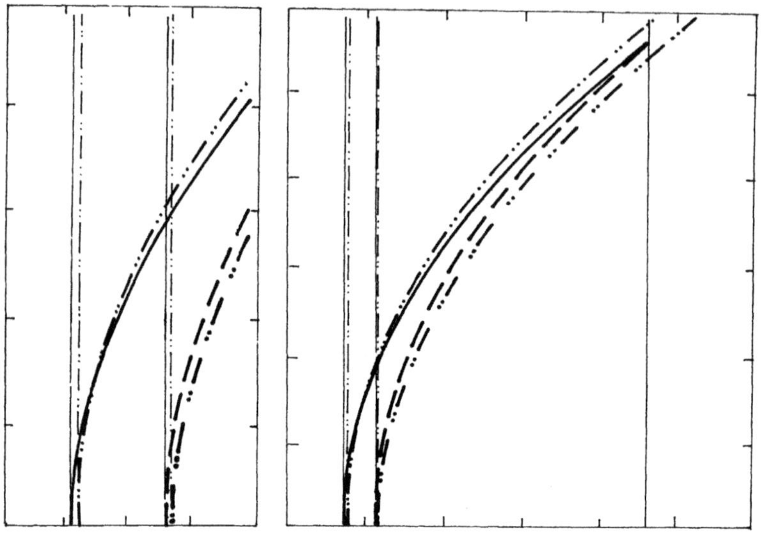

Figure 2 A. and B.

Stationary frequency-response curves associated with a principal parametric resonance are plotted in Figure 2. When the steady-state amplitude a is a function of the detuning parameter s, it is known that a positive cubic nonlinearity bends the two response curves to the right for a hard spring effect. Moreover, a stability analysis [8] has shown that the upper branch is stable and experimentally observable while the lower branch is unstable and physically unrealizable. In this figure, as well as in Figure 1, the characteristics of the three regions and the phenomena associated with principal parametric resonance were clearly explained in [2].

Figure 2(a) illustrates the nontrivial solutions at the onset of principal parametric resonance. When the response is near the resonance $\lambda \approx 2\omega$, it can be seen that the differences between the two methods are not so important. As shown in Figure 2(b), however, it makes an important difference between the two methods far from the resonance condition. Numerical calculations carried out on equation (6), analyzed by the asymptotic method, have resulted in the convergence of the stable and unstable solutions. Meanwhile, the two stable and unstable branches analyzed by the method of multiple scales diverge to infinity. In practice, this cannot happen [2]. Hence, it can be said that the method of multiple time scales is valid only in the region of parametric instability, or when the response is close to a principal parametric resonance.

For multiple resonances, for simplicity, results concerning simultaneous resonances will be presented in the form of the effect of an internal resonance on a principal parametric resonance. Therefore, the interactions between the internal resonance $3\omega_1 \approx \omega_2$ and the principal parametric resonances $\lambda \approx 2\omega_1$ and $\lambda \approx 2\omega_2$ on the frequency-response curves are illustrated in Figure 3. In the figure, the bar over an amplitude means that the amplitude is associated with simultaneous resonances, while the subscript i following the mth mode shape indicates that the amplitude is possible due to internal resonance.

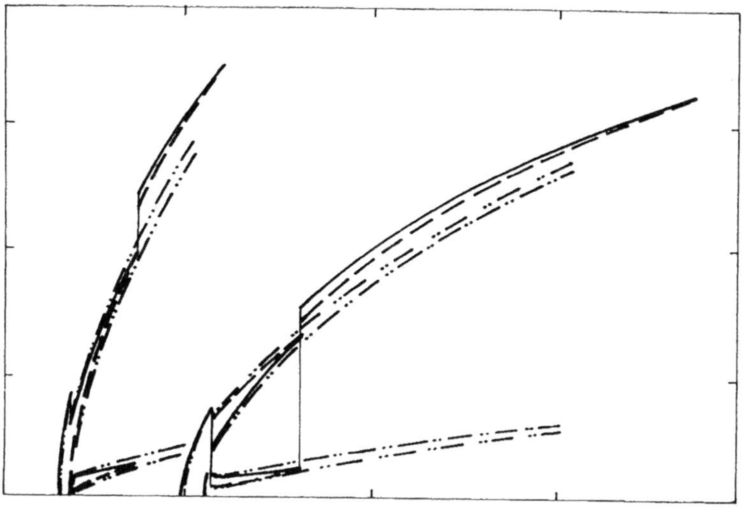

Figure 3

As can be seen, the parametric response of the first mode occurs when $\lambda \approx 2\omega_1$. At a certain frequency, however, a small part of energy from the first mode is transferred to the second mode, due to modal coupling between these two modes. Consequently, the amplitude of the first mode slightly decreases but remains larger than that for the second mode. The asymptotic solutions show that the energy transfer vanishes after a certain range; the amplitude of the second mode decays and the steady-state amplitude of the first mode regains its full strength. On the contrary, the multiple scales solutions will unrealistically go on forever.

In contrast to the previous case, the response of the system at $\lambda \approx 2\omega_2$ is particularly interesting. It can be observed that when the excitation frequency reaches the point where the first mode can be excited through internal resonance, the amplitude of the second mode, which is directly excited by the parametric excitation, drops drastically and becomes less than the amplitude of the first mode which is due to internal resonance. This implies that there is a significant transfer of energy from the second mode to the first. As before, the asymptotic solutions indicate that when the transfer of energy stops, the amplitude of the first mode vanishes and the second mode continues to be excited by the parametric excitation. In contrast to the previous method, for the multiple scales approximation, the transfer of energy never stops and the amplitudes of the two modes again go on forever.

5. Concluding Remarks

Two perturbation methods, the generalized asymptotic method and the method of multiple time scales, are used to solve weakly nonlinear temporal equations of motion. Such equations of motion are governed by second-order differential equations with periodic coefficients, cubic nonlinearities, and linear viscous damping.

For simple parametric resonance, the analyses predict the configuration of the principal parametric instability region and the frequency-response characteristics of the system. For the principal region of parametric instability, the results indicate that there is no important difference between the two perturbation methods when the magnitude of the excitation parameter is small and moderate. For stationary frequency-response curves associated with principal parametric resonance, there is no significant difference between the two methods in the instability zone, or when the response is near the resonance. The further the response is from the resonance condition, the greater the difference between the two methods. The validity of the results analyzed by the asymptotic technique has been ascertained experimentally [2].

For simultaneous resonances, the results obtained by the method of multiple time scales are not reasonable because the amplitudes of vibration unrealistically continue forever. On the contrary, the validity of the results obtained by the asymptotic method has once more been ascertained experimentally [3, 9].

6. References

1. Nguyen, H., and Ostiguy, G.L.: Effect of boundary conditions on the dynamic instability and nonlinear response of rectangular plates, Part I: Theory, *J. Sound and Vibration* **133** (1989), 381-400.
2. Nguyen, H., Ostiguy, G.L., and Samson, L.P.: Effect of boundary conditions on the dynamic instability and nonlinear response of rectangular plates, Part II: Experiment, *J. Sound and Vibration* **133** (1989), 401-422.
3. Nguyen, H.: *Effect of Boundary Conditions on the Dynamic Instability and Responses of Rectangular Plates*, Dr.Sc. Thesis, Ecole Polytechnique, Montreal, 1987.
4. Nguyen, H., and Ostiguy, G.L.: Recent developments on the dynamic stability and response of parametrically-excited rectangular plates. In *Applied Electromagnetism and Mechanics, Vol. 6* (K. Miya, M. Yamamoto and N.X. Hung, eds.), Japan Society of Applied Electromagnetics and Mechanics, 1998, pp. 102-111.
5. Mitropolskii, Yu.A.: *Problems of the Asymptotic Theory of Nonstationary Vibrations*, Daniel Davey, New York, 1965.
6. Nayfeh, A.H.: *Perturbation Methods*, Wiley-Interscience, New York, 1973.
7. Nayfeh, A.H., and Mook, D.T.: *Nonlinear Oscillations*, Wiley-Interscience, New York, 1979.
8. Nguyen, H.: *Résonances combinées d'une plaque rectangulaire soumise à une excitation paramétrique*, M.Sc. Thesis, Ecole Polytechnique, Montreal, 1982.
9. Ostiguy, G.L., Samson, L.P., and Nguyen, H.: On the occurrence of simultaneous resonances in parametrically-excited rectangular plates, *Trans. ASME, J. Vibration and Acoustics* **115** (1993), 344-352.

NEW METHOD CALCULATING THE VIBRATION OF OFFSHORE STRUCTURE SUBJECTED TO WAVE ACTION

NGUYEN XUAN HUNG, NGUYEN XUAN HOANG
Institute of Applied mechanics Vietnam(IAM)
291 Dien Bien Phu St., 3 Dist., Ho Chi Minh City.

1. Abstract

Usually, modal analysis is used for the vibration of simple offshore structures. For large structures the finite element method is applied to analyse dynamic deflections and stresses. Both methods are approximately; it is difficult to model the nonlinear component of the wave force due to the drag on each element and a common hydrodynamic damping is usually applied.

In this paper the vibration of offshore structures subject to harmonic and random waves is considered by using the dynamic stiffness matrix method (DSM method); the element drag force is obtained by considering the relative motion between the structure and sea water in detail.

2. The dynamic stiffness matrix of structure

Assume the structure is divided into N elements or substructures, the complex amplitude of which may be determined. We denote the complex amplitude of element i by $\hat{u}^{(i)}(\bar{x},\omega)$, where $\bar{x}$ is a coordinate of any point of element and ω is its frequency of vibration.

For element or substructure i we choose some points, through which the element or substructure is connected to other elements or substructures and denote the vector of complex amplitudes of generalized coordinates of the substructure at these connected points by

$$\bar{q}^{(i)} = \left(\hat{q}_1, \hat{q}_2, \ldots, \hat{q}_{n_i}\right)^T$$

The vector of complex amplitudes of generalized forces at the connected points is expressed by:

$$\bar{Q}^{(i)} = \left(\hat{Q}_1, \hat{Q}_2, \ldots, \hat{Q}_{n_i}\right)^T$$

where $\hat{Q}_i$ is complex amplitude of the generalized force corresponding to $\hat{q}_i$

N. Van Dao and E.J. Kreuzer (eds.),
IUTAM Symposium on Recent Developments in Non-linear Oscillations of Mechanical Systems, 175–185.

The dynamic stiffness matrix (DSM) $\overline{K}^{(i)}$ of element or substructure i is defined by the equation

$$\overline{Q}^{(i)} = \overline{K}^{(i)}\overline{q}^{(i)} \tag{1}$$

$\overline{K}^{(i)}$ is square symmetric of order n_i, the coefficients of which $\overline{k_{ij}}$ are functions of ω.

The *DSM* of the complete structure is determined from the dynamic stiffness matrices of its elements or substructures as follows:

$$K = \sum_{i=1}^{N} a_i^T K^{(i)} a_i = \sum_{i=1}^{N} \widetilde{K}^{(i)} \tag{2}$$

where $K^{(i)} = D_i^T \overline{K}^{(i)} D_i$ is the *DSM* of element i in the global coordinate system, D_i is the direction cosines matrix of element i; a_i is the Boolean matrix determining the location of element i in complete structure, and $\widetilde{K}^{(i)} = a_i^T K^{(i)} a_i$. K is a square symmetric matrix of order n, the coefficients k_{ij} of which are functions of frequency ω ; N is the number of elements in the structure; n is number of the generalized coordinates of the structure.

The inverse matrix of K is called the dynamic flexibility matrix (*DFM*) of the structure, or the matrix of the transfer functions of the structure H:

$$H = K^{-1} \tag{3}$$

In the following we consider the bar structure, the *DSM* of which may be calculated exactly.

3. The dynamic stiffness matrix of the bar structure

3.1 Governing equation

The differential equation of damped longitudinal vibration of the prismatic bar in viscous elastic field has the following form:

$$AE\left(\frac{\partial^2 u}{\partial x^2} + k_i \frac{\partial^3 u}{\partial x^2 \partial t}\right) - k_a^u \frac{\partial u}{\partial t} - c_a^u u - m\frac{\partial^2 u}{\partial t^2} = 0 \tag{4}$$

The differential equation of lateral vibration of the bar in a viscous elastic field, including shear deformation, rotatory inertia, and static axial force N_{st} is as follows:

$$EJ\left(\frac{\partial^4 v}{\partial x^4} + k_i \frac{\partial^4 v}{\partial x^4 \partial t}\right) + N\frac{\partial^2 v}{\partial x^2} + \left(c_a^v + k_a^v \frac{\partial v}{\partial x} + m\frac{\partial^2 v}{\partial t^2}\right)\left(1 - \frac{N_{st}}{k' AG}\right) - mr^2 \frac{\partial^4 v}{\partial x^4 \partial t^2} - k_a^\varphi \frac{\partial^3 v}{\partial t \partial x^2} -$$
$$c_a^\varphi v \frac{\partial^2 v}{\partial x^2} - \frac{EJ}{k' AG}\frac{\partial^2}{\partial x^2}\left(c_a^v v + k_a^v \frac{\partial v}{\partial t} + m\frac{\partial^2 v}{\partial t^2}\right) + \frac{mr^2}{k' AG}\frac{\partial^2}{\partial t^2}\left(c_a^v v + k_a^v \frac{\partial v}{\partial t} + m\frac{\partial^2 v}{\partial t^2}\right) +$$
$$\frac{k_a^\varphi}{k' AG}\frac{\partial}{\partial t}\left(c_a^v v + k_a^v \frac{\partial v}{\partial t} + m\frac{\partial^2 v}{\partial t^2}\right) + \frac{c_a^\varphi}{k' AG}\left(c_a^v v + k_a^v \frac{\partial v}{\partial t} + m\frac{\partial^2 v}{\partial t^2}\right) = 0 \tag{5}$$

In equations (4) and (5) $u(x, t)$, $v(x,t)$ are the longitudinal and lateral displacements of the bar, A and J are the area and the moment of inertia of the bar cross section, $r^2 = \frac{J}{A}$ is the radius of gyration of the bar cross section, m is the mass per unit length of the bar, $k_a^u, c_a^u, k_a^v, c_a^v$ and k_a^φ, c_a^φ are the elastic constants and damping constants of the field around the bar for longitudinal, lateral displacements, and rotation of the bar section respectively, coefficient k' relates to the non uniform distribution of shear force in the bar cross section (it is useful to recall that $k' = \frac{5}{6}$ for a rectangular section), E and G are the modulus of elasticity and shear modulus of the bar material, k_i is viscous damping coefficient.

If the bar harmonic vibrates with angular frequency Ω, then the displacements $u(x, t)$ and $v(x, t)$ may be taken in the following form:

$$u(x,t) = \hat{u}(x)e^{j\Omega t}$$
$$v(x,t) = \hat{v}(x)e^{j\Omega t} \tag{6}$$

where $j = \sqrt{-1}$

In general $\hat{u}(x)$ and $\hat{v}(x)$ are the complex functions of frequency Ω. They are called the complex amplitudes and can be expressed as follows:

$$\hat{u}(x) = |\hat{u}(x)|\, e^{\varphi_u(x)}$$
$$\hat{v}(x) = |\hat{v}(x)|\, e^{\varphi_v(x)} \tag{7}$$

in which $|\hat{u}(x)|$, $|\hat{v}(x)|$ and $\varphi_u(x)$, $\varphi_v(x)$ are the amplitudes and the phase angles of the displacements $u(x, t)$ and $v(x, t)$ at the section x respectively.

Substituting expressions (6) into equations (4) and (5) we obtain the following equations for $\hat{u}(x)$ and $\hat{v}(x)$:

$$\frac{d^2\hat{u}(x)}{dx^2} + \left(\frac{\lambda_l}{l}\right)^2 \hat{u}(x) = 0 \tag{8}$$

$$\frac{d^4\hat{v}}{dx^4} - \left(\frac{\lambda_b}{l}\right)^4 (1-\alpha_1\alpha_2)\hat{v} + \left(\frac{\lambda_b}{l}\right)^4 (\alpha_o+\alpha_2) l^2 \frac{d^2\hat{v}}{dx^2} + \lambda_b^4\left(\frac{\lambda_b}{l}\right)^4 \alpha_o\alpha_2\hat{v} + \frac{\alpha_1}{l^2}\frac{d^2\hat{v}}{dx^2} = 0 \tag{9}$$

where
$$\lambda_l = \left[\frac{m\Omega^2}{\hat{E}J}\left(1 - j\frac{k_a^l}{m\Omega} - \frac{c_a^l}{m\Omega^2}\right)\right]^{\frac{1}{2}} l;\; \lambda_b = \left[\frac{m\Omega^2}{\hat{E}J}\left(1 - j\frac{k_a^b}{m\Omega} - \frac{c_a^b}{m\Omega^2}\right)\right]^{\frac{1}{4}} l \tag{10}$$

$\hat{E} = E[1 + jk_i\Omega]$ is the complex modulus of elasticity,

$$\alpha_o = \frac{\left(1 - j\dfrac{k_a^\varphi}{mr^2}\dfrac{1}{\Omega} - \dfrac{c_a^\varphi}{mr^2}\dfrac{1}{\Omega^2}\right)}{\left(1 - j\dfrac{k_a^b}{m}\dfrac{1}{\Omega} - \dfrac{c_a^b}{m}\dfrac{1}{\Omega^2}\right)}\frac{r^2}{l^2};\quad \alpha_2 = \frac{1}{l^2}\frac{EJ}{k'AG} = \frac{2(1+\mu)}{k'}\frac{r^2}{l^2};\quad \alpha_1 = \frac{N_{st}l^2}{\hat{E}J} \tag{11}$$

We have assumed $E = 2(1+\mu)G$, where μ denotes Poisson's ratio.

By neglecting k_a^b, c_a^b, k_a^φ, c_a^φ, k_i we find that expression (11) becomes:

$$\alpha_o = \frac{r^2}{l^2};\qquad \alpha_2 = \frac{2(1+\mu)}{k'}\alpha_o;\qquad \alpha_1 = \frac{N_{st}l^2}{EJ} \tag{12}$$

3.2 Dynamic stiffness matrix (DSM) of the bar:

3.2.1 *DSM* for longitudinal vibration:

The solution of equation (8) is

$$u(x) = B_1 \sin\frac{\lambda_1}{l}x + B_2\cos\frac{\lambda_1}{l}x \tag{13}$$

The expression for complex amplitude of $\hat{N}(x)$ is

$$\hat{N} = A\hat{E}\frac{d\hat{u}}{dx} \tag{14}$$

Denoting $\hat{u}_1 = \hat{u}|_{x=0}$; $\hat{u}_2 = \hat{u}|_{x=l}$; $\hat{N}_1 = \hat{N}|_{x=0}$; $\hat{N}_2 = \hat{N}|_{x=l}$ for the displacements and the forces at the end nodes of the bar, from equation (13) and (14) we obtain the following matrix equation for $\hat{u}_i$ and $\hat{N}_i$:

$$\hat{N} = K_l \,.\, \hat{u} \tag{15}$$

where $\hat{N} = (\hat{N}_1 .\ \hat{N}_2)^T$; $\hat{u} = (\hat{u}_1 .\ \hat{u}_2)^T$

$$K_l = \begin{vmatrix} k_{11}^l & k_{12}^l \\ k_{21}^l & k_{22}^l \end{vmatrix} \tag{16}$$

$$k_{11}^l = k_{22}^l = \frac{A\hat{E}}{l}\frac{\lambda_l}{\tan\lambda_l} \;;\; k_{12}^l = k_{21}^l = -\frac{A\hat{E}}{l}\frac{\lambda_l}{\sin\lambda_l}$$

The matrix K_l is symmetric and is called the *DSM* of the bar, its coefficients k_{ij}^l $(i, j = 1,2..)$ are functions of frequency Ω.

Having the nodal displacements $\hat{u}_1$, $\hat{u}_2$ we determine the displacement of any cross section x from the equation:

$$\hat{u}(x) = \frac{\hat{u}_2 - \cos\lambda_l \hat{u}_1}{\sin\lambda_l}\sin\frac{\lambda_l}{l}x + \hat{u}_1\cos\frac{\lambda_l}{l}x. \tag{17}$$

3.2.2 DSM for lateral vibration:

The solution of equation (9) is as follows:

$$\hat{v}(x) = A_1\sin\left(\lambda_b\sqrt{q_1}\frac{x}{l}\right) + A_2\cos\left(\lambda_b\sqrt{q_1}\frac{x}{l}\right) + A_3\sinh\left(\lambda_b\sqrt{q_2}\frac{x}{l}\right) + A_4\cosh\left(\lambda_b\sqrt{q_2}\frac{x}{l}\right) \tag{18}$$

where

$$q_1 = \sqrt{1 - \alpha_1\alpha_2 + \left(\frac{\alpha_o - \alpha_2}{2}\right)^2\lambda_b^4 + \left(\frac{\alpha_o + \alpha_2}{2}\right)\alpha_1 + \frac{\alpha_1^2}{4\lambda_b^4}} + \frac{\alpha_o + \alpha_2}{2}\lambda_b^2 + \frac{\alpha_1}{2\lambda_b^2} \tag{19}$$

$$q_2 = \sqrt{1 - \alpha_1\alpha_2 + \left(\frac{\alpha_o - \alpha_2}{2}\right)^2\lambda_b^4 + \left(\frac{\alpha_o + \alpha_2}{2}\right)\alpha_1 + \frac{\alpha_1^2}{4\lambda_b^4}} - \frac{\alpha_o + \alpha_2}{2}\lambda_b^2 - \frac{\alpha_1}{2\lambda_b^2}$$

The equations for the complex amplitudes of rotation, shear force and bending moment of the bar are:

$$\hat{\varphi} = (1 + \alpha_2\rho_2)\frac{d\hat{v}}{dx} + \alpha_2\rho_1 l^2\frac{d^3\hat{v}}{dx^3}$$

$$\hat{Q} = \hat{E}J\left(\rho_1\frac{d^3\hat{v}}{dx^3} + \frac{\rho_2}{l^2}\frac{d\hat{v}}{dx}\right) \tag{20}$$

$$\hat{M} = \hat{E}J\left[(1 + \alpha_2\rho_2)\frac{d^2\hat{v}}{dx^2} + \alpha_2\rho_1 l^2\frac{d^4\hat{v}}{dx^4}\right]$$

where

$$\rho_1 = \frac{1}{1 - \alpha_o\alpha_2\lambda_b^4 - \alpha_1\alpha_2} \;; \qquad \rho_2 = \frac{(\alpha_o + \alpha_2)\lambda_b^4 + \alpha_1}{1 - \alpha_o\alpha_2\lambda_b^4 - \alpha_1\alpha_2} \tag{21}$$

Denoting the complex amplitude of nodal displacements, nodal rotations, nodal forces and nodal moments of the bar by:

$$\hat{v}_1 = \hat{v}|_{x=0}; \quad \hat{\varphi}_1 = -\hat{\varphi}|_{x=0}; \quad \hat{v}_2 = \hat{v}|_{x=l}; \quad \hat{\varphi}_2 = -\hat{\varphi}|_{x=l}$$

$$\hat{Q}_1 = \hat{Q}|_{x=0}; \quad \hat{M}_1 = \hat{M}|_{x=0}; \quad \hat{M}_2 = -\hat{M}|_{x=l}; \quad \hat{Q}_2 = -\hat{Q}|_{x=l}$$

from equations (18), (20) we obtain the following matrix equation for $\hat{v}_i$, $\hat{\varphi}_i$, $\hat{Q}_i$, $\hat{M}_i (i=1,2)$

$$\hat{Q} = K_b \cdot \hat{v}$$

where

$$\hat{v} = (\hat{v}_1, \hat{\varphi}_1, \hat{v}_2, \hat{\varphi}_2)^T$$

$$\hat{Q} = (\hat{Q}_1, \hat{M}_1, \hat{Q}_2, \hat{M}_2)^T$$

$$K_b = \frac{\hat{E}J}{l^3}\begin{bmatrix} \gamma & -\beta l & -\bar{\gamma} & -\bar{\beta} l \\ -\beta l & \alpha l^2 & \bar{\beta} l & \bar{\alpha} l^2 \\ -\bar{\gamma} & \bar{\beta} l & \gamma & \beta l \\ -\bar{\beta} l & \bar{\alpha} l^2 & \beta l & \alpha l^2 \end{bmatrix} \tag{22}$$

$$\gamma = \frac{\lambda_b^3}{d}(H_1 h_2 + H_2 h_1)(h_1 S_1 C_2 + h_2 S_2 C_1)$$

$$\beta = \frac{\lambda_b^2}{d}[(H_1 h_1 + H_2 h_2) S_1 S_2 + (H_2 h_1 - H_1 h_2)(1 - C_1 C_2)]$$

$$\alpha = \frac{\lambda_b}{d}(h_1\sqrt{q_1} + h_2\sqrt{q_2})(h_2 S_1 C_2 - h_1 C_1 S_2) \tag{23}$$

$$\bar{\gamma} = \frac{\lambda_b^3}{d}(H_1 h_2 + H_2 h_1)(h_1 S_1 + h_2 S_2) \quad ; \bar{\beta} = \frac{\lambda_b^2}{d}(h_1 H_2 + h_2 H_1)(C_2 - C_1)$$

$$\bar{\alpha} = \frac{\lambda_b}{d}(h_1\sqrt{q_1} + h_2\sqrt{q_2})(h_1 S_2 - h_2 S_1);$$

$$d = 2h_1 h_2(1 - C_1 C_2) + (h_2^2 - h_1^2) S_1 S_2$$

$$H_1 = \left(\rho_1 q_1 - \frac{\rho_2}{\lambda_b^2}\right)\sqrt{q_1} \quad ; \quad H_2 = \left(\rho_2 q_2 + \frac{\rho_2}{\lambda_b^2}\right)\sqrt{q_2} \tag{24}$$

$$h_1 = [(1 + \alpha_2 \rho_2) - \alpha_2 \rho_1 \lambda_b^2 q_1]\sqrt{q_1} \quad ; h_2 = [(1 + \alpha_2 \rho_2) + \alpha_2 \rho_1 \lambda_b^2 q_2]\sqrt{q_2}$$

$$S_1 = \sin\left(\lambda_b\sqrt{q_1}\right) \; ; S_2 = \sinh\left(\lambda_b\sqrt{q_2}\right); C_1 = \cos\left(\lambda_b\sqrt{q_1}\right) \; ; C_2 = \cosh\left(\lambda_b\sqrt{q_2}\right)$$

Having $\hat{v}_1, \hat{\varphi}_1, \hat{v}_2, \hat{\varphi}_2$ we determine the displacements of any section x from equation (18), in which $A_i (i = 1..4)$ will be obtained from

$$\begin{bmatrix} A_1 \\ A_2 \\ A_3 \\ A_4 \end{bmatrix} = \frac{1}{d'}\begin{bmatrix} -\frac{h_2 S_2 C_1 + h_1 S_1 C_2}{h_1} & \frac{l(h_1 S_1 S_2 + h_2 C_1 C_2 - h_2)}{\lambda_b h_1 h_2} & \frac{S_2 h_2 + S_1 h_1}{h_1} & \frac{l(C_2 - C_1)}{\lambda_b h_1} \\ \frac{h_1 + h_2 S_1 S_2 - h_1 C_1 C_2}{h_1} & \frac{l(h_1 C_1 S_2 - h_2 S_1 C_2)}{\lambda_b h_1 h_2} & C_1 - C_2 & \frac{l(h_2 S_1 - h_1 S_2)}{\lambda_b h_1 h_2} \\ \frac{h_2 C_1 S_2 + h_1 S_1 C_2}{h_2} & \frac{l(h_1 C_1 C_2 - h_2 S_1 S_2 - h_1)}{\lambda_b h_1 h_2} & -\frac{(h_2 S_2 + h_1 S_1)}{h_2} & \frac{l(C_1 - C_2)}{\lambda_b h_2} \\ \frac{h_2 - h_2 C_1 C_2 - h_1 S_1 S_2}{h_2} & \frac{l(h_2 S_1 C_2 - h_1 S_2 C_1)}{\lambda_b h_1 h_2} & C_2 - C_1 & \frac{l(h_1 S_2 - h_2 S_1)}{\lambda_b h_1 h_2} \end{bmatrix} \begin{bmatrix} \hat{v}_1 \\ \hat{\varphi}_1 \\ \hat{v}_2 \\ \hat{\varphi}_2 \end{bmatrix} \tag{25}$$

where $\quad d' = 2(1 - C_1 C_2) + \dfrac{1}{h_1 h_2}(h_2^2 - h_1^2) S_1 S_2$

For the case $q_2 < 0$ or $\lambda_b > \lambda_{cr} = \sqrt{1 - \alpha_1 \alpha_2}\sqrt[4]{\dfrac{1}{\alpha_1 \alpha_2 (1 - \alpha_1 \alpha_2)}}$ (26)

the equations (23), (24), (25) can be used after replacing $sinh(\lambda_b\sqrt{q_2})$ by $jsin(\lambda_b\sqrt{-q_2})$ and $\sqrt{q_2}$ by $j\sqrt{-q_2}$.

Expanding the coefficients of *DSM* as in power series and neglecting the higher powers of Ω we obtain the stiffness, mass and damping matrix of the finite element method [2].

4. Vibration equation of structure

Denoting the generalized coordinates and the corresponding generalized forces of the structure by q_i and Q_i $(i = 1, 2, 3, ..., n)$ respectively, we express the boundary conditions as follows

$$\hat{q}_i = 0\ (i = r_1, r_2, .., r_{ng}) \tag{27}$$

$$\hat{Q}_j = \hat{Q}_j(\omega)\quad (j = s_1, s_2, ..., s_{nf}) \tag{28}$$

with the condition $\quad i \neq j\ ;\ n_g + n_f = n$

The equation (27) expresses the geometrical constraints, and equation (28) expresses the force conditions on the structure.

The equation of motion of the constrained structure may be written in the form

$$\hat{Q}^* = K^*\hat{q}^* \tag{29}$$

$$\hat{q}^* = \left(\hat{q}_{s_1}, \hat{q}_{s_2}, ..., \hat{q}_{s_{nf}}\right)^T$$
$$\hat{Q}^* = \left(\hat{Q}_{s_1}, \hat{Q}_{s_2}, ..., \hat{Q}_{s_{nf}}\right)^T \tag{30}$$

K^* is the matrix of order n_f obtained by deleting the rows and columns $r_1, r_2, .., r_{ng}$ of matrix K and called the DSM of the constrained structure.

From equation (29) we have:

$$\hat{q}^* = (K^*)^{-1}\hat{Q}^* \tag{31}$$

The matrix

$$H^* = (K^*)^{-1} \tag{32}$$

is called the dynamic flexibility (DFM) of the constrained structure.

The complex amplitudes of the reaction forces of constrained are obtained by substituting the solution q^* and $q_i = 0$ $(i = r_1, r_2, .., r_{ng})$ into the equation

$$\hat{Q} = K\hat{q} \tag{33}$$

For a random external force with spectral density $S_{Q^*Q^*}(\omega)$, the spectral density of q^* is determined from following equation:

$$S_{q^*q^*}(\omega) = \overline{H}^* S_{Q^*Q^*}(\omega)\,(H^*)^T \tag{34}$$

where $\overline{H}^*$ is the complex conjugate matrix of H^*.

The natural frequencies of the complete structure are determined from the equation [2]:

$$\text{Det}\ |K^*(\omega)| = 0 \ \text{ or } \ \text{Det}\ |H^*(\omega)| \to \infty \tag{35}$$

The natural frequencies of each element of structure are determined from equation:

$$\text{Det}\ |H^*(\omega)| = 0 \ \text{ or } \ \text{Det}\ |K^*(\omega)| \to \infty \tag{36}$$

5. Vibration of the bar structure by wave action

Consider a bar structure, the elements of which are bars with constant cross section.

Denote the local coordinates of a bar element by $\bar{x},\bar{y},\bar{z}$; $\bar{x}$ is the bar axis, $\bar{y},\bar{z}$ are the principal inertia axes of the bar cross section. For a regular wave with frequency ω the complex amplitudes of the wave loading acting on a element of the structure are

$$\begin{aligned}\hat{\bar{q}}_{\bar{y}}(\omega,\bar{x}) &= C_{M\bar{y}}\rho A\hat{\ddot{\bar{u}}}_{o\bar{y}}(\bar{x}) + C_{h\bar{y}}(\bar{x})\hat{\dot{\bar{u}}}_{o\bar{y}} \\ \hat{\bar{q}}_{\bar{z}}(\omega,\bar{x}) &= C_{M\bar{z}}\rho A\hat{\ddot{\bar{u}}}_{o\bar{z}}(\bar{x}) + C_{h\bar{z}}(\bar{x})\hat{\dot{\bar{u}}}_{o\bar{z}}\end{aligned} \tag{37}$$

where ρ is the mass density of the sea water, A is the cross section area of element, $\hat{\dot{u}}_{oy},\hat{\dot{u}}_{oz},\hat{\ddot{u}}_{oy},\hat{\ddot{u}}_{oz}$ are the complex amplitude of the velocity and acceleration of the water particle in the $\bar{y},\bar{z}$ directions respectively, $C_{M\bar{y}},C_{M\bar{z}}$ are inertia coefficient of the section in the $\bar{y}$ and $\bar{z}$ directions respectively.

In equation (37) the drag term is linearized and the coefficients C_{hy},C_{hz} are determined from the following equations:

$$\begin{aligned}C_{h\bar{y}}(\bar{x}) &= \frac{4}{3\pi}\rho C_{D\bar{y}}D_{\bar{y}}\omega\left|\hat{\dot{\bar{u}}}_{r\bar{y}}\right| \\ C_{h\bar{z}}(\bar{x}) &= \frac{4}{3\pi}\rho C_{D\bar{z}}D_{\bar{z}}\omega\left|\hat{\dot{\bar{u}}}_{r\bar{z}}\right|\end{aligned} \tag{38}$$

where $\hat{\dot{\bar{u}}}_{r\bar{y}},\hat{\dot{\bar{u}}}_{r\bar{z}}$ are the complex amplitudes of velocity of the bar relative to the water, C_{Dy},C_{Dz} are the drag coefficients of the bar section in the $\bar{y}$ and $\bar{z}$ direction respectively.

The external damping coefficients of an element in the $\bar{y}$ and $\bar{z}$ direction are determined from the equations:

$$\begin{aligned}k_{ay} &= \frac{\int_o^l C_{h\bar{y}}(\bar{x})\left|g_{\bar{y}}(\bar{x})\right|^2 d\bar{x}}{\int_o^l \left|g_{\bar{y}}(\bar{x})\right|^2 d\bar{x}} \\ k_{az} &= \frac{\int_o^l C_{h\bar{z}}(\bar{x})\left|g_{\bar{z}}(\bar{x})\right|^2 d\bar{x}}{\int_o^l \left|g_{\bar{z}}(\bar{x})\right|^2 d\bar{x}}\end{aligned} \tag{39}$$

where $g_{\bar{y}},g_{\bar{z}}$ are the vibration forms of the bar in direction $\bar{y}$ and $\bar{z}$, which may be exactly determined by equations (18) and (25).

For a random wave the transfer functions of the loading acting on each bar element are

$$\begin{aligned}\bar{H}_{q_{\bar{y}}}(\omega,\bar{x}) &= C_{M\bar{y}}\rho A\bar{H}_{\ddot{u}_{o\bar{y}}}(\omega,\bar{x}) + C_{h\bar{y}}(\bar{x})\bar{H}_{\dot{u}_{o\bar{y}}}(\omega,\bar{x}) \\ \bar{H}_{q_{\bar{z}}}(\omega,\bar{x}) &= C_{M\bar{z}}\rho A\bar{H}_{\ddot{u}_{o\bar{z}}}(\omega,\bar{x}) + C_{h\bar{z}}(\bar{x})\bar{H}_{\dot{u}_{o\bar{z}}}(\omega,\bar{x})\end{aligned} \tag{40}$$

where $\bar{H}_{\dot{u}_{o\bar{y}}(z)}$, $\bar{H}_{\ddot{u}_{o\bar{y}}(\bar{z})}$ are the transfer functions of the velocity and acceleration in direction $\bar{y}(or\ \ \bar{z})$ with the wave height considered as the input.

In equation (40) the coefficients $C_{h\bar{y}},C_{h\bar{z}}$ are determined from:

$$\begin{aligned}C_{h\bar{y}}(\bar{x}) &= \sqrt{\frac{2}{\pi}}\rho C_{D\bar{y}}D_y\sigma_{\dot{\bar{u}}_{ry}} \\ C_{h\bar{z}}(\bar{x}) &= \sqrt{\frac{2}{\pi}}\rho C_{D\bar{z}}D_y\sigma_{\dot{\bar{u}}_{rz}}\end{aligned} \tag{41}$$

where $\sigma_{\dot{\bar{u}}_{r\bar{y}}},\sigma_{\dot{\bar{u}}_{r\bar{z}}}$ are the standard deviation of the relative velocity of the bar element to the water, they are determined from

$$\sigma_{\dot{\bar{u}}_{r\bar{y}}} = \sqrt{2\int_{o}^{\infty} S_{\dot{\bar{u}}_{r\bar{y}}\dot{\bar{u}}_{r\bar{y}}}\, d\omega}$$

$$\sigma_{\dot{\bar{u}}_{r\bar{z}}} = \sqrt{2\int_{o}^{\infty} S_{\dot{\bar{u}}_{r\bar{z}}\dot{\bar{u}}_{r\bar{z}}}\, d\omega} \tag{42}$$

In equation (42) $S_{\dot{\bar{u}}_{r\bar{y}}\dot{\bar{u}}_{r\bar{y}}}$, $S_{\dot{\bar{u}}_{r\bar{z}}\dot{\bar{u}}_{r\bar{z}}}$ are the spectral density of relative velocity of the bar element to the water in directions $\bar{y}$ and $\bar{z}$; they are determined from equations

$$S_{\dot{\bar{u}}_{r\bar{y}}\dot{\bar{u}}_{r\bar{y}}}(\bar{x},\omega) = \left|H_{\dot{\bar{u}}_{r\bar{y}}}(\bar{x},\omega)\right|^2 S_{\xi\xi}(\omega)$$

$$S_{\dot{\bar{u}}_{r\bar{z}}\dot{\bar{u}}_{r\bar{z}}}(\bar{x},\omega) = \left|H_{\dot{\bar{u}}_{r\bar{z}}}(\bar{x},\omega)\right|^2 S_{\xi\xi}(\omega) \tag{43}$$

where $H_{\dot{\bar{u}}_{ry}}, H_{\dot{\bar{u}}_{ry}}$ are the transfer function of the relative velocity of the bar element to the water with the wave height considered as the input, they are determined from

$$\overline{H}_{\dot{\bar{u}}_{ry}} = H_{\dot{\bar{u}}_{o\bar{y}}} - j\omega H_{\bar{u}_{\bar{y}}}$$

$$\overline{H}_{\dot{\bar{u}}_{rz}} = H_{\dot{\bar{u}}_{o\bar{z}}} - j\omega H_{\bar{u}_{\bar{z}}} \tag{44}$$

$\overline{H}_{u_y}$ and $\overline{H}_{u}$ in Eq(44) are the transfer functions of the displacement of bar element with the wave height considered as input, they may be determined from the transfer function of the structure [see Eq.(32)] as follows:

$$H = (K^*)^{-1}.H_F \tag{45}$$

where K^* is the *DSM* of structure and H_F is the transfer function of the wave loading with wave height considered as the input.

Having the loading, we may determine the vibration of the structure, the internal force and the stress in any point of the structure from equations (31), (34), (33), (14), (20).

Since the drag force depends on the vibration of structure, we must use the calculation interactively.

6. Example

Consider the structure shown in Fig 1 a with following input data[4]:

Columns: $D_1 = 9144mm$, $t_1 = 88{,}9mm$.

Braces: $D_2 = 914mm$, $t_2 = 38{,}1mm$.

The lowest horizontal brace $D_3 = 7316mm$, $t_3 = 51mm$.

The height of the first floor = $36{,}57m$

The height of the other floors = $30{,}48m$

The width of structure $b = 67{,}056m$

The water depth $d = 167{,}64m$

The concentrated mass on platform 42000t

Modulus of elasticity $E = 2{,}1.10^8$kN/m^2.

Damping coefficient $k_i = 0{,}02$ ($\hat{E} = E(1 + 0.02j)$)

The structure is modelled by the plane frame as shown in Fig 1b.

The first frequencies of the structure are given in following table:

Table 1 The natural frequencies of the structure is Fig 1.

$\omega\,(s^{-1})$	Elementary case	By including effect of axial force	By including effect of axial force, shear deformation, and rotatory inertia
ω_1	1, 34506	1, 34390	1, 34044
ω_2	5,00853	5,00569	4,93900
ω_3	7,52246	-	-
ω_4	8,85267	8,85311	8,87701
ω_5	9,17911	-	-

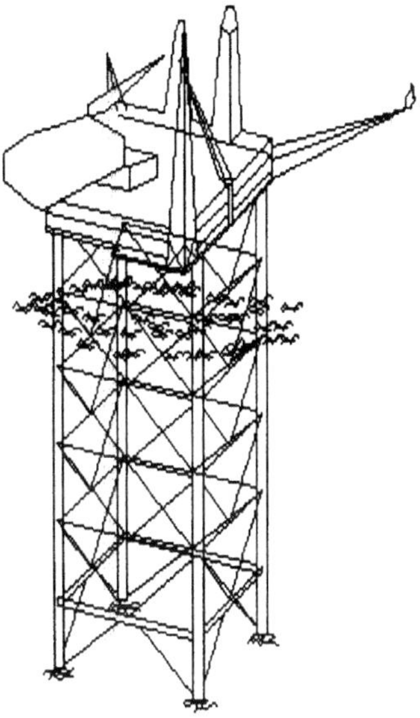

Figure 1a

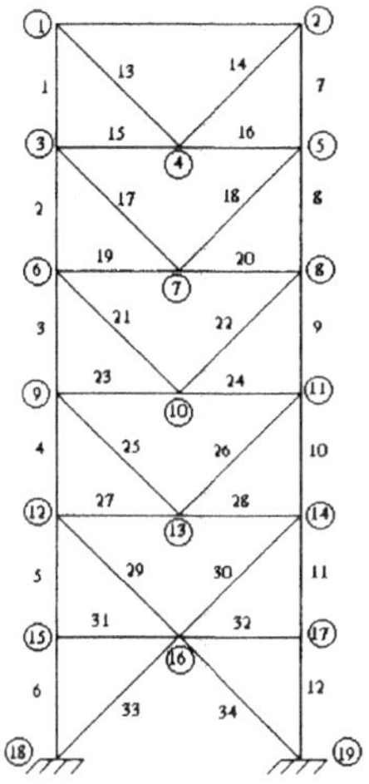

Figure 1b

By assuming that the relationship between the height and period of wave is expressed in the form:

$$h = AT^{B}$$

where $A = 6{,}91.10^{-4}$, $B - 4{,}095$, we find the computated results of the dynamic displacement and stress of structure are given as in table 2.

Table 3 shows the random vibration of the structure by using the wave spectral density function of Pierson Moskowitz[4]:

$$S_{\xi\xi}(\omega) = \frac{1}{2}\frac{A}{\omega^5}\exp\left(-\frac{B}{\omega^4}\right)$$

where $$A = 4\pi^3\frac{h_s^2}{T_o^4},\quad B = \frac{16\pi^3}{T_o^4}$$

In the table 3 $\sigma_{v_x^{(2)}}, \sigma_{\dot{v}_x^{(2)}}, \sigma_{\ddot{v}_x^{(2)}}$ are the standard deviations of the displacement, velocity and acceleration of the node 2 in horizontal direction respectively, $\sigma_{\sigma_{max}}$ is the standard deviation of the maximum stress in the structure.

Table 2 Harmonic vibration of structure in Fig 1 by a regular wave

Wave height $h_s(m)$	Wave frequency $\omega(s^{-1})$	Wave period T(s)	Wave length (m)	Maximum hydrodynamic damping coefficient k_{amax}(kNs/m^2)	Maximum amplitude of displacement (m)	Maximum amplitude of stress (kN/m^2)	The location of maximum stress
0,2	1,574	4,00	24,9	0,055	0,006	2900	node 1 element 21
0,387 (resonance)	1,340	4,69	34,3	1,5	0,5113	234690	node 1 element 22
0,664	1,174	5,35	44,7	0,187	0,0006	2597	node 2 element 15
1,5	0,962	6,53	66,5	0,507	0,0616	29261	node 1 element 22
2,5	0,85	7,40	85,4	0,726	0,0631	30307	node 1 element 22
4,5	0,736	8,54	113,8	1,33	0,0331	17688	node 2 element 16
6,3	0,678	9,27	113,8	2,00	0,0036	22158	node 1 element 15
9,5	0,613	10,25	164	3,1	0,0588	33426	node 1 element 15
13,5	0,563	11,16	194,6	5,01	0,1278	65769	node 2 element 30
16,65	0,536	11,73	214,6	6,4	0,178	94160	node 2 element 30

Table 3 Random vibration of structure in Fig 1

	Spectral density function of Pierson Moskowitz					
	$h_s = 10m;$ $T = 9(s)$	$h_s = 0,5m;$ $T = 5,3(s)$	$h_s = 1m;$ $T = 5(s)$	$h_s = 1m;$ $T = 5,5(s)$	$h_s = 1,5m;$ $T = 5,5(s)$	$h_s = 1m;$ $T = 7(s)$
$\sigma_{v_x^{(2)}}(m)$	0,6091	0,1035	0,2161	0,1881	0,2722	0,1254
$\sigma_{\dot{v}_x^{(2)}}(m/s)$	0,8129	0,1388	0,2899	0,2521	0,3648	0,1679
$\sigma_{\ddot{v}_x^{(2)}}(m/s^2)$	1,0908	0,1864	0,3894	0,3385	0,4900	0,2253
$\sigma_{\sigma_{max}}\left(kN/s^2\right)$	279873	47527	99193	86326	12492	57547
$k_{a\max}\left(kNs/m^2\right)$	4,69	0,61	1,31	1,69	1,695	0,77
Period of peak (s)	4,68	4,67	4,68	4,68	4,68	4,68
Spectral width	0,1039	0,053	0,051	0,061	0,067	0,066

7. References

1. NGUYEN XUAN HUNG .
   Dynamics of offshore structure - *Science - Technical* Pub Hanoi 1998.

2. NGUYEN XUAN HUNG .
   Dynamics of Structure and its Application in structure identification - *IAM – NCST* Vietnam.

3. NGUYEN XUAN HUNG .
   The dynamic stiffness matrix method for Bar structures.
   Proceedings of the Vietnam – Japan Symposium on advances in Applied Electromagnetics and Mechanics.
   January 19-21, 1998 Ho Chi Minh city Vietnam.

4. KARL HEIRZ HAPEL
   Festigkeisanalyse Dynamicsch Beanspruchter Offshore Konstructionen, Vieweg Verlag 1990.

5. BREBBIA C.A, WALKER S.L
   Dynamic Analysis of Offshore Structures, Newnes – Butterworths 1979.

6. PENZIEN J. CLOUGH RW
   Dynamic of Structures, Mc Graw Hill 1975.

7. NATKE H.G.
   Einführung in Theorie und Praxis der Zeitreihen und Modal Analyse , Wieveg Verlag 1992.

ON THE TRANSVERSE VIBRATIONS OF BEAM-BRIDGES UNDER THE ACTION OF SOME MOVING BODIES

NGUYEN VAN KHANG, HOANG HA, VU VAN KHIEM
and DO XUAN THO
Hanoi University of Technology

1. Introduction

Transverse vibration of continuous beam under the action of a moving body has been mentioned in works such as [1,2,3,4]. Currently transverse vibrations of the beam on many intermediate elastic supports are attracting increased attention in cable-stayed bridges. In this work, we use the method of substructures to derive transverse vibration equations of continuous beam with intermediate elastic supports under the action of moving bodies. An algorithm is built to solve the vibration equations. From this algorithm, a computer program is created using Turbo Pascal language.

2. Derivation of vibration equations using method of substructures

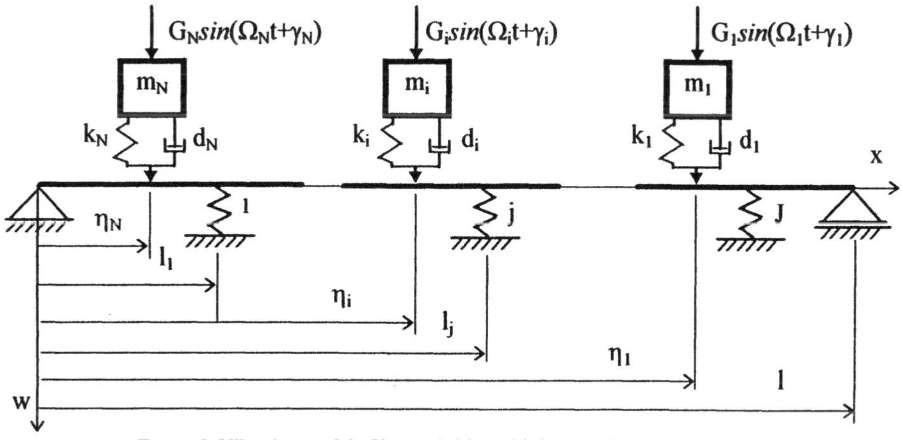

Figure 1. Vibration model of beam- bridge with intermediate elastic supports

Consider a continuous Euler-Bernoulli beam with J intermediate elastic supports and length l (figure 1). Suppose that its mass of length unit is μ ($\mu=\rho A$), and bending rigidity EI is constant across of its length where ρ is mass density, A- cross sectional area, E-elastic modulus, I- centroidal moment of inertia. c_j and l_j ($j=1,...,J$) respectively represent

N. Van Dao and E.J. Kreuzer (eds.),
IUTAM Symposium on Recent Developments in Non-linear Oscillations of Mechanical Systems, 187–195.

the rigidity and the coordinates of intermediate elastic support j. The i-body (i=1,...,N) consists of mass m_i attached to the spring system with rigidity k_i and damper d_i directly proportional to the velocity. The i-body moves with the velocity v_i and is subjected to the action of force $G_i sin(\Omega_i t+\gamma_i)$ caused by the unequilibrated mass which rotates with angled speed Ω_i. Here G_i is the amplitude of force.

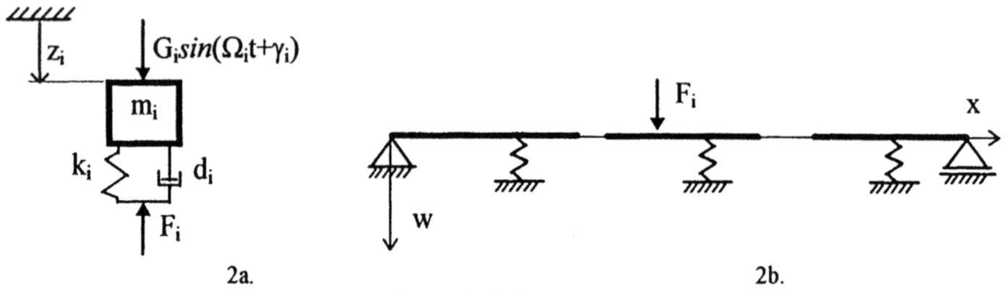

Figure 2. Substructures

Using the method of substructures to derive vibration equations of the beam and the bodies, we divide the system into N+1 substructures: beam and N bodies (figure 2).
The position of the i-body can be determined by

$$\eta_i = v_i\,(t-\tau_i),\quad t \geq \tau_i \tag{1}$$

where τ_i denotes the instant time when the i-body starts moving along the beam with the constant velocity v_i.
Additionally it is supposed that during the motion, the i-body is not separated from the beam and its velocity v_i satisfies the condition of non impact as follows

$$\eta_i > \eta_j \qquad\qquad (i<j)$$

Substructure 2b is considered as a simple beam subject to following forces:
- Pressure $p_t(x,z,t)$ of bodies on beam

$$p_t(x,z,t) = \sum_{i=1}^{N} L_i(t)[m_i g + G_i sin\varphi_i - m_i\,\ddot{z}_i\,]\,\delta(x-\eta_i) \tag{2}$$

with $\varphi_i = \Omega_i t + \gamma_i$
- Reactions of elastic supports

$$p_g(x,z,t) = -\sum_{j=1}^{J} c_j w(l_j,t)\,\delta(x-l_j) \tag{3}$$

Here we apply the Dirac-function $\delta(x-a)$ and the logic-function $L_i(t)$, which are defined by the following relations

$$L_i(t) = \begin{cases} 1 & \text{when} \quad \tau_i \le t \le T_i + \tau_i \\ 0 & \text{when} \quad t < \tau_i \text{ and } t > \tau_i \end{cases} \tag{4}$$

with $T_i = \dfrac{1}{v_i}$,

$$\delta(x\text{-}a) = \lim_{\varepsilon \to 0} \delta_\varepsilon(x\text{-}a) \tag{5}$$

with

$$\delta_\varepsilon(x-a) = \begin{cases} 1/2\varepsilon & \text{when} \quad |x-a| \le \varepsilon \\ 0 & \text{when} \quad |x-a| > \varepsilon \end{cases} \tag{6}$$

Vibrational differential equations for substructures can be obtained by applying the basic principles of dynamics. The equation describing transverse vibration of beam including internal friction is

$$EI\left(\frac{\partial^4 w}{\partial x^4} + \alpha \frac{\partial^5 w}{\partial x^4 \partial t}\right) + \mu\left(\frac{\partial^2 w}{\partial t^2} + \beta \frac{\partial w}{\partial t}\right) = p(x,z,t) \tag{7}$$

$$p(x,z,t) = p_g(x,z,t) + p_t(x,z,t)$$

in which α and β are damping constants.
The equation describing the vibration of i-body has the following form:

$$L_i(t)(m_i \ddot{z}_i + d_i \dot{z}_i + k_i z_i) = L_i(t)(m_i g + G_i \sin\varphi_i + d_i \dot{w}_{\eta_i} + k_i w_{\eta_i}) \tag{8}$$

$(i = 1,\ldots N)$

in which $w_{\eta_i} = w(t, \eta_i)$; $\dot{w}_{\eta_i} = \dfrac{\partial w(t, \eta_i)}{\partial t}$

The equations of motion (7) and (8) are a mixture of ordinary and partial differential equations. Four boundary conditions, two at x=0, two at x=l, and initial conditions must be specified for the solution of the equations.
The boundary conditions have the following forms:

$$x = 0: \qquad w(0,t) = \frac{\partial^2 w(0,t)}{\partial x^2} = 0 \tag{9}$$

$$x = l: \qquad w(l,t) = \frac{\partial^2 w(l,t)}{\partial x^2} = 0 \tag{10}$$

The initial and transformation conditions have expressions of the form

$$t=\tau_i: \qquad w(t,\tau_i) = f_i^{(1)}(x), \quad \frac{\partial w(t,\tau_i)}{\partial t} = f_i^{(2)}(x) \tag{11}$$

$$z_j(\tau_i) = z_j^{(0)}(\tau_i), \quad \dot{z}_j(\tau_i) = \dot{z}_j^{(0)}(\tau_i) \tag{12}$$

$$(i,j=1,\dots,N)$$

When we consider the non-linearity of intermediate elastic supports, we find the reaction of the elastic supports has the following form

$$p_g(x,z,t) = -\sum_{j=1}^{J} [c_j^{(1)} w(l_j,t) + c_j^{(3)} w^3(l_j,t)]\,\delta(x-l_j) \tag{13}$$

Thus, the equation describing non-linear transverse vibrations of beam is

$$EI\left(\frac{\partial^4 w}{\partial x^4} + \alpha\frac{\partial^5 w}{\partial x^4 \partial t}\right) + \mu\left(\frac{\partial^2 w}{\partial t^2} + \beta\frac{\partial w}{\partial t}\right) = p(x,z,t) \tag{14}$$

with

$$p(x,z,t) = \sum_{i=1}^{N} L_i(t)[m_i g + G_i \sin\varphi_i - m_i \ddot{z}_i]\,\delta(x-\eta_i) - \sum_{j=1}^{J} [c_j^{(1)} w(l_j,t) + c_j^{(3)} w^3(l_j,t)]\,\delta(x-l_j) \tag{15}$$

3. Transformation into ordinary differential equations

Assume the solution of (7) and (8) with the boundary conditions (9) and (10) to be of the form

$$w(x,t) = \sum_{r=1}^{n} q_r(t) \sin\frac{r\pi x}{l} \tag{16}$$

in which $q_r(t)$ $(r=1,\dots,n)$ are generalized coordinates to be determined. Substituting solution (16) into equations (7) and (8), we find

$$\ddot{q}_s = -\sum_{r=1}^{n}\{\delta_r^s[\frac{EI\alpha}{\mu}(\frac{\pi}{l})^4 s^4 + \beta] + \frac{2}{\mu l}\sum_{i=1}^{N} L_i(t)(d_i \sin\frac{s\pi\eta_i}{l}\sin\frac{r\pi\eta_i}{l})\}\,\dot{q}_r$$

$$+\frac{2}{\mu l}\sum_{i=1}^{N} L_i(t)(d_i \sin\frac{s\pi\eta_i}{l})\,\dot{z}_i - \sum_{r=1}^{n}\{\delta_r^s[\frac{EI}{\mu}(\frac{\pi}{l})^4 s^4 +$$

$$+\frac{2}{\mu l}\sum_{i=1}^{N} L_i(t)(d_i\frac{r\pi v_i}{l}\cos\frac{r\pi\eta_i}{l} + k_i \sin\frac{r\pi\eta_i}{l})\sin\frac{s\pi\eta_i}{l}\}\,q_r$$

$$-\frac{2}{\mu l}\sum_{r=1}^{n}(\sum_{k=1}^{J} c_k \sin\frac{s\pi l_k}{l}\sin\frac{r\pi l_k}{l})\,q_r + \frac{2}{\mu l}\sum_{i=1}^{N} L_i(t)(k_i \sin\frac{s\pi\eta_i}{l})\,z_i$$

$$(s=1,\ldots,n) \qquad (17)$$

and

$$L_i(t)\,\ddot{z}_i = L_i(t)\{\sum_{r=1}^{n}(\frac{d_i}{m_i}\sin\frac{r\pi\eta_i}{l})\,\dot{q}_r - \frac{d_i}{m_i}\,\dot{z}_i$$

$$+\sum_{r=1}^{n}(\frac{k_i}{m_i}\sin\frac{r\pi\eta_i}{l} + \frac{d_i r\pi v_i}{l m_i}\cos\frac{r\pi\eta_i}{l})\,q_r - \frac{k_i}{m_i}z_i + g + \frac{G_i}{m_i}\sin\varphi_i\}; \qquad (18)$$

$$(i=1,\ldots,N)$$

in which

$$\delta_r^s = \begin{cases} 1 & r = s \\ 0 & r \neq s \end{cases}$$

The differential equations (17) and (18) can be written in the following matrix form

$$\ddot{\mathbf{q}} = \mathbf{B}(t)\,\dot{\mathbf{q}} + \mathbf{C}(t)\mathbf{q} + \mathbf{f}(t) \qquad (19)$$

in which **B**(t), **C**(t) are square matrices of (n+N) degree, **q**, **f** are column vectors with (n+N) elements. Where

$$\mathbf{q} = [q_1, \ldots, q_n, L_1(t)z_1, \ldots, L_N(t)z_N]^T$$

$$\mathbf{f} = [0, \ldots, 0, f_{n+1}, \ldots, f_{n+N}]^T$$

The elements of the vector **f**, matrices **B** and **C** have the forms

$$f_{n+i} = L_i(t)[g + \frac{G_i}{m_i}\sin\varphi_i];\ (i = 1,\ldots,N)$$

$$b_{sr} = -\ \delta_r^s[\frac{EI\alpha}{\mu}(\frac{\pi}{l})^4 s^4 + \beta] - \frac{2}{\mu l}\sum_{i=1}^{N} L_i(t) d_i \sin\frac{s\pi\eta_i}{l}\sin\frac{r\pi\eta_i}{l};$$

$$b_{s,n+i} = \frac{2}{\mu l} L_i(t) d_i \sin\frac{s\pi\eta_i}{l}; \quad b_{n+i,r} = L_i(t)\frac{d_i}{m_i}\sin\frac{r\pi\eta_i}{l}; \quad b_{n+i,n+j} = -\delta_j^i L_i(t)\frac{d_i}{m_i};$$

$$(s,r = 1,\ldots,n;\ i,j = 1,\ldots,N)$$

$$c_{sr} = -\delta_r^s \frac{EI}{\mu}\left(\frac{\pi}{l}\right)^4 s^4 - \frac{2}{\mu l}\sum_{k=1}^{J} c_k \sin\frac{s\pi l_k}{l}\sin\frac{r\pi l_k}{l} -$$

$$\frac{2}{\mu l}\sum_{i=1}^{N} L_i(t)\left(d_i \frac{r\pi v_i}{l}\cos\frac{r\pi\eta_i}{l} + k_i \sin\frac{r\pi\eta_i}{l}\right)\sin\frac{s\pi\eta_i}{l};$$

$$c_{s,n+i} = \frac{2}{\mu l} L_i(t)\, k_i \sin\frac{s\pi\eta_i}{l}; \quad c_{n+i,r} = L_i(t)\left[\frac{k_i}{m_i}\sin\frac{r\pi\eta_i}{l} + \frac{d_i r\pi v_i}{l m_i}\cos\frac{r\pi\eta_i}{l}\right];$$

$$c_{n+i,n+j} = -\delta_j^i L_i(t)\frac{k_i}{m_i};$$

$$(s,r = 1,\ldots,n;\ i,j = 1,\ldots,N)$$

4. The transverse vibration of Dakrong cable-stayed-bridge in Quangtri province

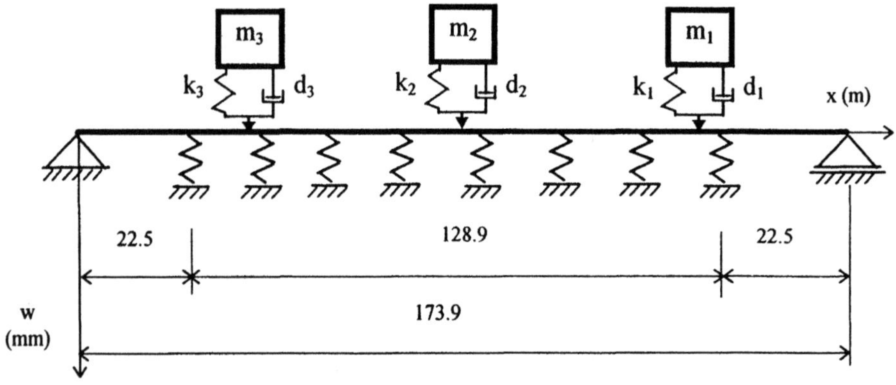

Figure 3. The calculated model of Dakrong cable-stayed-bridge in Quangtri province

The calculated model of Dakrong cable-stayed-bridge in Quangtri province in Vietnam influenced by the action of moving bodies is shown in figure 3. The comparison of calculated results with the experimental results is represented in figures 4 and 5. In figure 4 the curve of the transverse deflection of the cross section $x = 0.07l$ is demonstrated when there are 3 bodies with masses $m_1 = m_2 = m_3 = 13030$ kg, velocities $v_1 = 20$ km/h, $v_2 = 15$ km/h and $v_3 = 12$ km/h. The dotted curve is the experimental result, the full curve is the calculated result. Figure 5 shows the graph of the transverse deflection of the cross section in the middle of the beam when there is one body with mass m = 13030 kg and velocity v = 25 km/h. The dotted curve is experimental, the full curve is the calculated result. Figure 6 shows the curve the transverse deflection of the cross section in the middle of beam with one moving body. The maximum transverse deflection of the beam corresponding to different values of the velocity of moving bodies on the bridge is demonstrated in figure 7.

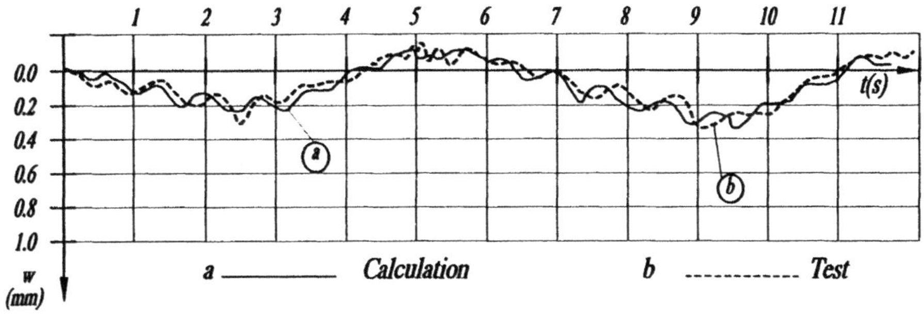

Figure 4. The transverse deflection of the cross section x = 0.07l, for three moving bodies with masses $m_1 = m_2 = m_3 = 13030$ kg, velocities $v_1 = 20$ km/h, $v_2 = 15$ km/h, $v_3 = 12$ km/h.

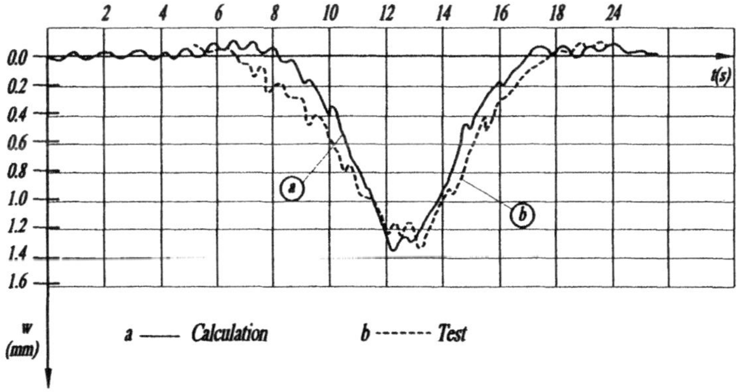

Figure 5. The transverse deflection of the cross section in the middle of beam, for one moving body with mass m = 13030 kg and velocity v = 25 km/h

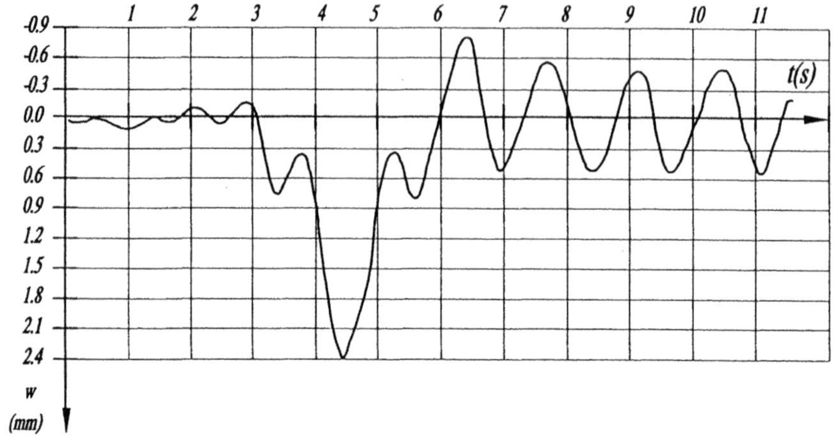

Figure 6. The transverse deflection of the cross section in the middle of the beam, for one moving body with mass m = 13030 kg and velocity v = 105 km/h

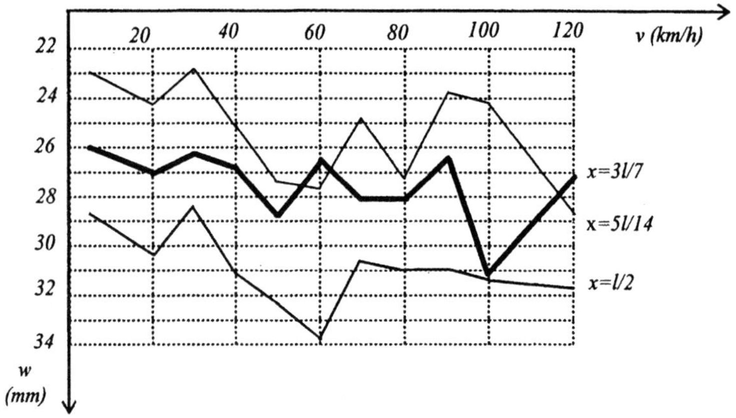

Figure 7. The dependence of the maximum transverse deflection on the velocity of moving bodies in three cross sections, for two bodies with masses $m_1 = m_2 = 13030$ kg and velocities $v_1 = v_2 = v$

5. Conclusion

A system of vibration differential equations of a continuous beam with elastic intermediate supports under the action of moving bodies has been constructed by applying the substructure method. An algorithm and a computer program (VIBEAM)

for the calculation of vibrations of beams have been created. The VIBEAM program can be used for calculations in the design of bridges that will bear moving bodies.

This paper was completed with the financial support of the Vietnam Basic Research Program in Natural Science.

References

1. A.P. Filippov, X.X. Kokhmaniuk, IU.X. Borobuv: *Action of Dynamically Loads on the Elements of Construction* (in Russian). Naukova dumka, Kiev (1974).
2. K. Popp, W. Schiehlen: *Fahrzeugdynamik*. B.G. Teubner, Stuttgart (1993).
3. Do Xuan Tho: *Calculation of the Transverse Vibration of Continuous Beams under Action of a Moving Body*. Ph.D Thesis, Hanoi University of Technology, Hanoi (1996).
4. Nguyen Van Khang, Hoang Ha: Determining parameters of the transverse vibration model for some beam bridges on the highway (in Vietnamese). *Vietnamese Journal of Science and Technology*, Hanoi 36, 1(1998), pp. 22-30.

PILE SOIL SYSTEM IDENTIFICATION BY A MODAL TESTING APPROACH

NGUYEN TIEN KHIEM, DAO NHU MAI,
NGUYEN VIET KHOA, LE VAN ANH
Department of Technical Diagnosis,
Institute of Mechanics
264 Doi Can Str., Hanoi, Vietnam

1. Introduction

In current construction methods the common practice is to use pile foundations, especially in areas of soft soil conditions. Once a pile has been driven into the ground, the task of checking its bearing capacity and integrity is required. For a long time the static test has been the only way to evaluate the pile foundation capacity. However, its use is restricted by the cost of implementation and the small amount of information resulting from the test. It is often necessary to know more specific parameters of the pile and the properties of soil, for use in further analysis of the overall structure to be constructed. In addition, the pile might be damaged during its installation and any information about the pile integrity could be very useful for evaluating its capacity. These requirements can be provided only by using a non-destructive technique which measures the vibration response of the structure to obtain its modal parameters; this is called the modal testing. This paper is devoted to the problem of parameter identification of the pile soil system by the modal testing method.

The pile is usually treated as a bar; its free axial vibrations described by an one-dimensional wave equation have been studied comprehensively in the literature. R. D. Adams et al. were among the first authors to study the nature of a damaged bar. In [1] they suggested modelling a damage located at a position in the bar by an axial spring, and investigated the damage detection problem based on receptance analysis. Further studies in this topic were carried out by Y. Narkis in [2]. The local flexibility model of damage is an useful method for solving the problem of damage detection, however it does not cover the case of the distributed damage. The aim of our study is to introduce damage

N. Van Dao and E.J. Kreuzer (eds.),
IUTAM Symposium on Recent Developments in Non-linear Oscillations of Mechanical Systems, 197–206.

length into the damage detection problem for a pile-soil system. The soil is modelled by a spring and dashpot. Our specific contribution is to distinguish between the stiffness of the soil surrounding the pile at the lower pile edge, and the general soil stiffness.

In section 2, a vibration model of the pile-soil system will be established. In the following sections two problems are considered: a system parameter identification problem when the pile has no defect (section 3); a pile damage detection problem when soil parameters are assumed to be given (section 4). Procedures for solving these two problems will be illustrated by numerical examples.

2. Vibration Model of System

2.1 DAMAGED PILE MODEL

Suppose that a bar of the total length L, cross sectional area F, density ρ and Young's modulus E has a defect in the interval (x_1, x_2). The length of the defect $\Delta x = x_2 - x_1$ is assumed to be small enough so that the defect bar specimen described by the parameters E_x, F_x, ρ_x can be considered as an axial spring of stiffness K. Let u_1, u_2 be the axial displacements of the bar at the sections x_1, x_2 respectively, then

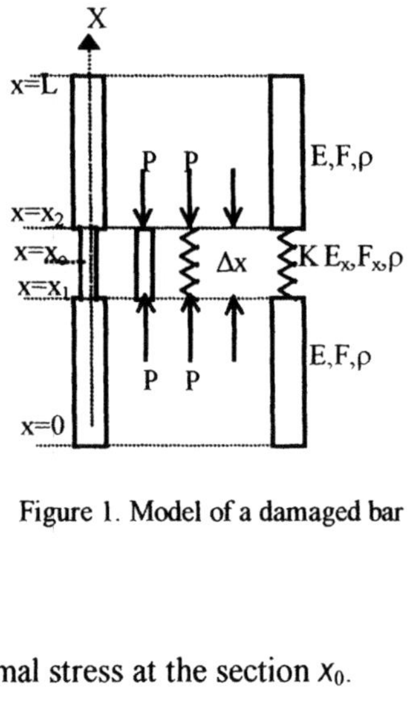

Figure 1. Model of a damaged bar

$$P = F_x\sigma = E_xF_x\frac{u_2 - u_1}{\Delta x} = K(u_2 - u_1)$$

where P the tension force along the axial, σ - the normal stress at the section x_0. From this relationship the stiffness K can be found as

$$K = \frac{E_xF_x}{\Delta x};$$

Introduce a dimensionless stiffness k by

$$k = \frac{K}{EF} = \frac{E_xF_x}{EF}\cdot\frac{1}{\Delta x} = \frac{\gamma}{\Delta x}.$$

There are some specific cases:

a) Undamaged pile: $E_x=E$, $F_x=F$, $\Delta x\to 0 \Rightarrow k\to\infty$

b) Change in cross section area: $E_x=E$, $\Delta x\neq 0 \Rightarrow \quad \gamma_F = \dfrac{F_x}{F} = k\Delta x$ (1)

c) Change in elasticity modulus: $F_x=F$, $\Delta x\neq 0 \Rightarrow \quad \gamma_E = \dfrac{E_x}{E} = k\Delta x$ (2)

If $E_x/E \to \infty$, then $k=\infty$ and one gets an absolutely rigid specimen of length $\Delta x \neq 0$; on the other hand, when $E_x/E \to 0$, $k=0$ we have a gap of length Δx (in the sense of stiffness).

d) Cracked pile: $E_xF_x \to 0$, $\Delta x \to 0$, but

$$\lim_{\Delta x\to 0} \frac{E_xF_x}{\Delta x} = k_c \Rightarrow k = \frac{k_c}{EF} \tag{3}$$

as considered in [1] by Adams R. D. et al. For instant, according to [3], the coefficient k_c can be found by the formulas

$$\frac{1}{k_c} = \frac{EF}{k} = 2\pi b\gamma^2\left(0.7442 - 0.8463\gamma + 1.376\gamma^2 - 0.7540\gamma^3 + 0.5476\gamma^4\right) \tag{4}$$

where $\gamma = \dfrac{2d}{b}$; d - crack depth, b - beam high. This formula is obtained in conjunction with the stress intensity factor as equivalent spring constant for axial tension.

2.2 VIBRATION MODEL OF PILE-SOIL SYSTEM

Consider the pile soil system modelled in Fig. 2. Impedance characteristics of the soil surrounding the pile are elastic coefficient K_x, viscous damping coefficient C_x. For the soil at the edge of the pile only the elastic coefficient K_0 is considered. Free vibration of the system is given by the equation

$$\rho F\frac{\partial^2 u}{\partial t^2} + C_x\frac{\partial u}{\partial t} + K_x u - EF\frac{\partial^2 u}{\partial x^2} = 0 \tag{5}$$

Boundary conditions at $x=0$, $x=L$ and at sections $x=x_1$, $x=x_2$ have forms

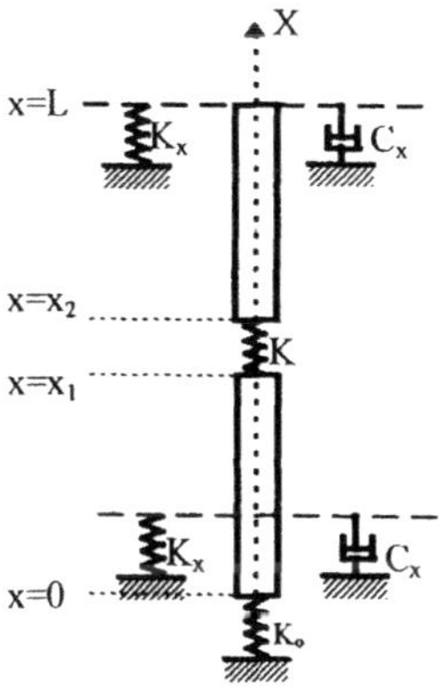

Figure 2

$$\left(EF\frac{\partial u}{\partial x}-K_o u\right)\bigg|_{x=0}=0;\quad EF\frac{\partial u}{\partial x}\bigg|_{x=L}=0; \tag{6}$$

$$\left(EF\frac{\partial u}{\partial x}+Ku\right)\bigg|_{x_1}=Ku\big|_{x_2};\quad \left(EF\frac{\partial u}{\partial x}-Ku\right)\bigg|_{x_2}=-Ku\big|_{x_1}. \tag{7}$$

Introducing the notations

$$a^2=\frac{E}{\rho};\quad h=\frac{C_x}{2\rho F};\quad k_x=\frac{K_x}{EF};\quad k_0=\frac{K_0}{EF};\quad k=\frac{K}{EF} \tag{8}$$

and assuming the well known form of free vibration

$$u(x,t)=\Phi(x)\cdot\exp\{(-h+i\omega)t\}$$

we find that equation (5) becomes

$$\Phi''(x)+\lambda^2\Phi(x)=0; \tag{9a}$$

and conditions (6), (7) reduce to

$$\begin{aligned}&\Phi'(L)=0 \qquad \Phi'(0)-k_o\Phi(0)=0\\ &\Phi'(x_1)+k[\Phi(x_1)-\Phi(x_2)]=0\\ &\Phi'(x_2)-k[\Phi(x_2)-\Phi(x_1)]=0\end{aligned} \tag{9b}$$

where $$\lambda^2=\frac{\omega^2+h^2}{a^2}-k_x. \tag{9c}$$

The appropriate solution of equation (9a) is

$$\Phi(x)=\begin{cases}C_1L_1(x) & 0\le x\le x_1\\ C_2L_2(x) & x_2\le x\le L\end{cases}\quad \text{where}\quad \begin{aligned}L_1&=\lambda\cos\lambda x+k_o\sin\lambda x\\ L_2&=\cos\lambda(x-L)\end{aligned}$$

functions L_1, L_2 obviously satisfy two first conditions in (9b). C_1 and C_2 are arbitrary constants that will be determined from the two last conditions in (9b). Thus, substituting the expression of Φ into (9b) leads to a system of linear equations for C_1 and C_2. The condition for existence of non-trivial solutions with respect to C_1, C_2 is

$$L_1'(x_1)\cdot L_2'(x_2)-k[L_1'(x_1)\cdot L_2(x_2)-L_1(x_1)\cdot L_2'(x_2)]=0$$

or

$$\frac{1}{k}+\frac{k_0\tan(\lambda x_1)+\lambda}{\lambda(k_0-\lambda\tan(\lambda x_1))}+\frac{1}{\lambda}\cot(\lambda(x_2-L))=0. \tag{10}$$

This is the frequency (characteristic) equation of the system. We distinguish particular cases

a) Undamaged pile $x_1=x_2$; $k=\infty$:

$$\lambda\tan(\lambda L)=k_0 \tag{11a}$$

b) Pile with a rigid specimen of length $\Delta x\neq 0$; $k=\infty$:

$$\lambda\tan(\lambda(L-\Delta x))=k_0 \tag{11b}$$

Equation (11b) differs from (11a) by the pile length reduced from L to $L'=L-\Delta x$

c) The case investigated by Adams R. D. et al. [1]: $\Delta x\to 0$, $k_0=0$:

$$\frac{1}{k}-\frac{1}{\lambda}\left[\cot(\lambda x_0)+\cot(\lambda(L-x_0))\right]=0 \tag{11c}$$

d) If $k_0\to\infty$, equation (10) takes the form

$$\frac{1}{k}+\frac{1}{\lambda}\tan(\lambda x_1)+\frac{1}{\lambda}\cot(\lambda(x_2-L))=0 \tag{11d}$$

3. Pile-Soil System Parameter Identification

In this section, we consider the problem of determining the system parameters: the pile length (L) and the soil properties such as C_x, K_x, K_0, when there are measured natural frequencies $\omega_1^*,\cdots,\omega_M^*$ and structural damping ratios $h_1^*,\cdots,h_M^*$. This problem includes the undamaged pile-soil system (11a) and the case of pile with a rigid specimen (11b). Thus, the basic equations for solving the problem are

$$\lambda_j^2=\frac{\omega_j^{*2}+h_j^{*2}}{a^2}-k_x,\qquad C_x=2\rho F h_j^*,$$

$$\lambda_j\tan(\lambda_j L)=k_0;\qquad j=1,\ldots,M \tag{12}$$

It is evident that C_x may be calculated approximately by

$$C_x=\frac{2\rho F}{M}\sum_1^M h_j^*$$

Three unknowns L, k_x, k_0 must be determined, therefore three modes should be measured. We simplify by using the following change in notations:

$$\beta_j = \lambda_j L\,;\; Lk_0 = z\,;\; \varpi_j^2 = \omega_j^{*2} + h_j^{*2}\,;\; p = \frac{L^2}{a^2}\,;\; q = L^2 k_x$$

equations (12) can be written as

$$\beta_j^2 - \varpi_j^2 p + q = 0\,; \qquad \beta_j \tan\beta_j = z\,; \qquad j=1,..3 \tag{13}$$

From the first equations in (13) one gets

$$p = \frac{\beta_2^2 - \beta_1^2}{\varpi_2^2 - \varpi_1^2}\,; \quad q = \frac{\varpi_1^2\beta_2^2 - \varpi_2^2\beta_1^2}{\varpi_2^2 - \varpi_1^2}\,; \tag{14}$$

and

$$\left(\varpi_1^2 - \varpi_2^2\right)\cdot\beta_3^2 + \left(\varpi_2^2 - \varpi_3^2\right)\cdot\beta_1^2 + \left(\varpi_3^2 - \varpi_1^2\right)\cdot\beta_2^2 = 0\,. \tag{15}$$

The last equations in (13) imply that β_j, $j=1,..3$ are solutions of the equation $\beta\tan\beta = z$, therefore the β_j will be functions of z and can be found by numerical computation. Thus, we need to solve only one equation (15) with respect to z:

$$f(z) \equiv \left(\varpi_1^2 - \varpi_2^2\right)\cdot\beta_3^2(z) + \left(\varpi_2^2 - \varpi_3^2\right)\cdot\beta_1^2(z) + \left(\varpi_3^2 - \varpi_1^2\right)\cdot\beta_2^2(z) = 0 \tag{16}$$

where $\beta_j(z)$, $j=1, 2, 3$ are solutions of the equation $\beta\tan\beta = z$, corresponding to the measured frequencies ϖ_j. Suppose that z^* is a solution of (16), the parameters p and q can be calculated for the z^* by the formulas (14)

$$p^* = \frac{\beta_2^2(z^*) - \beta_1^2(z^*)}{\varpi_2^2 - \varpi_1^2}\,; \qquad q^* = \frac{\varpi_1^2\beta_2^2(z^*) - \varpi_2^2\beta_1^2(z^*)}{\varpi_2^2 - \varpi_1^2}$$

Furthermore, having z^*, p^* and q^*, we can determine the unknowns L, k_0, k_x from

$$L = a\sqrt{p^*}\,,\; k = \frac{Z^*}{a\sqrt{p^*}}\,, k_x = \frac{q^*}{a^2 p^*} \text{ or } K_0 = \frac{maz^*}{\sqrt{p^*}}\,, K_x = \frac{mq^*}{p^*}$$

where $m = \rho F$.

A program has been written to carry out the proposed algorithm, and a numerical example has been constructed to verify the program. Inputs for the program are: a, m and ω_1^*, ω_2^*, ω_3^*, h_1^*, h_2^*, h_3^*. In the results there are three parameters: L, K_0, K_x. In the example, three cases have been studied, in each case one of the parameters takes different values and two others remain unchanged. Table 1 shows results of the identification compared with the initial values of that parameters. The results show that the method is accurate.

Table 1

Given			Identified		
L (m)	K_0 (10^7 N/m)	K_x (10^6 N/m^2)	L^*(m)	K_0^* (10^7 N/m)	K_x^* (10^6 N/m^2)
10	2.94	2.94	9.99997	2.94955	2.94022
20	2.94	2.94	20.0002	2.94021	2.94002
30	2.94	2.94	29.9999	2.94010	2.94002
40	2.94	2.94	40.0006	2.94067	2.93999
50	2.94	2.94	50.0006	2.94070	2.93998
20	1.47	2.94	19.99992	1.46975	2.94006
20	2.94	2.94	20.00017	2.94024	2.94002
20	4.41	2.94	19.99990	4.41001	2.94005
20	5.88	2.94	20.00030	5.88132	2.93995
20	7.35	2.94	20.00029	7.35182	2.93986
20	1.47	2.94	19.99994	1.46977	2.94006
20	1.47	5.88	19.99994	1.46976	5.88006
20	1.47	8.82	19.99990	1.46976	8.82006
20	1.47	11.76	19.99990	1.46978	11.76006
20	1.47	14.70	19.99990	1.46977	14.70005

4. Damage detection of pile

Consider the damaged pile-soil system as shown in Fig. 2 with the parameters $a = \sqrt{\frac{E}{\rho}}$, $m=\rho F$, L, K_x, C_x, K_0 assumed to be previously determined ($K_0 = \infty$). Furthermore, let ω_j^*, h_j^*, $j = 1,..3$ be measured; the damage detection problem is to find three parameters: K, $\Delta x = x_2 - x_1$, $x_0 = \frac{1}{2}(x_1 + x_2)$ denoting the damage degree, the length of the damage and its position, respectively. For this purpose, the frequency equation (11.d) is transformed into the form

$$\frac{1}{\overline{k}} = \frac{1}{\beta}\left\{\frac{2\cos\beta(1-\overline{\Delta x})}{\sin\beta(1-\overline{\Delta x}) + \sin\beta(1-2\overline{x}_0)}\right\} \tag{17}$$

where $\bar{k} = kL$, $\beta = \lambda L$, $\overline{\Delta x} = \Delta x/L$, $\bar{x}_0 = x_0/L$ .Using the formulas (10.c) with the given parameters one has

$$\beta_j^* = L\sqrt{\frac{\omega_j^{*2} + h_j^{*2}}{a^2} - k_x}, j = 1,2,3 \tag{18}$$

as given data for the damage detection. Thus, $\overline{\Delta x}, \bar{x}_0$ can be found from the two equations

$$f_1(\bar{x}_0, \overline{\Delta x}) = f_2(\bar{x}_0, \overline{\Delta x}) = f_3(\bar{x}_0, \overline{\Delta x}) \tag{19}$$

where

$$f_j(\bar{x}_0, \overline{\Delta x}) = \frac{\cos\beta_j^*(1 - \overline{\Delta x})}{\beta_j^*\left[\sin\beta_j^*(1 - \overline{\Delta x}) + \sin\beta_j^*(1 - 2\bar{x}_0)\right]}, j = 1,2,3$$

Thereafter, the damage degree $\bar{k}$ is calculated by (17) using the solution $\overline{\Delta x}^*$, $\bar{x}_0^*$ of the equations (19). The system of equations (19) may be solved by one of the well known numerical methods. However, because of the error in either measurements or computation the system of equations (19) may have no solution. To overcome the later case we must chose another approach that will be given below.

Introducing following notations

$$\begin{aligned}
&\vec{a} = (a_1, a_2, a_3)^T \quad \vec{b} = (b_1, b_2, b_3)^T \\
&|\vec{a}|^2 = \sum_{j=1,}^{3} a_j^2 \quad |\vec{b}|^2 = \sum_{j=1,}^{3} b_j^2, \\
&(\vec{a}, \vec{b}) = \sum_{j=1}^{3} a_j b_j \\
&a_j = \cos\beta_j^*(1 - \overline{\Delta x}), \\
&b_j = \beta_j^*\left[\sin\beta_j^*(1 - \overline{\Delta x}) + \sin\beta_j^*(1 - 2\bar{x}_0)\right] \; j = 1,2,3
\end{aligned} \tag{20}$$

system (19) implies existence of a scalar α such that $\vec{a} = \alpha\vec{b}$. This condition leads to the equality $E(\bar{x}_0, \overline{\Delta x}) \equiv |\vec{a}|^2.|\vec{b}|^2 - (\vec{a}, \vec{b})^2 = 0$, that must be satisfied by the solution of equations (19). Of course, due to measurement error the function $E(\bar{x}_0, \overline{\Delta x}) > 0$. Therefore, we can find an approximate solution of the system (19) by solving the optimisation problem

$$\min E(\overline{x}_0, \overline{\Delta x});\ \forall(\overline{x}_0, \overline{\Delta x}) \in G = \{0 \le \overline{\Delta x} < 1; 0 < \overline{x}_0 < 1\} \tag{21}$$

For the obtained approximate solution $\overline{\Delta x}^*, \overline{x}_o^*$ the damage magnitude will be found by the formula

$$d = \overline{k}^{-1} = 2(\vec{a}^*, \vec{b}^*).|\vec{b}|^{-2}. \tag{22}$$

This approximation solution will be the exact one if measurement is free of noises. An example has been investigated to illustrate the proposed procedure; results of the damage detection problem are given in Table 2. The error function $E(\overline{x}_0, \overline{\Delta x})$ is plotted versus $\overline{x}_0$ for $\overline{\Delta x} = 0.05$ and versus $\overline{\Delta x}$ for $\overline{x}_0 = 0.5$ and given in Fig. 3, 4. A minimum of the function is clearly defined as shown in the graphics and it is very close to zero because in this particular case no measurement error has been assumed. This solution is exact one.

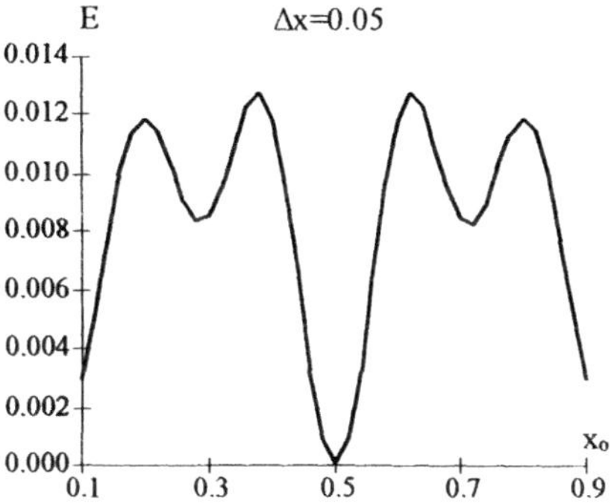

Figure 3

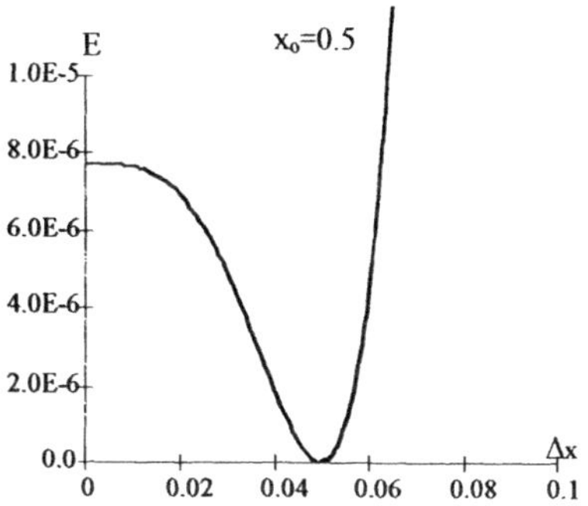

Figure 4

Table 2.

Given			Identified		
$1/\overline{k}$	$\overline{\Delta x}$	$\overline{x}_0$	$1/\overline{k}$	$\overline{\Delta x}$	$\overline{x}_0$
0.1	0.05	0.5	0.09998	0.05000	0.50000
0.2	0.07	0.2	0.20000	0.07000	0.20000
0.3	0.09	0.6	0.30001	0.09000	0.60000
0.4	0.03	0.8	0.40001	0.03000	0.80000
0.5	0.01	0.3	0.49995	0.01000	0.30000

5. Conclusion

This report presented some results in the parameter identification and damage detection for a pile-soil system. First, a general model of the pile-soil system was introduced to describe not only the pile and the soil, but also the damage of the pile. The damage was modelled by an axial spring, and the length of the damage is also an unknown parameter. For the soil a distinction was made between the stiffness of the soil surrounding the pile and the soil at the lower pile edge. Next, by dynamic testing approach two problems were solved for the pile-soil system: the model parameters identification for undamaged pile; and the damage detection when soil parameters are given.

The research results have been implemented in a project for assessing and evaluating the technical condition of the girders and pillars of a series of bridges in Thaibinh province, Vietnam.

Acknowledgement: The work has completed with the support from the Vietnam National Program for Basic Research in Natural Sciences.

References

1. Adams R. D., Cawley P., Pye C. J. and Stone B. J.: A vibration technique for non-destructively assessing the integrity of structures, *J. Mechanical Engineering Science*, Vol. 20, No. 2 (1978), 93-100.

2. Narkis Y.: Identification of crack location in vibrating simply supported beams, *J. of Sound and Vibration*, Vol. 172, No 4 (1994), 549-558.

3. Haisty B. S. and Springer W. T.: A general beam element for use in damage assessment of complex structures, *J. of Vibration, Acoustics, Stress and Reliability in Design*, Vol. 110, July 1988, 389-394.

ON NUMERICAL METHODS FOR CONSTRAINED MECHANICAL SYSTEMS

NGUYEN NHAT LE, DINH VAN PHONG, DO SANH
Department of Applied Mechanics, Hanoi University of Technology

1. Introduction

The dynamics of complex mechanical system is becoming increasingly important. Detailed descriptions are needed for physical processes occurring in machines, robots and other multibody systems, see [11], [14], [15]. Accuracy must be maintained as computational volume increases. Many mechanical systems are controlled by programs. The controlling quantities are reaction forces corresponding to constraints. Numerical errors in computations may lead to violation of these constraints; the level of violation depends on the complexity of the system and the nature of the constraints.

The equations of motion of complex mechanical systems may be expressed in many ways, but only some of them lead to accurate modelling of the constraint reactions. This article reviews various alternative formulations of the problem and illustrates results derived from them.

2. The motion equation of a mechanical system

Consider the motion of a mechanical system. The configuration of the system is described by n Lagrangian coordinates q_i , i = 1,...,n or in the matrix form by the vector $\mathbf{q} = \lfloor q_i \rfloor^T$ (the symbol ^{T} denotes transpose) . The kinetic energy of this mechanical system can be expressed as:

$$T = \frac{1}{2} \dot{\mathbf{q}}^T \mathbf{A} \dot{\mathbf{q}} \qquad (1)$$

where **A** is a nxn matrix, symmetric and nonsingular, depending only on the coordinates q_i, it means: $\mathbf{A} = \mathbf{A}(\mathbf{q})$. $\dot{\mathbf{q}}$ is the n-vector of generalized velocities: $\dot{\mathbf{q}} = [\dot{q}_i]^T$. The generalized forces, corresponding to generalized coordinates are denoted by f_i, i=1,...,n or in the matrix form by the vector $\mathbf{f} = [f_i]^T$.

Assume that we have s constraints applied to the system and they have the matrix forms:

$$\mathbf{g}(\mathbf{q}, \dot{\mathbf{q}}, t) = \mathbf{0} \qquad (2)$$

where **g** is an sx1 matrix. The constraints (2) could be holonomic or nonholonomic.

N. Van Dao and E.J. Kreuzer (eds.),
IUTAM Symposium on Recent Developments in Non-linear Oscillations of Mechanical Systems, 207–216.

Firstly, we can write the equations of motion of system with the well-known Lagrange multipliers. Assume that the constraints (2) can be expressed in the following form:

$$\mathbf{B} \,.\, \ddot{\mathbf{q}} + \mathbf{b}_o = \mathbf{0} \tag{3}$$

where **B** is an sxn matrix and $\mathbf{b}_o$ is an sx1 matrix. The elements of these matrices are the functions of the generalized coordinates and generalized velocities. The equations of motion of system with multipliers can be written in the form , see e.g. [10], [13]:

$$\frac{d}{dt}\frac{\partial T}{\partial \dot{\mathbf{q}}} - \frac{\partial T}{\partial \mathbf{q}} = \mathbf{f} + \mathbf{B}^T \boldsymbol{\lambda} \tag{4}$$

where $\boldsymbol{\lambda}$ is the column vector of unknown Lagrange multipliers $\lambda = [\lambda_1, \lambda_2, \dots, \lambda_s]^T$.

The equations (3) and (4) create a system of (n+s) equations. The (n+s) unknowns are q_i, i=1,...,n and λ_j, j=1, ..., s .

The great advantage of this system of equations of motion is the fact that they can be written easily. But on the other hand the solution of the whole system of (n+s) equations is not so easy and moreover the initial values of λ_j , j=1,...s are unknown. Other difficulties, e.g. the problem of violation of constraint equations, appear when investigating large mechanical system or systems with many equations.

Another possibility is the principle of compatibility [6]. Using this method we first write the system of equations for determining the constraint reactions:

$$\mathbf{G} \,.\, \mathbf{r} + \mathbf{b} = \mathbf{0} \tag{5}$$

$$\mathbf{D} \,.\, \mathbf{r} = \mathbf{0} \tag{6}$$

In these equations $\mathbf{r} = [r_1, r_2, \dots, r_n]^T$ is the vector of reaction forces. **G** is an sxn matrix, **b** is an sx1 matrix.They are calculated according to the formulas:

$$\mathbf{G} = \mathbf{B} \,.\, \mathbf{A}^{-1}, \quad \mathbf{b} = \mathbf{G} \,.\, \mathbf{f} + \mathbf{B} \,.\, \boldsymbol{\psi} \tag{7}$$

where $\boldsymbol{\psi}$ is the nx1 column matrix of Christoffel symbols of the first kind (3 indices). In equation (6), **D** is a (n-s)xn matrix of coefficients relating the generalized accelerations with independent accelerations, [6].

After completely determining **r** from (5) and (6), the equations of motion of the mechanical system are represented in the form:

$$\mathbf{A} \, \ddot{\mathbf{q}} = \mathbf{h} + \mathbf{f} + \mathbf{r} \tag{8}$$

in which **h** is a nx1 matrix and its coefficients are determined from the inertia matrix **A**.

In this method we can separate the investigation of the system motion into two steps, and in each step the number of equations is only n. From (5) and (6) one sometimes obtains the analytical formulas for reaction forces:

$$\mathbf{r} = \mathbf{r}\,(t, \mathbf{q}, \dot{\mathbf{q}}) \tag{9}$$

especially when using the symbolic computation. This can help us to understand the structure of the reaction forces. The numerical alternative of this method is described in [4]. The system of equations (5), (6), (8) is solved simultaneously by using numerical methods. Moreover the components of reaction forces corresponding to each constraint can be also calculated.

The drawback of using (7) is the matrix inversion. For large mechanical systems this operation may lead to unacceptable numerical error, and as in the use of

Lagrangian multipliers, the violation of constraints. In order to overcome this problem we can proceed as follows, [5], [7], [8].

Write the motion equation of system in the form

$$\mathbf{D}_o\,\ddot{\mathbf{q}} = \mathbf{h}_0 + \mathbf{f}_o \tag{10}$$

where $\mathbf{D}_o$ is a (n-s)xn matrix, $\mathbf{h}_o$ and $\mathbf{f}_o$ are (n-s)x1 matrices. They are calculated according to the formulas:

$$\mathbf{D}_o = \mathbf{D}\,.\,\mathbf{A}\,,\quad \mathbf{h}_o = \mathbf{D}\,.\,\mathbf{h}\,,\quad \mathbf{f}_o = \mathbf{D}\,.\,\mathbf{f} \tag{11}$$

The equations (10) and (2) create a system of differential-algebraic equations. This system enables us to determine the time history of the system motion from predetermined initial conditions:

$$\mathbf{q} = \mathbf{q}(t),\ \dot{\mathbf{q}} = \dot{\mathbf{q}}\,(t),\ \ddot{\mathbf{q}} = \ddot{\mathbf{q}}\,(t) \tag{12}$$

The reaction forces of constraints **r** are determined from the equation:

$$\mathbf{r}(t) = \mathbf{A}(t)\,.\,\ddot{\mathbf{q}}\,(t) - \mathbf{h}(t) - \mathbf{f}(t) \tag{13}$$

The advantage of this procedure is that the equation of motion does not include the reaction forces explicitly; they are evaluated from (13).Therefore the initial values of reaction forces are not required. Note also that the mixed system of differential-algebraic equations (10) and (2) has only n equations; this makes the integrating process easier and more accurate. The system of differential equations (10) includes only a minimal set of (n-s) equations, i.e. the number of degree of freedom of the system under consideration. On the other hand these equations require the determination of the coefficient matrices **h** and **D**. Some algorithms for their construction are described in [3], [5], [7], [8]. The algorithm in [3] is based on the solution of a linear system of algebraic equations. The processing is purely numerical so all operations with matrices are carried out repeatedly for each time step. Another algorithm suitable even for symbolical processing is developed in [5]. The advantage of this technique is that the matrix **D** can be derived symbolically only once at the beginning of the integration process; the equation of motion of the system has an exact analytical form which is simply evaluated at each time step. The process of finding $\mathbf{D}^T = [\mathbf{d}_1, \mathbf{d}_2, \ldots, \mathbf{d}_m]$, (m=n-s), consists of two steps. In the first the Gram-Schmidt orthogonalization is realized, and in the next the orthonormal vector system is created and nx1 column matrices $\mathbf{d}_i$, i=1,...,m, are found.

3. Computer implementation

The first problem is the choice of integrating methods. Our system of equations of motion is a system of differential-algebraic equations (DAE's) that differs in significant ways from ordinary differential equations (ODE's). Applying ODE methods such as Runge-Kutta to DAE’s can have drastic consequences.

In conventional procedures one uses the derivative forms of constraint equations, in which accelerations appear explicitly, e.g. (3). So first, generalized accelerations $\ddot{\mathbf{q}}$ are calculated from other quantities, e.g. Lagrange multipliers. Then an explicit ODE's scheme can be applied to provide the integration. Clearly, the violation of original constraints can occur and only derivative forms of constraint conditions are satisfied. Therefore some constraint violation stabilization methods should be used to achieve the stability of the integration process, [14].

An alternative technique considers the DAE system as a stiff system and uses a special integration scheme for stiff systems. Many examples have been tested and show good results with the formulas known as Gear algorithms. The great drawback of the formulas is that they are not, as other multistep algorithms, self starting, and always requires starting values

In contrast, one-step formulas of Runge-Kutta's implicit method need only function values at one previous node, so are self-starting. Many implicit Runge-Kutta methods may be found in [1]. Many of them were tested and implemented in our program. For the simplest example, Gauss-Legendre quadrature q=2, order 4, gives stable results. For more complex problems, higher order algorithms are more convenient, e.g.Gauss-Legendre quadrature q=3, order 6 or another quadrature technique, Radau IIa with q=3, order 5 etc.

The computer implementation of various methods of different orders involves the problems of step size, convergence and stability. So far it has been implicitly assumed that for given initial-values, a numerical integration algorithm of a certain order is selected and this order remains fixed during the entire integration process. Under this assumption the time step is calculated so as to obtain an acceptable numerical error, but the algorithm remains numerically stable. But from a practical point of view, changing the order (or algorithm) requires less effort than changing the time step. Increasing the order would require an increase in the number of coefficients; this corresponds to a small increase of storage space and computing time. Consequently, it is often more efficient to vary both the order and the step size during each time step. Moreover one specific algorithm may tally with some particular technical problem but not another. So it is reasonable to consider the possibility of changing the integration methods and step size automatically in the program.

Defining the initial values is another question that deserves particular investigation. For n unknowns in a first-order ODE system we have just n initial values. The algebraic equations in the system reduce the number of initial values; at the start of integration process we need only the number of initial conditions that correspond to the degrees of freedom. On the other hand the numerical scheme requires initial values for all variables. So special preprocessing should be provided before the integration process begins. The algorithm implemented in our program is described in [2]. It is assumed that we solve the system of algebraic equations consisting of equations of motion and constraint equations. The initial values of coordinates, i.e. $\mathbf{q}$ and $\dot{\mathbf{q}}$, are derived from constraint equations and/or their derivative forms. The values of generalized reaction forces $\mathbf{r}$, if needed, are calculated at the same time as the accelerations from the equations of motion and the derivative forms of the constraint equations in which accelerations appear explicitly.

From this point of view the most advantageous concept is the modification of the principle of compatibility (10). No initial values of reaction forces are required, but these quantities can be evaluated easily whenever needed during the integrating process, according to (13).

Finally, we discuss symbolic versus numerical computation. Some algorithms for deriving the equations of motion, as shown in section 2, can be processed symbolically. This could avoid the numerical errors and save much computer time, since the particular forms are derived once at the beginning of integration process and then evaluated as needed for each time step. On the other hand, if the symbolic algorithm is weak, the final symbolic formulas may be too complicated, and the advantages disappear due to rounding errors.

In the Department of Applied Mechanics of Hanoi University of Technology the software is based on the module DAESOL. The programming languages are C++ and FORTRAN 77.

4. Example illustration

4.1 EXAMPLE 1

Consider the steering motion of an automobile in a horizontal plane, Figure 1a. The mass, the moment of inertia about the center, and the length are m, J and l (l=a+b) respectively. The force F acts along the lengthwise axis of the automobile. Choose the generalized coordinates x, y,φ, i.e. the coordinates of the mass center and the angle between the lengthwise and x axes. The controlling parameter is θ, i.e. the inclined angle of the front wheel plane with respect to the lengthwise axis.

With 3 redundant coordinates, the kinetic energy of the system can be written easily:

$$T = \frac{1}{2} m \left(\dot{x}^2 + \dot{y}^2 \right) + \frac{1}{2} J \dot{\varphi}^2 \tag{14}$$

The system is coupled by two nonholonomic constraint conditions:

$$\dot{x} \sin\varphi - \dot{y} \cos\varphi + b\, \dot{\varphi} = 0 \tag{15}$$

$$\dot{x} \sin(\varphi + \theta) - \dot{y} \cos(\varphi + \theta) - a \cos\theta\, \dot{\varphi} = 0 \tag{16}$$

In [9] the example is treated analytically with Lagrange multipliers. The equations of motions are presented in the forms:

$$m\, \ddot{x} = F \cos\varphi + \lambda_1 \sin\varphi + \lambda_2 \sin(\varphi + \theta) \tag{17}$$

$$m\, \ddot{y} = F \sin\varphi - \lambda_1 \cos\varphi - \lambda_2 \cos(\varphi + \theta) \tag{18}$$

$$J\, \ddot{\varphi} = \lambda_1 b - \lambda_2 a \cos\theta \tag{19}$$

The solution of the system is complicated due to need for analytical forms of λ_1 and λ_2. Numerical processing is more convenient but requires initial values of λ_1 and λ_2.

Now consider the principle of compatibility. As shown earlier, this techniquc includes the reactions forces, i.e. $\mathbf{r} = [r_x , r_y , r_\varphi]$, in the equations of motion. One of the key points is the determination of coefficient matrix **D** in equation (6). Two algorithms can be applied when the coefficient matrix **B** is given, [3], [5]. Here we extract the results:

$$\mathbf{B} = \begin{bmatrix} \sin\varphi & -\cos\varphi & b \\ \sin(\varphi+\theta) & -\cos(\varphi+\theta) & -a\cos\theta \end{bmatrix} \tag{20}$$

$$\mathbf{D} = [l\cot\theta\cos\varphi - b\sin\varphi \qquad l\cot\theta\sin\varphi + b\cos\varphi \qquad 1] \tag{21}$$

This gives the equations of motion according to the formulas (5), (6) and (8):

$$\sin\varphi\, r_x - \cos\varphi\, r_y + \frac{b\,m}{J}\, r_\varphi + m(\cos\varphi\,\dot{x} + \sin\varphi\,\dot{y})\,\dot{\varphi} = 0 \tag{22}$$

$$\sin(\varphi+\theta) r_x - \cos(\varphi+\theta)\, r_y - \frac{a\,m}{J}\cos\theta\, r_\varphi + F\sin\theta + m\{[\cos(\varphi+\theta)\dot{x} + \sin(\varphi+\theta)\dot{y}](\dot{\varphi} + \dot{\theta})$$

$$- a\sin\theta\,\dot{\varphi}\,\dot{\theta}\} = 0 \tag{23}$$

$$(l\cot\theta\cos\varphi - b\sin\varphi)\, r_x + (l\cot\theta\sin\varphi + b\cos\varphi)\, r_y + r_\varphi = 0 \tag{24}$$

$$m\,\ddot{x} = F\cos\varphi + r_x \tag{25}$$

$$m\,\ddot{y} = F\sin\varphi + r_y \tag{26}$$

$$J\,\ddot{\varphi} = r_\varphi \tag{27}$$

This system of differential-algebraic equations has 6 unknowns x, y, φ, r_x, r_y, r_φ. The reaction forces appear explicitly. Since one can show the relation of these forces to the physical contact forces, they could be used for understanding the behaviour of the automobile.

The disadvantage of this form, as in the case of equations with Lagrange multipliers, is need of determinate of initial values of x, y, φ, $\dot{x}$, $\dot{y}$, $\dot{\varphi}$, but also r_x, r_y, r_φ. One must solve a system of nonlinear algebraic equations, in which the unknowns are r_x, r_y, r_φ, $\ddot{x}$, $\ddot{y}$, $\ddot{\varphi}$ at the beginning of the integrating process.

In order to reduce the number of equations and to avoid the determination of initial values of reaction forces, we write the equations of motion as in the modification of the principle of compatibility (10). Since the matrix **h** is

$$\mathbf{h} = \mathbf{0} \tag{28}$$

and **D** is known, one easily gets from (10) and (11) the following equation:

$$m\,(l\cot\theta\cos\varphi - b\sin\varphi)\,\ddot{x} + m\,(l\cot\theta\sin\varphi + b\cos\varphi)\,\ddot{y} + J\ddot{\varphi} - Fl\cot\theta = 0 \tag{29}$$

Together with the constraint equations (15) and (16) we have 3 differential-algebraic equations for 3 unknowns x, y, φ. Note that for the Lagrange equations with multipliers the number of equations is 5, and for the original form of the principle of compatibility this number is 6.

The system of equations (15), (16), (29) can be solved with DAESOL without any problem concerning initial conditions. The reaction forces are easily evaluated since (13) yields

$$r_x = m\,\ddot{x} - F\cos\varphi, \quad r_y = m\,\ddot{y} - F\sin\varphi, \quad r_\varphi = J\,\ddot{\varphi} \tag{30}$$

Now, we can simulate the automobile manoeuvres for various values of controlling parameter θ. Some interesting cases occur:

- θ is constant;

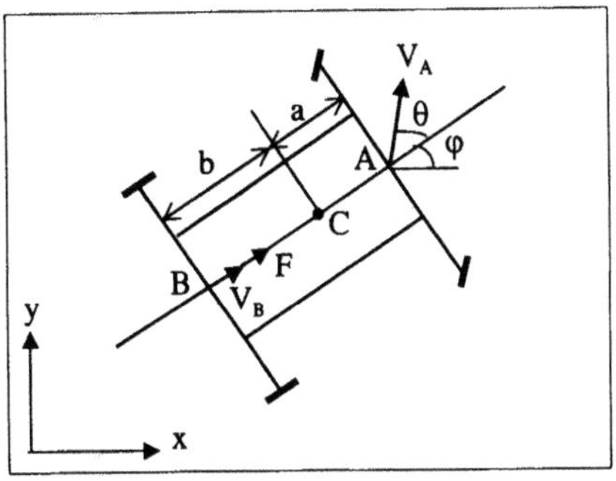

Figure 1a. Dynamical model of example 1

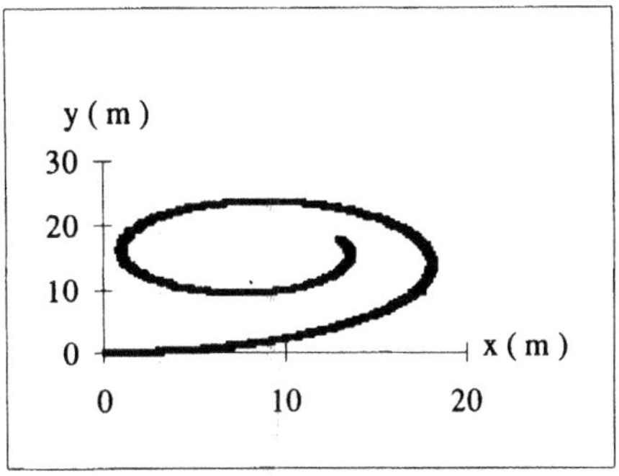

Figure 1b. Trajectory of mass center C for $\dot{\theta}$ = const = 0.2 rad/s

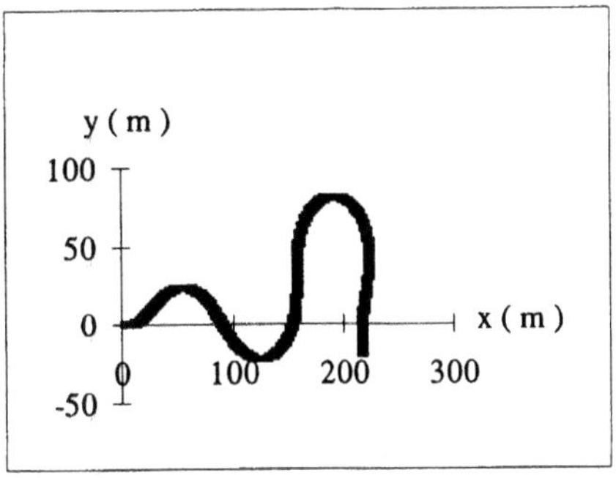

Figure 1c. Trajectory of mass center C for θ(t) = 0.1543sin(1.5708t+0.6008)

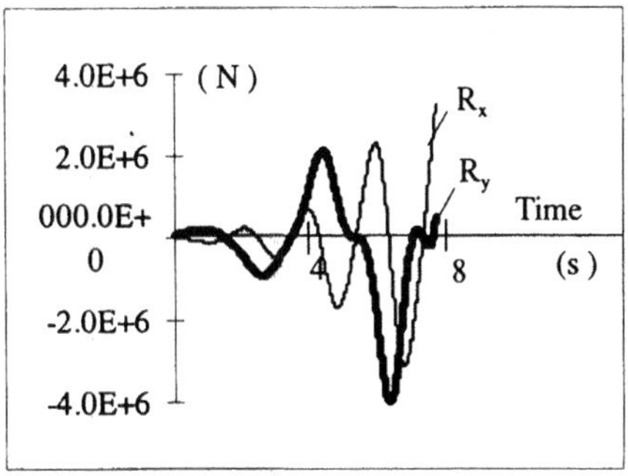

Figure 1d. Time-history of R_x and R_y for θ(t) = 0.1543sin(1.5708t+0.6008)

Figure 1. Parameters of example 1 : $m = 20.10^3$ kg; $J = 5.10^3$ kgm^2; $F = 200.10^3$ N; a = 3 m; b = 1.5 m. Initial conditions : x = y = 0; φ = 0; θ = 0.0872 rad; $\dot{x}$ = 10 m/s; $\dot{y}$ = 0.2926 m/s; $\dot{\varphi}$ =0.1944 rad/s; $\dot{\theta}$ = 0.2 rad/s

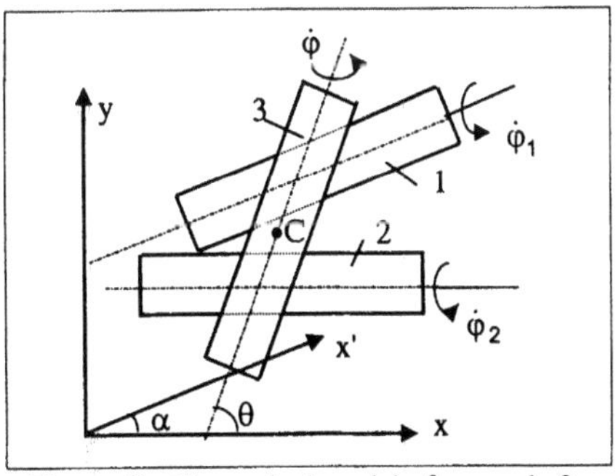

Figure 2a. Dynamical model of example 2

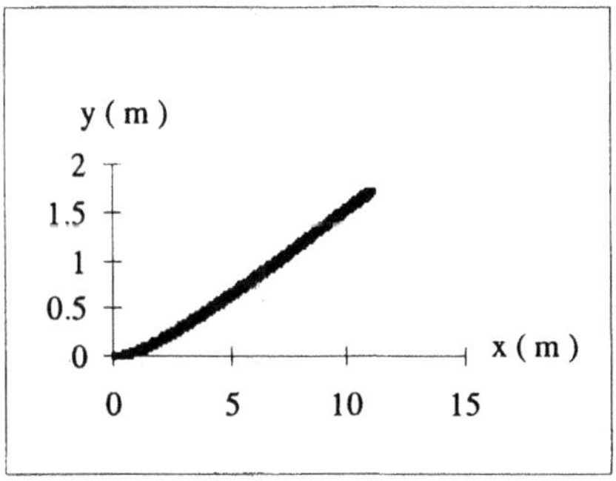

Figure 2b. Trajectory of mass center C

Figure 2. Parameters of example 2: m = 20 kg; J = 1.6 kgm^2; $J_1 = J_2 = 0.8$ kgm^2; K = 48 kgm^2; α = 0.524 rad. Initial conditions: x = y = 0; $\varphi = \varphi_1 = 0$; $\varphi_2 = 10\pi$ rad; θ = 1.047 rad; $\dot{x}$ = 2.5 m/s; $\dot{y}$ = 0; $\dot{\varphi} = \dot{\varphi}_1$ = 7.217 rad/s; $\dot{\varphi}_2$ = 12.5 rad/s; $\dot{\theta}$ =0.2 rad/s

- $\dot{\theta}$ is constant;
- θ is a sinusoidal function.

The numerical results of the two first cases coincide with the analytical invetigation in [9] and [11]. Figures 1b, 1c, 1d show some results of our simulation.

4.2 EXAMPLE 2

Consider two cylinders, rolling without slipping on a fixed plane. The cylinders have the same radius a and their moments of inertia about generating lines are J_1 and J_2 respectively. A third cylinder rolls without slipping on the cylinders 1 and 2. The mass, radius and principal central moments of inertia of the third cylinder are m, b, K and J. The angle α between the axes of cylinders 1 and 2 is constant due to their rolling without slipping, Figure 2a.

To investigate the motion of this system we choose 6 coordinates: x, y, θ, φ, φ_1, φ_2. The coordinates x and y are those of the center of the third cylinder on the fixed plane, θ is the angle between its axis and the x-axis, φ is the angle of its rolling motion. φ_1 and φ_2 are the angles of rolling motions of the cylinders 1 respectively 2. The coordinates are dependent, and the condition of rolling without slipping yields 4 nonholonomic constraints:

$$\dot{x} - b\sin\theta\,\dot{\varphi} - (a\,\varphi_1 - y)\,\dot{\theta} = 0 \tag{31}$$

$$\dot{y} + a\cos\theta\,\dot{\varphi} + (a\,\varphi_1 - y)\cot\theta\,\dot{\theta} - 2a\,\dot{\varphi}_1 = 0 \tag{32}$$

$$\sin(\theta-\alpha)\,\dot{x} - b\sin\theta\sin(\theta-\alpha)\,\dot{\varphi} + 2a\sin\alpha\sin(\theta-\alpha)\,\dot{\varphi}_2 - (a\varphi_2 + x\sin\alpha - y\cos\alpha)\sin\theta\,\dot{\theta} = 0 \tag{33}$$

$$\sin(\theta-\alpha)\,\dot{y} - b\cos\theta\sin(\theta-\alpha)\,\dot{\varphi} - 2a\cos\alpha\sin(\theta-\alpha)\,\dot{\varphi}_2 + (a\varphi_2 + x\sin\alpha - y\cos\alpha)\cos\theta\,\dot{\theta} = 0 \tag{34}$$

Again, the kinetic energy can be written very easily with redundant coordinates:

$$T = \frac{1}{2}\left[m(\dot{x}^2 + \dot{y}^2) + K\dot{\theta}^2 + J\dot{\varphi}^2 + J_1\dot{\varphi}_1^2 + J_2\dot{\varphi}_2^2 \right] \tag{35}$$

The equations of motion can be derived with Lagrange multipliers but they comprise a complicated system of 10 differential-algebraic equations, see [13]. Moreover the problem with the initial values of the multipliers reoccurs and the numerical solution of the system is difficult. We modify the principle of compatibility. From equation (10) one gets

$$\begin{aligned} & m(a\varphi_1 - y)\,\ddot{x} - \frac{m}{\sin\alpha}\left[(a\,\varphi_1 - y)\cos\alpha - (a\varphi_2 + x\sin\alpha - y\cos\alpha)\right]\ddot{y} + K\ddot{\theta} \\ & + \frac{J_1}{2\,r\sin\alpha}\left[(a\varphi_2 + x\sin\alpha - y\cos\alpha) - \frac{(a\,\varphi_1 - y)\sin(\theta-\alpha)}{\sin\theta}\right]\ddot{\varphi}_1 \\ & + \frac{J_2}{2a\sin\alpha\sin(\theta-\alpha)}\left[(a\varphi_2 + x\sin\alpha - y\cos\alpha)\sin\theta - (a\varphi_1 - y)\sin(\theta-\alpha)\right]\ddot{\varphi}_2 = 0 \end{aligned} \tag{36}$$

$$m\,b\sin\theta\,\ddot{x} - m\,b\cos\theta\,\ddot{y} + J\ddot{\varphi} = 0 \tag{37}$$

Together with (31)-(34) we have the mixed system of 6 differential-algebraic equations for 6 unknowns that can be solved numerically. The reactions forces, corresponding to generalized coordinated can be evaluated at each time step according to the equation (13). An example of numerical solution, giving the trajectory of the mass center of the third cylinder, is depicted in figure 2b.

5. Conclusion

Formerly it was necessary to apply Lagrange equations with multipliers, or other techniques such was as Chaplyghin's equations or Woronetz equations etc. [9], [11], [13], when the expression of kinetic energy and constraints were written in quasicoordinates. But many problems occur for large systems.

This article deals with various forms of the equations of motion, their implementation and solution. The concept of a system of differential-algebraic equations, including the algebraic constraint equations, appears very convenient for use in a stable and effective numerical integration scheme.

The principle of compatibility in different forms enables a flexible realization on the computer. If needed, we can reduce the number of differential equations to the minimum and remove all reaction forces from the first direct stage of integration. In this manner we avoid the problem of initial conditions, and the integration is more efficient, and the reaction forces can be found. This has a great advantage over the conventional way which uses Lagrange equations. It should be emphasized that the equations of motion are written in the matrix form basing on only two quantities: the inertia matrix and the vector of generalized forces.

A software package based on this method is being developed in the Department of Applied Mechanics of HUT.

The publication is completed with financial support from the National Basic Research Program in Natural Sciences.

References

1. Butcher, J.C.: *The Numerical Analysis of Ordinary Differential Equations*. John Willey & sons, 1989.
2. Dinh Van Phong: Direct using constraint equation of mechanical systems. *Proceedings of the International Conference on Applied Dynamics*, Hanoi, 1995, p.79-85.
3. Dinh Van Phong: Principle of compatibility and criteria of ideality in study of a constrained mechanical system. *Journal of Czech and Slovak Mechanical Engineering*, **47**(1996) 2-11.
4. Dinh Van Phong: An algorithm for calculating reaction forces in constrained mechanical systems. *Engineering Mechanics* (IM), **5**(1998) 291-297.
5. Dinh Van Phong: An algorithm for deriving equations of motion of constrained mechanical system. *Journal of Mechanics, NCNST of Vietnam*, **1**(1999) 1-10.
6. Do Sanh: *On the Motion of Constrained Mechanical Systems*. The thesis of doctor of science, Hanoi University of Technology, Hanoi, 1984, (in Vietnamese).
7. Do Sanh: A form of equations of motion of a mechanical system. *Journal of Mechanics, NCNST of Vietnam*, **3**(1995)45-48.

8. Do Sanh: A new form of equation of motion of a mechanical system. *Journal of Mechanics, NCNST of Vietnam*, **3**(1997) 47-52.
9. Dobronravov, V.V.: *Bases of Nonholonomic Mechanics*. Vyska Skola, Moscow, 1976, (in Russian).
10. Ferrers, M.: Extension of Lagrange's equations. *Quart. Journ.*, No.45, V.12, b.4, 1872.
11. Lobas, L.G.: *Nonholonomic Models of a Wheeled Vehicle*. Kiev Naukova Dumka, 1986, (in Russian).
12. Morring B.W., Genin J.: Comparison of slip angle and nonholonomic constraint vehicle models. *J. Sound and Vibration*, **65**(1979) 85-95.
13. Nejmark, J.I., Fafajew N.A: *Dynamics of Nonholonomic System*. Nauka, Moscow,1984, (in Russian).
14. Nikravesh, P.E.: *Computer-Aided Analysis of Mechanical Systems*. Prentice-Hall, Englewood, Cliffs,1988.
15. Nikravesh , P.E., Affifi, H.E.: Construction of the equations of motion of multibody dynamics using point and joint coordinates. In: *Computer-Aided Analysis of Rigid and Flexible Mechanical Systems*, Edited by Manuel F.O. Seabra Pereira and Jorge A.C. Ambrosio. Kluwer Academic Publishers, Dodrecht/Boston/London, 1994, p.31-63.

A TECHNIQUE FOR SOLVING NON-LINEAR SYSTEMS SUBJECT TO RANDOM EXCITATION

NGUYEN DONG ANH - NINH QUANG HAI

Institute of Mechanics
224 Doi Can, Hanoi Vietnam

A solution technique based on the representation of the response of the non-linear system by a polynomial in the response of the linearized system is presented. Extended moment equations are developed and their closed set is to be solved to determine unknowns. For the Duffing oscillator subject to white noise excitation, the technique gives good approximation to the response moments as well as the probability density function and the power spectral density of the system response.

1. Introduction

Since all real engineering systems are, more or less, non-linear and for those systems the exact solutions are known only for a number of special cases, it is necessary to develop approximate techniques to determine the response of non-linear systems under excitations. Several books examine approximate techniques for solving deterministic and / or random vibration problems, for instance, see [1-2] and see [3-4], respectively. This paper presents a solution technique based on the representation of the response of a non-linear system by a polynomial of the response of the linearized system. Extended moment equations are developed and their closed set is to be solved to determine unknowns. For the Duffing oscillator subject to white noise excitation the technique gives good approximation to the response moments as well as the probability density function and the power spectral density of the system response.

2. Representation of the Response of Non-Linear System by a Function of the Response of Linear System

It is well-known that a T-periodic continuous function z(t) can be approximately expressed by a finite summation of its Fourier series (see [5]).

$$z(t) = \sum_{n=1}^{N} (c_n \cos nkt + s_n \sin nkt) \tag{1}$$

N. Van Dao and E.J. Kreuzer (eds.),
IUTAM Symposium on Recent Developments in Non-linear Oscillations of Mechanical Systems, 217–226.

where k = 2π/T and c_n, s_n are Fourier coefficients. Using the formulas

$$\begin{aligned}\cos nkt &= \cos^n kt - C_n^2 \cos^{n-2} kt \sin^2 kt + C_n^4 \cos^{n-4} kt \sin^4 kt \mp \ldots \\ \sin nkt &= C_n^1 \cos^{n-1} kt \sin kt - C_n^3 \cos^{n-3} kt \sin^3 kt \pm \ldots\end{aligned} \tag{2}$$

one can represent z(t) in a polynomial form of $x(t) = \cos kt$ *and* $\dot{x}(t) = -k \sin kt$:

$$z(t) = \sum_{n=0}^{N} \sum_{m=0}^{N} \alpha_{nm} x^n(t) \dot{x}^m(t) \tag{3}$$

and α_{nm} are coefficients depending on k, c_1, s_1, c_2, s_2,... and dots denote time differentiation. Thus, a periodic function can be approximately represented by a polynomial of its first harmonics. Generalizing this property one can propose a technique for solving non-linear systems, namely, *to represent the response of the non-linear system by a polynomial of the response of the linearized system.*

In deterministic vibration problems this representation may be used for determining approximate periodic solutions, for example, of the differential equation of motion

$$\ddot{z} + f(z, \dot{z}) = 0 \tag{4}$$

Namely, one can find z(t) in the form (3) where x(t) is a periodic solution of the corresponding linearized equation

$$\ddot{x} + k^2 x = 0 \tag{5}$$

Example: Consider the Duffing oscillator

$$\ddot{z} + z + \varepsilon z^3 = 0 \tag{6}$$

An approximate periodic solution of (6) takes the form (see [2])

$$z = a \cos\varphi + \frac{\varepsilon}{32} a^3 \cos 3\varphi + \ldots \quad , \quad \varphi = kt + \theta \tag{7}$$

where a, θ are constants. One can rewrite (7) in the cubic polynomial as follows:

$$z = (1 - \frac{3\varepsilon}{32} a^2) x + \frac{\varepsilon}{8} x^3 + \ldots \quad , \quad x = a \cos\varphi \tag{8}$$

Analogously, in the field of random vibration one can propose a technique for solving random vibration systems using a polynomial of Gaussian process. It should be noted that the use of polynomials in non-linear vibration problems has been considered in the literature (see, for example [6-8]).

3. Extended Moment Equations and Polynomial Form for the System Response

Consider the equation of motion of a single - degree - of - freedom mechanical system

$$\ddot{z} + 2h\dot{z} + \beta z + f(z,\dot{z}) = \sigma\dot{\xi}(t) \tag{9}$$

where h is a positive constant, β is a constant, $f(z,\dot{z})$ is a polynomial function or approximated as a polynomial function of $z, \dot{z}$; ξ (t) is a zero mean white noise with the autocorrelation and spectral density given, respectively, by

$$R(\tau) = <\dot{\xi}(t)\dot{\xi}(t+\tau)> = \sigma^2\delta(\tau) \qquad and \qquad S(\omega) = \frac{\sigma^2}{2\pi}$$

where $\delta(\tau)$ is Dirac delta function, < > denotes the expectation operator. For sake of simplicity we suppose <z> = 0. Together with (9) one considers the corresponding linearized equation

$$\ddot{x} + 2\bar{h}\dot{x} + \bar{\omega}_0^2 x = \sigma\dot{\xi}(t) \tag{10}$$

For an arbitrary up-to-second-order differentiable function $\varphi(z,\dot{z},x,\dot{x})$ the extended moment equation corresponding to a stationary solution of (9), (10) has the form:

$$\begin{aligned} &<\frac{\partial\varphi}{\partial z}\dot{z}> - <\frac{\partial\varphi}{\partial \dot{z}}(2h\dot{z} + \beta z + f(z,\dot{z}))> + <\frac{\partial\varphi}{\partial x}\dot{x}> - <\frac{\partial\varphi}{\partial \dot{x}}(2\bar{h}\dot{x} + \bar{\omega}_0^2 x)> + \\ &+\frac{\sigma^2}{2}\left\{<\frac{\partial^2\varphi}{\partial\dot{z}^2}> +2<\frac{\partial^2\varphi}{\partial\dot{z}\partial\dot{x}}> + <\frac{\partial^2\varphi}{\partial\dot{x}^2}>\right\} = 0 \end{aligned} \tag{11}$$

The equation (11) is derived from Fokker-Planck equation (e.g. see [3]).

Using (11) one can get, in the non-linear case, i.e., f ≠ 0, a so - called infinite hierarchy of linear algebraic equations for the response moments in the sense that all finite sets of moment equations contain a number of involved unknown moments more than the number of equations (see [9-12]). Particularly, taking the 8 "lowest" polynomial functions $\varphi(z,\dot{z},x,\dot{x})$, and using the well-known definition:

$$m_{ij} = <z^i\dot{z}^j> = E[z^i\dot{z}^j]$$

one has following equations:

$$for\ \varphi = \frac{1}{2}\dot{z}^2 : \qquad -2hm_{02} - <f(z,\dot{z})\dot{z}> + \frac{\sigma^2}{2} = 0 \tag{12}$$

$$for\ \varphi = z\dot{z}\ :\qquad m_{02} - \beta m_{20} - \langle f(z,\dot{z})z \rangle = 0 \qquad (13)$$

$$for\ \varphi = \frac{1}{2}\dot{x}^2\ :\qquad 2\bar{h} \langle \dot{x}^2 \rangle - \frac{\sigma^2}{2} = 0 \qquad (14)$$

$$for\ \varphi = x\dot{x}\ :\qquad \bar{\omega}_0^2 \langle x^2 \rangle - \langle \dot{x}^2 \rangle = 0 \qquad (15)$$

$$for\ \varphi = zx\ :\qquad \langle x\dot{z} \rangle + \langle \dot{x}z \rangle = 0 \qquad (16)$$

$$for\ \varphi = \dot{x}z\ :\qquad \langle \dot{x}\dot{z} \rangle - 2\bar{h} \langle \dot{x}z \rangle - \bar{\omega}_0^2 \langle zx \rangle = 0 \qquad (17)$$

$$for\ \varphi = x\dot{z}\ :\qquad \langle \dot{x}\dot{z} \rangle - 2h \langle x\dot{z} \rangle - \beta \langle zx \rangle - \langle xf(z,\dot{z}) \rangle = 0 \qquad (18)$$

$$for\ \varphi = \dot{x}\dot{z}:\quad \sigma^2 - 2h \langle \dot{x}\dot{z} \rangle - \beta \langle \dot{x}z \rangle - \langle \dot{x}f(z,\dot{z}) \rangle - 2\bar{h} \langle \dot{x}\dot{z} \rangle - \bar{\omega}_0^2 \langle \dot{z}x \rangle = 0 \quad (19)$$

The moment equations (12-15) are conventional ones for separated moments of $z, \dot{z}$ and $x, \dot{x}$, while the moment equations (16-19) contain mixed moments of $z, \dot{z}$, $x, \dot{x}$. The equations (14), (15) are used separately to determine linearized coefficients:

$$\bar{h} = \frac{\sigma^2}{4 \langle \dot{x}^2 \rangle} \quad ; \quad \bar{\omega}_0^2 = \frac{\langle \dot{x}^2 \rangle}{\langle x^2 \rangle} \qquad (20)$$

It is seen that moment equations (12 - 19) have higher order moments. To close a set of moment equations one needs some additional relationships between moments. For instance, in Gaussian closure one puts z(t) = x(t) and only 2 moment equations (12), (13) can be satisfied, namely, they are used to find $\langle z^2 \rangle$ and $\langle \dot{z}^2 \rangle$. Thus, in Gaussian closure, in principle, only 2 equations from the hierarchy of moment equations can be satisfied. On the other hand, *one might suppose that the accuracy of a closure technique would be better if more moment equations could be satisfied.*

In the paper, we consider the following form of the system response

$$z(t) = x(t) + \alpha_1 x^3(t) + \alpha_2 x^5(t) \qquad (21)$$

Substituting (21) into (16-19) shows that the equations (16), (17) are satisfied for any α. Thus, finally one has 4 equations (12), (13), (18), (19) for determining 4 unknowns $\langle x^2 \rangle, \langle \dot{x}^2 \rangle, \alpha_1, \alpha_2$. It is noted that, the solution of the problem must satisfy not only the moment equations but also must preserve moment properties such as non-negativeness of even-order moments and satisfaction of Schwartz' inequality. The considered moment equations may yield a unique solution, multiple solutions or no solution. This matter should be investigated in other research. Particularly, for Gaussian closure we may see[13]. The probabilistic characteristics of z(t) can be obtained from (21) if x(t), α_1, α_2 are known.

4. Probability Density Function of Response (PDF)

Using the expression (21) one can easily calculate the PDF of the response process z(t). In fact, the PDF of x is known as:

$$p_x = \exp\left\{-x^2/2\sigma_x^2\right\}/\sqrt{2\pi\sigma_x^2} \tag{22}$$

The probability distribution function of z is defined as

$$F_z(y) = P\{-\infty \prec z \prec y\} \tag{23}$$

where P{E} denotes the probability of E. The PDF of z can be found as follows

$$w_z(y) = \lim_{\Delta y\to 0}\frac{F_z(y+\Delta y)-F_z(y)}{\Delta y} = \lim_{\Delta y\to 0}\frac{P\{y \prec z \prec y+\Delta y\}}{\Delta y} = \lim_{\Delta y\to 0}\frac{1}{\Delta y}\int_d p_x dx \tag{24}$$

where the integration domain d takes the form

$$d = \{x : y < z < y+\Delta y\} = \{x : y < x+\alpha_1 x^3 + \alpha_2 x^5 < y+\Delta y\} \tag{25}$$

Thus, a numerical procedure to determine this function can be established.

5. Power Spectral Density of Response (PSD)

The PSD is an important statistical characteristic of the stationary process. There are many investigations concerning with the PSD of the stationary response of non-linear stochastic systems (see [14-15]). Here using the expression (21) one can find the PSD of z(t). The higher moments of a Gaussian random process are related to its second-order moments by known equations (see [16]). Applying the properties of Gaussian process we can express the second order autocorrelation function of z(t) in terms of $R_x(\tau)$:

$$\begin{aligned} R_z(\tau) &= \langle z(t)z(t+\tau)\rangle = \langle[x(t)+\alpha_1 x^3(t)+\alpha_2 x^5(t)] \\ &\quad [x(t+\tau)+\alpha_1 x^3(t+\tau)+\alpha_2 x^5(t+\tau)]\rangle = \Psi(R_x(\tau)) \end{aligned} \tag{26}$$

in which $R_x(\tau)$ (the second order autocorrelation function of x(t)) can be defined once the linearized equation (10) has been known (see[3]). The PSD of z(t) is then defined as:

$$S_z(\omega) = \frac{1}{\pi}\int_0^{+\infty} R_z(\tau)\cos\omega\tau\, d\tau \tag{27}$$

Based on the above-mentioned formulas, a numerical procedure for determining the PDS of z(t) can be established.

6. Duffing Oscillator under White Noise Excitation

To elucidate the proposed technique, consider the Duffing oscillator under white noise

$$\ddot{z} + 2h\dot{z} + \beta z + \varepsilon z^3 = \sigma\xi(t) \tag{28}$$

6.1. MOMENT EQUATIONS AND POLYNOMIAL FORM FOR SOLUTION

For sake of simplicity and due to the limited capacity of our computer to determine $<x^2>$, $<\dot{x}^2>, \alpha_1$, the solution of the equation (28) is taken in the form:

$$z(t) = x(t) + \alpha x^3(t) \tag{29}$$

We consider 3 equations (12), (13), (18) which now, respectively, become

$$2hm_{02} - \frac{\sigma^2}{2} = 0 \tag{30}$$

$$m_{02} - \beta m_{20} - \varepsilon m_{40} = 0 \tag{31}$$

$$<\dot{x}^2> - \beta <zx> - \varepsilon <z^3x> + 3\alpha <x^2><\dot{x}^2> = 0 \tag{32}$$

From (29), (30) in the case where $\sigma^2/4h = 1$, after simple calculation one gets:

$$<\dot{x}^2> = 1/[1 + 6\alpha <x^2> + 27\alpha^2 <x^2>^2] \tag{33}$$

Denoting $y =< x^2 >$, using (29), (33) after some calculation, it follows from (31), (32):

$$1-\beta y-6\beta\alpha y^2-3\varepsilon y^2-15\beta\alpha^2 y^3-60\varepsilon\alpha y^3-630\varepsilon\alpha^2 y^4-3780\varepsilon\alpha^3 y^5-10395\varepsilon\alpha^4 y^6 = 0 \tag{34}$$

$$1+3\alpha y-\beta y-9\beta\alpha y^2-3\varepsilon y^2-45\beta\alpha^2 y^3-63\varepsilon\alpha y^3-81\beta\alpha^3 y^4-666\varepsilon\alpha^2 y^4-4050\varepsilon\alpha^3 y^5- -14175\varepsilon\alpha^4 y^6-25515\varepsilon\alpha^5 y^7 = 0 \tag{35}$$

for 2 unknowns: y and α. For these equations, the moment properties such as non-negativeness of even-order moments and satisfaction of Schwartz' inequality are used to exclude some extraneous solutions. Finally, the second moment m_{20} (or $<z^2>$) can be calculated from (29). The results $<z^2>$ obtained by the proposed procedure and that obtained by Gaussian closure with the values $\beta = 1$ and $\beta = -1$ are compared in Table 1 and Table 2 for $\sigma^2 = 4h$ and different values of ε, respectively. It is seen that the proposed solutions $<z^2>$ are much closer to the exact solutions $<z^2>_E$ than the solutions obtained by Gaussian closure $<x^2>_G$ except the case of small ε ($\varepsilon = 0.1$) and $\beta = 1$.

TABLE 1. Approximate mean squares of response of the Duffing equation with $\beta = 1$ (Unimodal case).

N	ε	$<z^2>_E$	$<x^2>_G$	$<z^2>$	α	$<x^2>$
1	0.1	0.8176	0.8054 (-1.49%)	0.6059 (-26%)	-0.1236	1.5034
2	1	0.4679	0.4343 (-7.19%)	0.4466 (-4.54%)	-0.1424	0.9846
3	10	0.1889	0.1667 (-11.8%)	0.1909 (1.03%)	-0.3207	0.4000
4	100	0.0650	0.0561 (-13.6%)	0.0663 (1.96%)	-0.9135	0.1368

TABLE 2. Approximate mean squares of response of the Duffing equation with $\beta = -1$ (Bimodal case).

N	ε	$<z^2>_E$	$<x^2>_G$	$<z^2>$	α	$<x^2>$
1	0.1	8.7136	4.1387 (-52.5%)	6.2556 (-28.2%)	-0.0089	11.543
2	1	1.0418	0.7676 (-26.3%)	1.031 (-1.04%)	-0.0562	2.0026
3	10	0.2435	0.2 (-17.9%)	0.2497 (2.53%)	-0.2379	0.5020
4	100	0.0704	0.0594 (-15.6%)	0.0722 (2.5%)	-0.8311	0.1470

6.2. PROBABILITY DENSITY FUNCTION OF RESPONSE

The PDF of the response of the bimodal Duffing oscillator is investigated, for example, in [17]. Herein, the PDF is calculated from (24) where the formula (25) takes the form:

$$d = \{x : y < z = x + \alpha x^3 < y + \Delta y\} \tag{36}$$

and $\sigma_x^2 = < x^2 >$ is found from (34-35).

The graphs of exact probability density function, of that obtained by the Gaussian closure and by the proposed procedure are shown in Figures 1-a, 1-b, 2-a, 2-b. It is seen from these Figures that in comparison with the PDF of the Gaussian closure, the graphs of PDF of the proposed method are closer to the exact ones. In the bimodal case the graphs of PDF obtained by the proposed technique do not have two extremes. That is a shortcoming of the procedure. However, as is seen from Figures 2-a, 2-b, the proposed method gives a better prediction for PDF than the Gaussian closure does.

6.3. POWER SPECTRAL DENSITY

The power spectral density of the Duffing oscillator to white noise excitation is investigated in [15] and the third harmonic peak is shown using other approach. Here, using (29) we have

$$\begin{aligned} R_z(\tau) &= < z(t)z(t+\tau) >=< [x(t) + \alpha x^3(t)][x(t+\tau) + \alpha x^3(t+\tau)] >= \\ &= R_x^{(2)}(\tau) + \alpha R_x^{(4)}(\tau,\tau,\tau) + \alpha R_x^{(4)}(0,0,\tau) + \alpha^2 R_x^{(6)}(0,0,\tau,\tau,\tau) \end{aligned} \tag{37}$$

Unimodal Case:

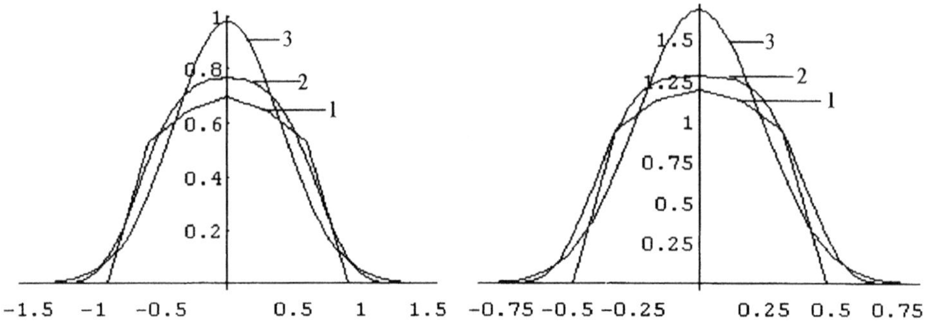

Figure 1-a $\beta = 1,\ \varepsilon = 10.$ *Figure 1-b* $\beta = 1,\ \varepsilon = 100.$
1-The PDF of proposed method, 2-The exact PDF, 3-The PDF of Gaussian closure

Bimodal Case

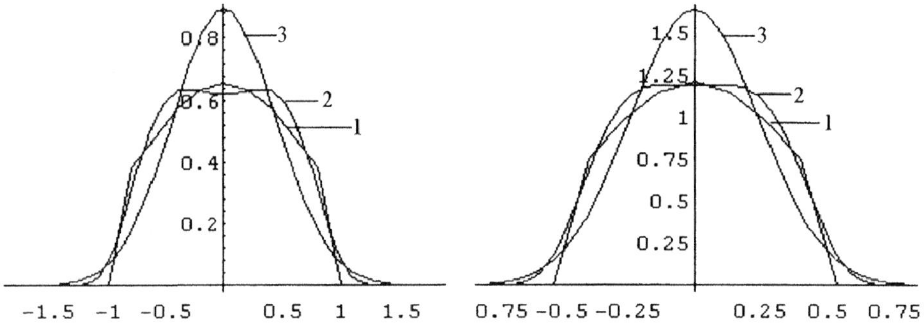

Figure 2-a $\beta = -1,\ \varepsilon = 10.$ *Figure 2-b* $\beta = -1,\ \varepsilon = 100.$
1-The PDF of proposed method, 2-The exact PDF, 3-The PDF of Gaussian closure

Applying the properties of Gaussian process yields

$$R_z(\tau) = R_x(\tau)\left[1 + 6\alpha < x^2 > +9\alpha^2 < x^2 >^2\right] + 6\alpha^2\left[R_x(\tau)\right]^3 \tag{38}$$

It follows:

$$S_z(\omega) = (1 + 6\alpha < x^2 > \ +9\alpha^2 < x^2 >^2)S_x(\omega) + \frac{6\alpha^2}{\pi}\int_0^{+\infty}\left[R_x(\tau)\right]^3 \cos\omega\tau d\tau \tag{39}$$

where $R_x(\tau)$ and $S_x(\omega)$ are formulated (see [3]).

Finally, from (20), (33) one gets the coefficients of the corresponding linearized equation $\overline{h}$ *and* $\overline{\omega}_0^2$, in the case $\sigma^2 = 1$, as follows:

$$\bar{h} = \frac{1+6\alpha < x^2 > +27\alpha^2 < x^2 >^2}{4} \; ; \; \bar{\omega}_0^2 = \frac{1}{< x^2 > [1+6\alpha < x^2 > +27\alpha^2 < x^2 >^2]} \tag{40}$$

The graphs of the PSD are shown in Figs. 3-a, 3-b and Figs.4 -a, 4-b, respectively.

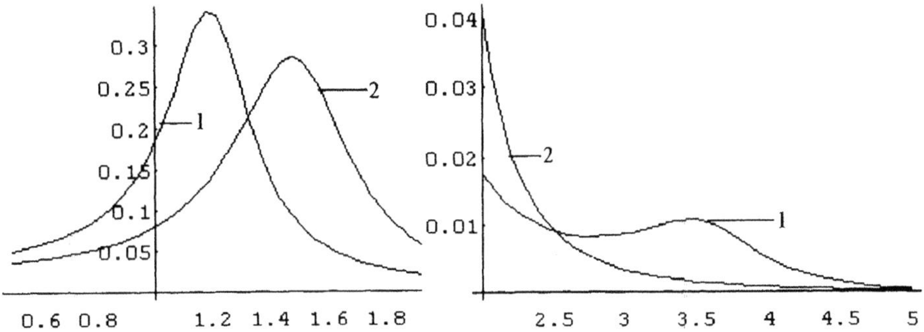

Figure 3-a $\beta = 1$, $\varepsilon = 1$. *The first harmonic peak of Duffing oscillator.* *Figure 3-b* $\beta = 1$, $\varepsilon = 1$. *The third harmonic peak of Duffing oscillator.*
1-The PSD of the proposed method, 2-The PSD of the Gaussian closure.

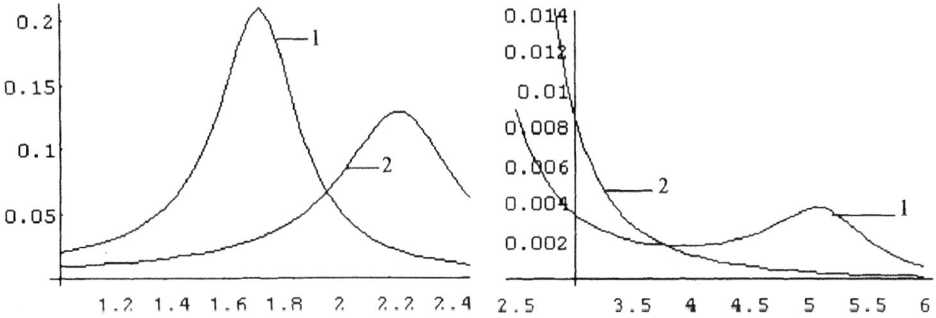

Figure 4-a $\beta = -1$, $\varepsilon = 10$. *The first harmonic peak of Duffing oscillator.* *Figure 4-b* $\beta = -1$, $\varepsilon = 10$. *The third harmonic peak of Duffing oscillator.*
1-The PSD of the proposed method, 2-The PSD of the Gaussian closure.

It is seen from Fig. 3-b and Fig. 4-b that the proposed technique can also give the third harmonic peak, whereas the Gaussian closure can not.

6. Conclusion

The main idea of the proposed technique is to represent the system response by a polynomial of Gaussian process and the application is then simplified by using the so-called extended moment equations involving mixed moments of the original non-linear and linearized systems. Thus, a possible way to determine the polynomial coefficients and the Gaussian process can be derived. The technique is quite simple since it can use properties of the Gaussian process although the calculations are more complicated than the Gaussian closure. However, the technique should be tested for other non-linear systems and some related questions may occur: Which set of extended moment equations and which form of polynomials should be chosen to get a better approximate solution?

Acknowledgement. The research has been supported by a grant of the Fundamental Research Program in Natural Sciences.

7. Reference

1. Bogoliubov, N.N. and Mitroposkii, Iu. A. (1961) *Asymptotic Methods in the Theory of Nonlinear Oscillations,* Nauka, Moscow, (Russian).
2. Nayfeh, A. H. (1973) *Perturbation Methods.* John Wiley & sons.
3. Roberts, J. B. and Spanos, P.D. (1990) *Random Vibration and Stochastic Linearization*, John Wiley and Sons .
4 Lin, Y.K., Cai, G. Q. (1995) *Probabilistic Structural Dynamics. Advanced theory and Applications,* Mc Graw-Hill, Inc.
5. Korn, G. A., Korn, T. M. (1968) *Mathematical Handbook,* Mc Graw-Hill.
6. Winterstain, S. R. (1988) "Non-linear vibration models for extremes and fatigue", *Journal of engineering Mechanics*, ASCE, vol. **114**, No. 10. pp. 1772-1790.
7. Grigoriu, M. (1995) *Applied Non-Gaussian Processes*, PTR Prentice Hall, Englewood Cliffs, NJ.
8. Lutes, L. D. and Sarkani, S. (1997) *Stochastic Analysis of Structures and Mechanics*, Prentice-Hall, Englewood Cliffs, NJ.
9. Crandall, S.H. (1980) Non-Gaussian closure for random of non-linear oscillators. *Int. J. Non-linear Mech.* , **15**, 303 - 313.
10. Ibrahim, R.A. ,Soundararajan, A. and Heo, H. (1985) Stochastic response of non- linear dynamic systems based on non - Gaussian closure. *Trans. Of the ASME, J. of Appl. Mech.* , (52), 965 - 970.
11. Wojtkiewicz, S. F. , Spencer, Jr., B.F. , Bergman L, A. (1996) New insights on the application of moment closure methods to non-linear stochastic systems. *In Proc. of IUTAM Symposium on Non-Linear Stochastic Dynamics, eds Naess A. and S. Krenk,Kluwer* , 479 - 488.
12. Anh N.D., Schiehlen, W. (1994) An approach to the problem of closure in the non- linear stochastic Mechanics. *Int. J. of Mechanics*, **29,** 109 - 123.
13. Spanos, P.D. and Iwan, W.D. (1978) " on the Existence and Uniqueness of solutions Generated by Equivalent linearization," *International Journal of Nonlinear Mechanics*, **13**, pp. 71 - 78.
14. Miles, R. N. (1989) An approximate solution for the spectral response of Duffing's oscillator with random input. *Journal of Sound and Vibration* 132 (1), 43-49.
15. Bouc, R. (1994). The power spectral density of response for strongly non-linear random oscillator. *Journal of Sound and Vibration*, **175** (3), 317-331.
16. Robson., J. D. (1981) A simplified quasi-Gaussian random process model based on non-linearity. *Journal of Sound and Vibration* 76 (2), 169-177.
17. Alaoui Ismaili, M. & Bernard, P. (1997). Asymptotic analysis and linearization of the randomly perturbed Two-Well Duffing oscillator. *Prob. Eng. Mech.* Vol. 12, No 3, pp. 171-178.

ON THE IDENTIFICATION OF NONLINEAR SYSTEMS

K. POPP AND J.-U. BRUNS
Institute of Mechanics, University of Hannover
Appelstrasse 11, 30167 Hannover

1. Introduction

In many engineering disciplines the identification of dynamical systems from measured signals is an important task in the modelling process. If information on the structure of the system is available, the task is reduced to the identification of parameters. Often, however, either such information is not available or the system structure is known but is very complex and the relevant part of the structure is not easily discerned. As a first step in the identification process two questions will then have to be addressed:

- How complex does the model have to be? How many state variables are required to describe the system's behaviour?
- Is a simple linear model sufficient or do complex nonlinear models have to be considered?

In this paper the second question will be dealt with.

2. Detection of nonlinearities

Every real dynamical system contains nonlinear elements to a certain extent. The modelling process should therefore be preceded by a (non-)linearity test to answer the question whether the system can be adequately described by a simple linear model or not.

In its simplest form such a test only examines whether the hypotheses of a linear model can be accepted or has to be rejected. If the system under investigation is found to be nonlinear, the test results may also give additional information on the type of nonlinearity reflected in the system dynamics.

Table 1 gives an overview on a selection of linearity tests that can be found in the literature. The overview is by no means complete and the reader is referred to [6],[12],[1] and [5] for further references.

N. Van Dao and E.J. Kreuzer (eds.),
IUTAM Symposium on Recent Developments in Non-linear Oscillations of Mechanical Systems, 227–234.

TABLE 1. Overview on linearity tests

	Output signal	Input/Output signals
Time domain	• Testing for stochastically excited linear system (Keenan/Tsay) [8, 13] • Method of Internal Harmonics Cross-Correlation (Dimentberg/Sokolov/Haenisch) [4, 3]	• Superposition of input signals: [6] $u_1 \to y_1$, $u_2 \to y_2$ $\Rightarrow u_1 + u_2 \to y_1 + y_2$ • Identification of ARMAX-models before and after filtering [9]
Frequency domain	• Bispectrum test [11, 7]	• Hilbert transform [10] $G = \mathcal{H}(\mathcal{G})$ • Total harmonic distortion $THD = \sqrt{\frac{a_2^2+a_3^2+\ldots}{a_1^2+a_2^2+a_3^2+\ldots}}$ • Linear spectral density [2] $\lvert S_{uy}\rvert^2 = S_{uu} * S_{yy}$

The tests have been classified with respect to two aspects of the testing procedure:

- Does the test require input/output signals or just output signals?
- Is the test performed in the time or frequency domain?

The test involving the Hilbert transform and the Method of Internal Harmonics Cross-Correlation are considered in the following sections. The underlying ideas as well as simulation results will be given.

2.1. HILBERT TRANSFORM OF FREQUENCY RESPONSE FUNCTION

The application of the Hilbert transform to characterize nonlinearities in mechanical structures is described in [10], the initial suggestion that the Hilbert transform might be useful in detecting nonlinearity was given by Vinh.

The transfer function, $G(s)$, of a linear system can be considered as an analytic function in the complex s-plane. Cauchy's formula allows one to express its value at $s = \sigma + i\omega$ by evaluating the integral along a closed contour, Γ, as long as s is inside Γ,

$$G(s) = \frac{1}{2\pi i} \oint_{\Gamma} \frac{G(s^*)}{s^* - s} ds^*. \tag{1}$$

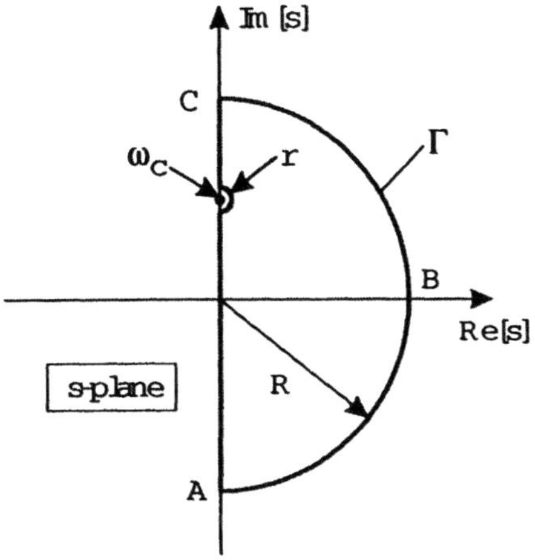

Figure 1. **Integration path Γ used to evaluate the value of the transfer function $G(s)$ at $s = i\omega_c$ using Cauchy's formula. The poles of $G(s)$ in the left half plane and the extra pole at $s = i\omega_c$ are avoided.**

With $G(s)$ representing a stable system all the poles of $G(s)$ are in the left half of the complex s-plane and $G(s)$ is therefore analytic on the right half plane. Figure 1 shows a contour Γ that avoids the poles in the left half plane and the extra pole introduced by evaluating (1) at $s = i\omega_c$.

Dividing the integration path Γ into two parts,

$$\oint_\Gamma = \oint_{ABC} + \oint_{CA}, \tag{2}$$

leads to a significant simplification; as the radius, R, of the semicircle ABC tends to infinity, the first integral on the right hand side of (2) tends to zero. This leaves only the integration along the imaginary axis of the s-plane, the ω-axis,

$$G(i\omega_c) = -\frac{1}{\pi i}\text{P.V.}\int_{-\infty}^{\infty} \frac{G(i\omega)}{\omega - \omega_c} d\omega, \tag{3}$$

where P.V. indicates Cauchy's Principal Value of the integral. Because of the similarity between (3) and the definition of the Hilbert transform of a real function, $f(x)$,

$$\mathcal{H}\{f(x)\} = \frac{1}{\pi}\int_{-\infty}^{+\infty} \frac{f(a)}{x-a} da, \tag{4}$$

the Hilbert transform of a transfer function has been defined as, cp. [10]:

$$\mathcal{H}\{G(i\omega_c)\} = H(i\omega_c) = -\frac{1}{\pi i}\text{P.V.}\int_{-\infty}^{\infty} \frac{G(i\omega)}{\omega - \omega_c} d\omega. \tag{5}$$

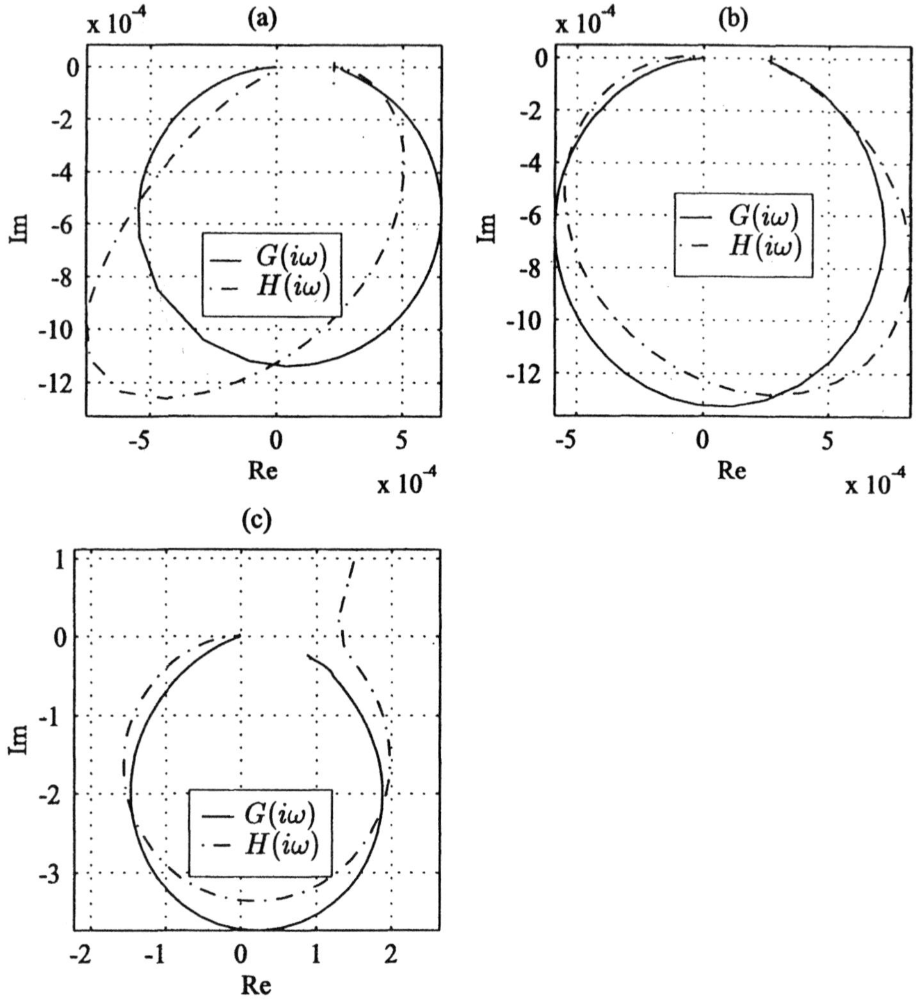

Figure 2. Simulated frequency response functions $G(i\omega)$ of nonlinear oscillators and their Hilbert transforms $H(i\omega)$, (a) progressive spring stiffness (b) degressive spring stiffness (c) Coulomb friction

Applying the Hilbert transform as defined in (5) to the transfer function $G(s)$ of a linear system does not change $G(s)$. Thus, the identity of $G(s)$ and its Hilbert transform $H(s)$ can be used to detect nonlinearities from frequency response data.

By splitting (5) into its real and imaginary part and considering the symmetries of transfer functions,

$$\begin{aligned} \mathrm{Re}\{\mathrm{G}(-\mathrm{i}\omega)\} &= \mathrm{Re}\{\mathrm{G}(\mathrm{i}\omega)\} \\ \mathrm{Im}\{\mathrm{G}(-\mathrm{i}\omega)\} &= -\mathrm{Im}\{\mathrm{G}(\mathrm{i}\omega)\} \end{aligned} \tag{6}$$

the integration can be limited to the positive frequency axis:

$$\mathrm{Re}\{\mathrm{G}(\mathrm{i}\omega_c)\} = -\frac{2}{\pi}\mathrm{P.V.}\int_0^{+\infty} \frac{\omega \mathrm{Im}\{\mathrm{G}(\mathrm{i}\omega)\}}{\omega^2 - \omega_c^2}\, d\omega \tag{7}$$

$$\mathrm{Im}\{\mathrm{G}(\mathrm{i}\omega_c)\} = \frac{2\omega_c}{\pi}\mathrm{P.V.}\int_0^{+\infty} \frac{\mathrm{Re}\{\mathrm{G}(\mathrm{i}\omega)\}}{\omega^2 - \omega_c^2}\, d\omega.$$

Obviously, the frequency response function of a real system can in general only be measured in a frequency range $\omega_l \leq \omega \leq \omega_u$, but a good approximation of (7) can often be calculated based on a broad enough frequency range.

Additional correction terms to compensate for the lower and upper frequency ranges that cannot be measured are described in [10]. The correction terms are based on modal parameters that can be identified from the measured frequency response data.

Figure 2 shows the results of the Hilbert transform applied to simple damped nonlinear oscillators containing different nonlinear elements:

- Progressive/degressive spring stiffness
- Coulomb friction.

For these types of nonlinearity, the Hilbert transform H of the simulated frequency response functions shows characteristic deviations of the frequency response function G.

2.2. METHOD OF INTERNAL HARMONICS CROSS-CORRELATION

The Method of Internal Harmonics Cross-Correlation is described by Dimentberg in [4, 3]. It is assumed that the system under study is excited by a broadband signal and that only the system's output can be measured. The particular system considered is a one degree of freedom mechanical system

$$\ddot{x} + 2\alpha\dot{x} + f(x) = \zeta(t); \quad f(x) = \omega_0^2 x + g(x) \tag{8}$$

where $f(x)$ is a possibly nonlinear restoring force and $\zeta(t)$ is Gaussian white noise. If the system is linear, the frequency components in the output signal are independent because the frequency components at the input are independent (Gaussian white noise). In a nonlinear system oscillations contain additional frequency components at the higher harmonics of the fundamental frequency. The output of a nonlinear system that is excited by a white noise signal should therefore show correlation between the slowly varying amplitudes of components at the fundamental frequency of the system and its higher harmonics. The procedure outlined in [4] involves filtering the

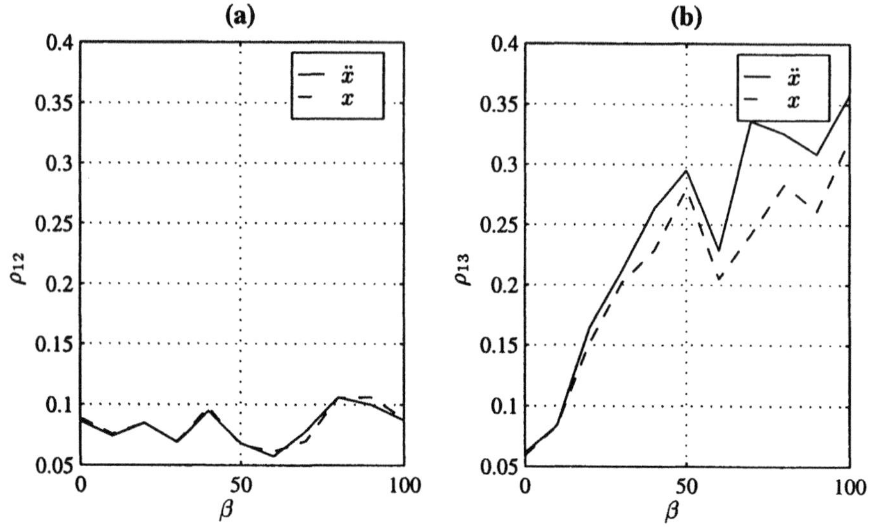

Figure 3. Normalized cross-correlation factors ρ_{12} (a) and ρ_{13} (b) for an oscillator with progressive spring-stiffness, ρ_{13} significantly increases with parameter β of nonlinearity

output signal $x(t)$ by a pair of band-pass filters that are tuned to main frequency ω_0 and a higher harmonic $k\omega_0$, respectively. Then, the slowly varying amplitudes (A_1 and A_k) of the filtered signals are extracted and their zero-mean parts cross-correlated. The normalized cross-correlation factor ρ_{1k} is then defined as

$$\rho_{1k} = \frac{\langle (A_1 - \langle A_1 \rangle)\,(A_k - \langle A_k \rangle) \rangle}{\left[\left\langle (A_1 - \langle A_1 \rangle)^2 \right\rangle \left\langle (A_k - \langle A_k \rangle)^2 \right\rangle \right]^{1/2}}. \tag{9}$$

Here, the mathematical expectation is denoted by $\langle \cdot \rangle$. Figure 3 shows ρ_{12} and ρ_{13} as calculated for a simulated oscillator with a progressive spring-stiffness

$$\ddot{x} + 2D\omega_0\dot{x} + (\omega_0^2 + \beta x^2)x = \zeta(t) \tag{10}$$

with $D = 0.01$, $\omega_0 = 2\pi$ and $\zeta(t)$ being Gaussian white noise of variance $\sigma_\zeta^2 = 1$. The relative filter-bandwidth (filter-bandwidth divided by center frequency) used was $\delta = 0.1$ and the main frequency, which depends on β, was selected automatically from the FFT of the output signal. For this type of nonlinearity only odd harmonics occur which is reflected in a significant amount of correlation between the main and the third harmonic , ρ_{13}.

The second example was given in [4] and describes a system with a "flapping" crack represented by a piecewise linear restoring force $f(x)$ in

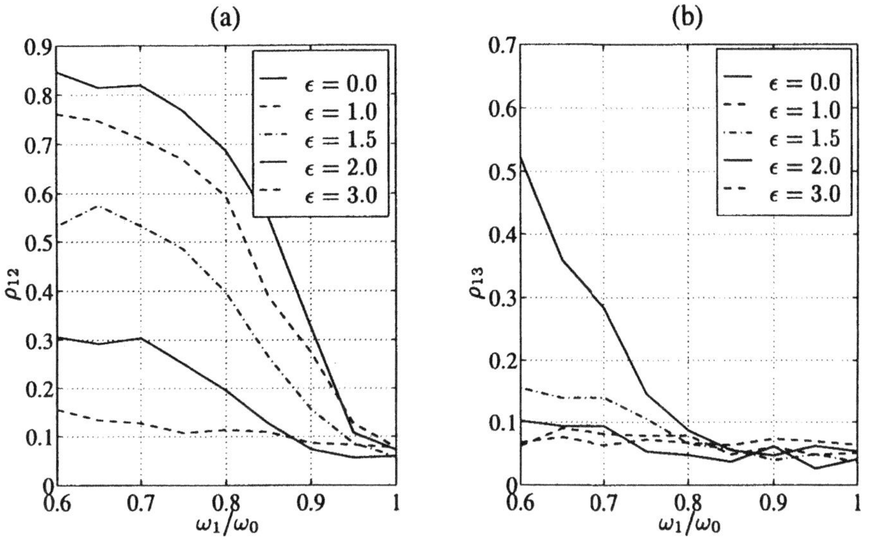

Figure 4. Normalized cross-correlation factors ρ_{12} (a) and ρ_{13} (b) for a mechanical oscillator with crack-type nonlinearity.

the model described by equation (8):

$$f(x) = \begin{cases} \omega_0^2 x & \text{for} \quad x \leq x_0 \\ \omega_1^2 x + (\omega_0^2 - \omega_1^2)\, x_0 & \text{for} \quad x > x_0 \end{cases}, \quad \omega_0 > \omega_1 \tag{11}$$

The difference $1 - \omega_1/\omega_0$ is a measure of the nonlinearity of the system and may be related to the crack depth whereas x_0 is the threshold value at which transition from the "hard" to the "soft" spring occurs. Increasing this value reduces the nonlinear character of the system. The crack-type restoring force (11) contains odd and even components, therefore odd and even higher harmonics occur.

Figure 4 shows the normalized cross-correlation factors ρ_{12} and ρ_{13} versus ω_1/ω_0. As this ratio changes from 0.6 to 1.0 the system becomes more linear which is reflected in the decreasing value of ρ_{12}. The curves in the diagram belong to different values of the normalized threshold $\epsilon = x_0/\sigma_{lin}$. Here σ_{lin} denotes the standard deviation of the linear system excited by Gaussian white noise. As expected, increasing the threshold ϵ reduces the systems nonlinearity because the timespan in which the "soft" spring is active is reduced.

The factor ρ_{13} is smaller than ρ_{12} and indicates significant nonlinearity only for $\epsilon = 0$. The results agree qualitatively with the results presented for this system in [4] although a different implementation of the algorithm was used to produce the results.

3. Conclusions

After a brief overview on nonlinearity tests in the literature two methods have been considered and numerical examples have been shown. Both tests require only experimental data that is readily available in most experimental set-ups. The test based on the Hilbert transform requires frequency response data from a frequency sweep while the test based on cross-correlation of higher harmonics relies on output measurements of stochastically excited systems. The second test is suitable for a wider range of applications because the system's input does not necessarily have to be controlled. This is e.g. important in the investigation of flow induced vibrations. The preliminary results show that certain types of nonlinearity can be detected and the Hilbert transform is also able to distinguish between them.

Further investigations considering more complex models are necessary and the usefulness of the test has to be verified in real experiments.

References

1. Barnett, W. A., Gallant, A. R., Hinich, M. J., Jungeilges, J. A., Kaplan, D. T., Jensen, M. J. (1995) Robustness of nonlinearity and chaos tests to measurement error, inference method, and sample size, *Journal of economic behavior and organization*, **27**, pp. 301-320
2. Bendat, J. S., Piersol, A. G. (1980) *Engineering applications of correlation and spectral analysis.*John Wiley and Sons, New York.
3. Dimentberg, M. F., Haenisch, H. G. (1998) Method of Internal Harmonics Cross-Correlation for On-Line Detection of Structural Nonlinearities. In *Proceedings of the 12th Engineering Mechanics Conference.* La Jolla, California
4. Dimentberg, M. F., Sokolov, A. A. (1991) Identification of restoring force nonlinearity from a system's response to a white-noise excitation, *International Journal of Non-Linear Mechanics*, **26**, pp. 851-855
5. Gabr, M. M. (1992) Nonlinearity tests for bilinear systems, *Journal of computational and applied mathematics*, **40**, pp. 313-322
6. Haber, R. (1985) Nonlinearity tests for dynamic processes. In *IFAC Symp. on Identification and System Parameter Estimation.* York, UK
7. Hinich, M. (1982) Testing for Gaussianity and linearity of a stationary time series, *J. Time Ser. Anal.*, **3**, pp. 169-176
8. Keenan, D. M. (1985) A Tukey nonadditivity-type test for time series nonlinearity, *Biometrika*, **72**, pp. 39-44
9. Peyton Jones, J. C., Billings, S. A. (1988) *Testing for nonlinearity using a prediction error filter.* Research Report 332, Department of Control Engineering, University of Sheffield.
10. Simon, M., Tomlinson, G. R. (1984) Use of the Hilbert transform in modal analysis of linear and non-linear structures, *Journal of Sound and Vibration*, **96**, pp. 421-436
11. Subba Rao, T., Gabr, M. M. (1980) A test for linearity of stationary time series, *J. Time Ser. Anal.*, **1**, pp. 145-158
12. Tong, H. (1990) *Non-linear time series: a dynamical system approach.*Oxford University Press, Oxford.
13. Tsay, Ruey S. (1986) Nonlinearity tests for time series, *Biometrika*, **73**, pp. 461-466

REDUCTION METHODS FOR NONLINEAR VIBRATIONS OF SPATIALLY CONTINUOUS SYSTEMS WITH INITIAL CURVATURE

G. REGA AND W. LACARBONARA
Dipartimento di Ingegneria Strutturale e Geotecnica,
University of Rome La Sapienza,
via A. Gramsci, 53 Rome 00197, Italy

AND

A. H. NAYFEH
Department of Engineering Science and Mechanics, MC 0219
Virginia Polytechnic Institute and State University
Blacksburg, Virginia 24061 U.S.A.

Abstract. Reduction methods and the resulting models for studying nonlinear vibrations of shallow monodimensional continuous systems are discussed. Primary resonances of the first mode of buckled beams and suspended cables are investigated. The convergence of the relevant Galerkin-reduced models with variation of the nondimensional buckling level (buckled beam) and the elasto-geometric parameter (cable) is analyzed. For low values of the control parameter, one-dof models (with the first relevant linear eigenmode) are sufficiently accurate, whereas, for higher values of the parameter (above the first crossover), three- or higher-dof models (with the lowest relevant symmetric eigenmodes) are the minimum reduced-order models that can capture qualitatively the symmetric planar dynamics of the original systems. The major modification of the mode shapes of the lowest symmetric modes with respect to the initial nonlinear static shape, due to crossover and snap-through-type bifurcations, is highlighted as the key mechanism for the breakdown of low-dimensional models.

1. Introduction

The local dynamics of a shallow monodimensional self-adjoint system around its initial curved configuration are governed by a set of integral-partial-differential equations and associated boundary conditions with quadratic and cubic geometric nonlinearities. It has been shown (Nayfeh et *al.* [1]; Pakdemirli and Boyaci [2]; Lacarbonara [3]) that direct attack of the equations of motion with a reduction method, such as the method of multiple scales, and application of the same method

N. Van Dao and E.J. Kreuzer (eds.),
IUTAM Symposium on Recent Developments in Non-linear Oscillations of Mechanical Systems, 235–246.

to the associated ∞-dimensional Galerkin-discretized models (obtained with the relevant base of the eigenmodes of the linearized problem) lead to the same asymptotic dynamics. These results show that, in principle, all of the mode shapes from the pertinent eigenspectrum contribute to the nonlinear motions arising from a primary resonance of a mode that is not involved in internal resonances with other modes.

Lacarbonara et *al.* [4] showed that the direct approach yields, for high buckling levels, results in agreement with the experimental response of a fixed-fixed first-mode buckled beam to a primary resonance of its first mode, whereas low-order Galerkin-reduced models lead to qualitatively erroneous results. Further, Rega et *al.* [5] showed that the direct procedure and a four-mode discretization yield slightly different bifurcation patterns for multiple resonances in suspended cables. To overcome the shortcomings of finite-dimensional Galerkin discretizations, Nayfeh [6] developed a novel technique for constructing reduced-order models of nonlinear distributed-parameter systems.

Here the problem of convergence of Galerkin-reduced models for primary resonances of the first mode of fixed-fixed first-mode buckled beams and suspended cables is addressed. The associated issue of extracting appropriate low-dimensional models from the original systems is discussed. It is shown that one-dof models (with the first relevant directly excited eigenmode), contrary to the common intuition, breakdown in some ranges of the buckling level (buckled beam) or the elasto-geometric parameter (cable). In these ranges, three- or higher-dof models (with the lowest relevant symmetric eigenmodes) are the minimum reduced-order models that can capture, at least qualitatively, the correct planar dynamics of the original systems.

2. A Class of Shallow Elastic One-Dimensional Systems

Nonlinear vibrations of shallow elastic continuous systems around their initial static configurations are governed, in nondimensional form, by

$$\ddot{u} + \mathcal{L}u = \mathcal{G}_2(u,u) + \mathcal{G}_3(u,u,u) - c\dot{u} + F(s,t) \tag{1}$$

subject to the linear homogeneous boundary conditions $\mathcal{B}(u) = 0$. In Eq. (1), s is the coordinate along the horizontal projection of the centerline of the system; the overdot indicates differentiation with respect to dimensionless time t; $u(s,t)$ is, in general, a three-dimensional vector of dynamic displacements measured from the curved configuration; $\mathcal{L}$ is a linear, homogeneous, self-adjoint, positive-definite integral-differential operator; $\mathcal{G}_2$ and $\mathcal{G}_3$ are quadratic and cubic geometric operators; and $F(s,t)$ is the forcing function. The nonlinear operators are non-commutative; i.e., $\mathcal{G}_2(v,w) \neq \mathcal{G}_2(w,v)$.

By virtue of the symmetric nature of the linear unforced undamped problem, the eigenfunctions $\phi_m(s)$ are mutually orthogonal and we assume they have been normalized such that $\int_0^1 \phi_m(s)\phi_n(s)ds = \delta_{mn}$ and $\int_0^1 \phi_m \mathcal{L}\phi_n ds = \omega_n^2\delta_{mn}$ where

δ_{mn} is the Kronecker delta. The eigenvalue problem governing the frequencies and associated mode shapes is

$$\mathcal{E}[\phi;\omega] = (\mathcal{L} - \omega^2\mathcal{I})\phi = 0, \quad \mathcal{B}(\phi) = 0$$

where $\mathcal{I}$ is the identity operator.

For planar vibrations of a shallow arch with initial shape $w(s)$ and an end-load p,

$$\mathcal{L}v = v'''' + Pv'' - \psi'' \int_0^1 v'\psi' ds, \quad P = p + \frac{1}{2}\int_0^1 \left(w'^2 - \psi'^2\right) ds \tag{2}$$

$$\mathcal{G}_2(v,v) = v'' \int_0^1 v'\psi' ds + \frac{1}{2}\psi'' \int_0^1 v'^2 ds, \; \mathcal{G}_3(v,v,v) = \frac{1}{2}v'' \int_0^1 v'^2 ds \tag{3}$$

where v denotes the transverse dynamic deflection measured from the shallow static nonlinear configuration ψ; and the prime indicates differentiation with respect to s. For an initially straight nth-mode buckled beam, we put $P = P_c$ in Eq. (2), where P_c is the nth critical Euler load.

For a suspended homogeneous elastic cable with small sag-to-span ratios (Benedettini et *al.* [7]),

$$\mathcal{L}\mathbf{u} = -\frac{1}{\rho}\begin{bmatrix} u_1'' + \frac{1}{64}\lambda^2\psi'' \int_0^1 u_1'\psi' ds & 0 \\ 0 & u_2'' \end{bmatrix} \tag{4}$$

$$\mathcal{G}_2(\mathbf{u},\mathbf{u}) = 8\frac{Kb}{\rho}\left\{\begin{array}{c} u_1'' \int_0^1 u_1'\psi' ds + \frac{1}{2}\psi'' \int_0^1 (u_1'^2 + u_2'^2)\, ds \\ u_2'' \int_0^1 u_1'\psi' ds \end{array}\right\} \tag{5}$$

$$\mathcal{G}_3(\mathbf{u},\mathbf{u},\mathbf{u}) = 4\frac{Kb}{\rho}\left\{\begin{array}{c} u_1'' \int_0^1 (u_1'^2 + u_2'^2)\, ds \\ u_2'' \, (u_1'^2 + u_2'^2)\, ds \end{array}\right\} \tag{6}$$

$$\mathbf{C}\dot{\mathbf{u}} = \begin{bmatrix} c_1\dot{u}_1 & 0 \\ 0 & c_2\dot{u}_2 \end{bmatrix}, \quad \mathbf{F}(s,t) = \left\{\begin{array}{c} P_1(s)\cos\Omega t \\ P_2(s)\cos(\Omega t + \tau) \end{array}\right\} \tag{7}$$

where ρ is a nondimensional inertia term, $K = E/(mg\ell)$ is a nondimensional parameter depending on the Young modulus E, the volume mass density m, and the span ℓ; b is the sag-to-span ratio; $\lambda^2 = 512Kb^3$ is the square of the elasto-geometric parameter [8]; $\psi = 4bs(1-s)$ is the initial parabolic shape; and $\mathbf{u}^{\mathrm{T}} = \{u_1, u_2\}$ denotes the vector of in-plane (vertical) and out-of-plane (horizontal) displacement components.

3. Direct Approach, Full-basis, and Finite Galerkin Discretizations

Direct application of the method of multiple scales [9,10] to Eq. (1) and associated boundary conditions produces a second-order approximation of the response of the system to a primary resonance of the non-internally resonant nth mode in the form

$$\begin{aligned} u(s,t) = & \; a_n \cos(\Omega t - \gamma_n)\phi_n(s) \\ & + \frac{1}{2}a_n^2 \left[\cos 2\left(\Omega t - \gamma_n\right) \psi_{1n}(s) + \psi_{2n}(s) \right] + \ldots \end{aligned} \tag{8}$$

where the ψ_{jn} are solutions of the boundary-value problems

$$\mathcal{E}[\psi_{1n}; 2\omega_n] = \mathcal{G}_2(\phi_n, \phi_n), \qquad \mathcal{E}[\psi_{2n}; 0] = \mathcal{G}_2(\phi_n, \phi_n) \tag{9}$$

with the pertinent boundary conditions. The amplitude a_n and the frequency detuning σ are related by the following frequency-response equation:

$$\sigma = -\alpha_{nn} a_n^2 \pm \left(\frac{f_n^2}{4\omega_n^2 a_n^2} - \mu^2 \right)^{\frac{1}{2}} \tag{10}$$

where f_n is the nth modal projection of the force and the *effective nonlinearity coefficient* is given by

$$\alpha_{nn} = \frac{1}{8\omega_n}\left(S_{nn} + 3\gamma_{nnnn} \right), \qquad \gamma_{nnnn} = \int_0^1 \phi_n \mathcal{G}_3(\phi_n, \phi_n, \phi_n) ds, \tag{11}$$

$$S_{nn} = \int_0^1 \phi_n \left\{ \sum_{j=1}^{2} (1 + \delta_{2j}) \left[\mathcal{G}_2(\phi_n, \psi_{jn}) + \mathcal{G}_2(\psi_{jn}, \phi_n) \right] \right\} ds. \tag{12}$$

Performing a full-basis Galerkin discretization of Eq. (1) and the pertinent boundary conditions by using the complete set of eigenfunctions ϕ_n and subsequently applying the method of multiple scales to the resulting infinite set of ordinary-differential equations yields (Lacarbonara [3]) the same approximate solution as that obtained with the direct approach; that is, Eqs. (8) and (10)–(12). The remarkable formal difference is that the second-order shape functions and the softening term in the effective nonlinearity coefficient are modal realizations of their counterparts obtained with the direct approach. For planar vibrations of fixed-fixed buckled beams and suspended cables (they are governed by the same quadratic operators), the series for the softening term becomes

$$S_{nn}^{(\infty)} = \sum_{k=1}^{\infty} \Pi_k, \quad \Pi_k = 2\left(\frac{2}{\omega_k^2} + \frac{1}{\omega_k^2 - 4\omega_n^2} \right) \left[b I_{1n} I_{2k} + \frac{1}{2} b I_{1k} I_{2n} \right]^2 \tag{13}$$

where

$$I_{1k} = \int_0^1 \varphi' \phi'_k ds \quad \text{and} \quad I_{2k} = \int_0^1 \phi'_n \phi'_k ds, \text{ with } \varphi = \psi/b.$$

For suspended cables, Π_k is multiplied by $8Kb/\rho$. We note that, for primary resonances of a symmetric mode, Π_k is zero for k corresponding to antisymmetric eigenmodes.

Finite-dof models can be extracted from the ∞-dimensional model by retaining a finite number of modes. The effects of the quadratic nonlinearities are twofold. First, they make the spatial variation ψ_{jn} depend on all of the modes. Second, they modify the effective nonlinearity coefficient through ψ_{jn}. The latter coefficient consists of two terms: the term $3\gamma_{nnnn}/8\omega_n$, generated by the cubic nonlinearity, and the term $S_{nn}^{(\infty)}/8\omega_n$ ($S_{nn}^{(M)}/8\omega_n$), calculated with the direct procedure (M−dof model), generated by the quadratic nonlinearities. The first term is responsible for a hardening behavior and, to this order of approximation, does not depend on the method employed. The second term is responsible for a softening behavior; it depends on the order (i.e., number of discretizing modes) of the discretization procedure.

The key issue related to the order reduction is convergence. Because the amplitude and phase of the response depend quantitatively but, most importantly, qualitatively on the effective nonlinearity coefficient, we focus on the convergence of this coefficient rather than on that of the displacement field. Convergence of the series (13) should guarantee convergence of the function series $\hat{\psi}_{jn}$.

3.1 PRIMARY RESONANCE OF A FIXED-FIXED BUCKLED BEAM

In this section, the convergence results conducted on the effective nonlinearity coefficient of the first mode of a fixed-fixed first-mode buckled beam are summarized. The critical Euler buckling load is $P_c = 4\pi^2$ and the corresponding buckling mode shape is $\psi = (1/2)b(1 - \cos 2\pi s)$, with b denoting the buckling level nondimensionalized with respect to the radius of gyration of the cross-section of the beam. We note that (Lacarbonara et *al.* [4]) $\psi(s)$ is the exact solution for the nonlinear post-buckling shape. In this case, $n = 1$; hence, for even k, $\Pi_k = 0$. The coefficient α_{11} was computed with the direct approach, using Eq. (11), and with the low-order models with $M = 1, 2, 3, 4$, where M indicates the number of the retained lowest symmetric modes. We show the results of these computations in Fig. 1(a). For low buckling levels b, the results of the low-order models, including the one-mode model, are in good quantitative agreement with the results obtained with the direct approach. However, as the buckling level increases, quantitative and even qualitative differences occur. The effective nonlinearity coefficient obtained with the direct approach is positive for all b of interest (i.e., the first mode is always softening), whereas, when computed with the one- and two-mode models, the coefficient becomes negative above a threshold value b_c (close but below the second crossover level). On the

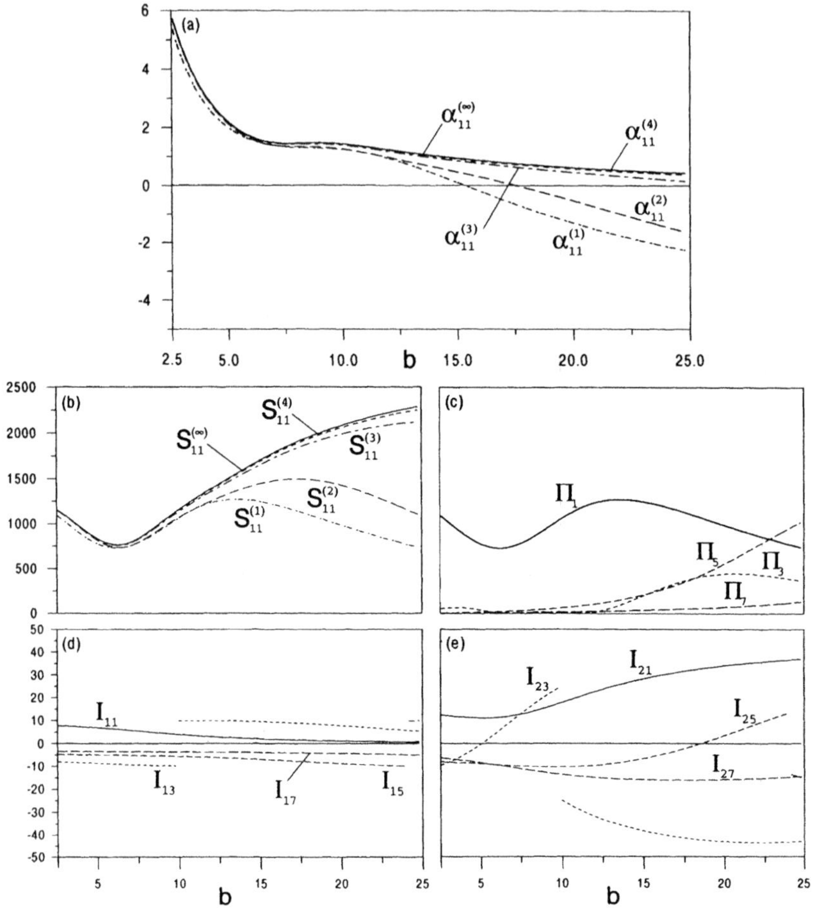

Figure 1. Variation of (a) the effective nonlinearity coefficient $\alpha_{11}^{(M)}$, (b) $S_{11}^{(M)}$, (c) Π_k, (d) I_{1k}, and (e) I_{2k}.

other hand, the three-mode and four-mode models predict a softening behavior for all b of interest in qualitative agreement with the direct model.

To investigate the convergence properties of the effective nonlinearity coefficient, we study convergence of the series given by Eqs. (13). In Fig. 1(b), we show the sum of the series $S_{11}^{(\infty)}$ and the partial sums of order one, two, three, and four. The softening term $S_{11}^{(\infty)}$, except for a minimum around $b = 6$, increases monotonically with increasing b. In contrast, the partial sums of order one and two increase until they attain a maximum. Above this maximum, they decrease with increasing b. Therefore, comparing Fig. 1(a) with Fig. 1(b), we observe that the breakdown of the one-mode and two-mode models, above b_c, is due to the underestimation of the softening effects.

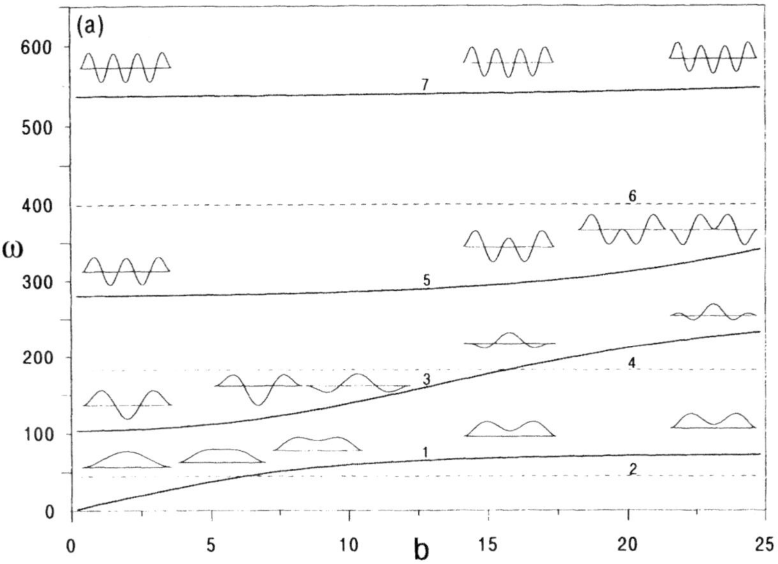

Figure 2. Variation with the control parameter of the lowest four symmetric frequencies and mode shapes of fixed-fixed buckled beams.

In Fig. 1(c), we show variation of each individual Π_k term with b. As expected, except for high values of b, the effect of the first mode is bigger than those of the higher modes. However, above a certain buckling level ($b \approx 13$), the effect of the first mode decreases monotonically with increasing b, and, for high values of b, it is lower than the effect of the fifth mode. In addition, the effect of the latter increases monotonically with b, and, except for two small intervals of b, it is higher than the effect of the third mode.

In Figs. 1(d) and 1(e), we show variation of the integrals I_{1k} and I_{2k}. In Fig. 1(d) we note that I_{11} is positive and decreases monotonically with increasing b, whereas I_{13} and I_{15} experience jumps from negative to positive values at $b \approx 9.82$ and $b \approx 24.02$, respectively. At these buckling levels, the third and fifth modes undergo snap-through-type bifurcations. Similarly, I_{23} and I_{25} experience jumps at the same b–values due to the above-mentioned bifurcations. The symmetric eigenmodes depend on the first-order component of the dynamic stretching force which, in turn, depends on the applied end-load. Increasing the buckling level (i.e., the end-load), these modes can undergo crossover and, for higher loads, snap-through-type bifurcations which entail major modifications of the structure of the associated shapes as it is shown in Fig. 2.

Observing the evolution of these mode shapes, above the first crossover, we are now able to interpret the behavior of I_{1k} and I_{2k}. For example, for the first mode, the presence of a hump in the midspan is responsible for the increased overall curvature of the mode shape, i.e. I_{21} (Fig. 1(e)); in addition, the resulting deviation between

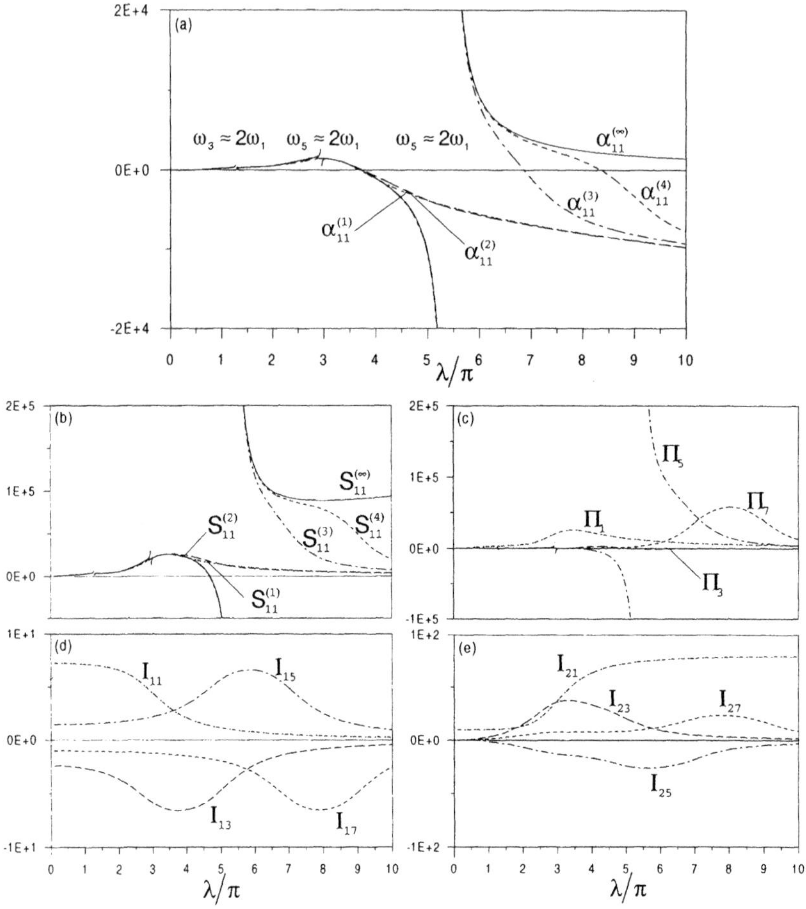

Figure 3. Variation of (a) the effective nonlinearity coefficient $\alpha_{11}^{(M)}$, (b) $S_{11}^{(M)}$, (c) Π_k, (d) I_{1k}, and (e) I_{2k} when $K = 588.581$.

the first mode shape and the post-buckling deflection of the beam is responsible for decreasing I_{11} (Fig. 1(d)). These two integrals have antagonist effects in the softening term; however, because $I_{11} < 1$ for high b, the product of I_{11} and I_{21} in Eq. (13) results in the underestimation of the softening effect in the one-mode model. At the same time, the evolution of the second and third symmetric mode shapes for high buckling levels seems to be responsible for the increasingly important contributions to the softening effects from the latter modes.

The implications of these multi-mode behaviors for high b on the underlying nonlinear dynamics of the system are such that a more complex spatial content is present in the dynamic deflections (i.e., 'spatial overtones').

3.2 PRIMARY RESONANCE OF A SUSPENDED CABLE

In this section, the results obtained for the effective nonlinearity coefficient of the first in-plane (vertical) mode of a suspended cable are summarized.

In Figs. 3(a) and 3(b), we show variation of the effective nonlinearity coefficient and of the softening term, respectively, with λ, as calculated with the direct approach and the one-, two-, three-, and four-dof models. The effective nonlinearity coefficient of the first mode is predicted to be of the hardening type for very low values of λ by all of the models (when λ approaches zero the cable tends to a string). As λ is increased, it becomes softening. As λ increases above $\lambda \approx 3.5\pi$, there is a further switch from softening to hardening behavior. In the range of high values of λ, the one- and two-mode models predict a hardening behavior, whereas the direct approach as well as the three- and higher-dof models predict a softening behavior. However, for higher values of λ, also the three- and four-dof models breakdown. It should be mentioned that for typical values of K (common technical cables), this range of λ corresponds to high sag-to-span ratios (i.e., slack cables) where the assumptions on the initial parabolic profile may become questionable. This does not occur for the experimental cable considered here (Rega et *al.* [11]). For this cable, characterized by a high value of K, typical values of b of a parabolic profile (up to one-eight) correspond to the overall considered range of λ.

We note that the discontinuous portions of Figs. 3(a) and 3(b) correspond to divergent solutions and are located in the neighborhoods of 2:1 internal resonances involving the first and second symmetric modes and the first and third symmetric modes. In these regions, the softening term (Fig. 3(b)) and the effective nonlinearity coefficient (Fig. 3(a)) calculated with the direct approach (or two- and higher-dof models) attain arbitrarily large values. This is due to the vanishing denominator $(\omega_j^2 - 4\omega_1^2)$ associated with the particular solution of the higher-frequency mode involved in the 2:1 internal resonance with the first mode. More specifically, the strongly divergent behavior for $\lambda \approx 5.5\pi$ is due to the largely detuned 2:1 internal resonance between the first and third symmetric planar modes. Evidently, the non-internally resonant expansion does not hold any more in these ranges. However, it is worth remarking that the breakdown of the expansion is naturally highlighted by the accomplished direct procedure.

Again, a close look at Figs. 3(b) and 3(c) leads to the conclusion that for high values of λ, the one- and two-dof models underestimate the softening effects of the quadratic nonlinearities. In Fig. 3(c) we note that (i) Π_1 attains a maximum between the first and second crossovers and then decreases with increasing λ; (ii) the term corresponding to the second symmetric mode is relatively small in the whole range of interest; and (iii) more interestingly, the terms corresponding to the third and fourth symmetric modes are higher than that corresponding to the first mode for high λ.

In Figs. 3(d) and 3(e), we show variation of the integrals I_{1k} and I_{2k}. We note

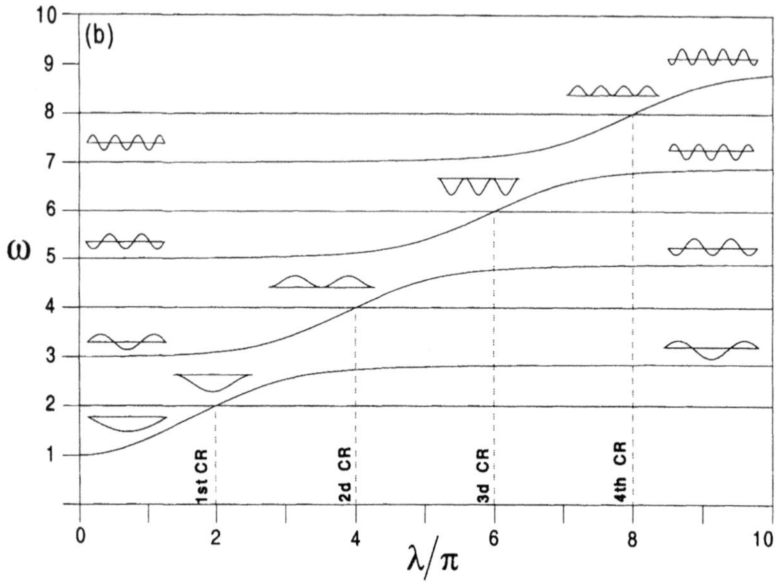

Figure 4. Variation with the control parameter of the lowest four symmetric frequencies and mode shapes of suspended cables.

that I_{11} is positive and decreases monotonically with increasing λ, as in buckled beams, and I_{21} has also a comparable behavior. On the other hand, I_{13}, I_{15}, and I_{17} are characterized by maxima attained at the crossover points of the associated modes (second, third, and fourth crossovers in Fig. 4). Again, observation of the evolution of the mode shapes, above the first crossover, allows us to interpret the behavior of I_{1k} and I_{2k}. Moreover, combination of the latter in Eq. (13) expounds their different contributions to the softening effects.

4. Conclusions

Convergence of Galerkin-discretized models for primary resonances of the first planar non-internally resonant mode of a fixed-fixed first-mode buckled beam and a suspended cable has been investigated. The used control parameter is the nondimensional buckling level for the buckled beam and the elasto-geometric parameter λ for the suspended cable. The limit sums of the investigated series have been obtained with direct application of the method of multiple scales to the relevant equations of motion. The governing equations of the two structural systems have the same geometric nonlinear operators, but have different linear stiffness operators. In addition, the initial normalized nonlinear shapes of both systems do not vary with the control parameter. Therefore, the difference between the dynamics of the beam and cable is the result of the variation of the eigenvalue problem with the control parameter.

By applying the method of multiple scales directly to the equations of motion, it is shown that one-dof models breakdown in some ranges of the control parameter (above the first crossover) due to underestimation of the softening effects. In these ranges, three- or higher-dof models are the minimum reduced-order models that can capture, at least qualitatively, the correct symmetric planar non-internally resonant dynamics of the original systems. The first-order dynamic stretching force is responsible for crossover bifurcations. In the case of buckled beams, it is likely to be responsible also for snap-through-type bifurcations of the symmetric modes. Both bifurcations entail major modifications of the associated mode shapes. Specifically, the variation of the shape of the first mode has two antagonist effects: (i) the increased overall curvature of the shape tends to emphasize the softening behavior; (ii) the deviation of the mode shape with respect to the initial nonlinear shape (i.e., presence of regions of opposite curvature) has a detrimental effect on the softening behavior. In both systems, this latter effect prevails and is responsible for the breakdown of one-dof model.

We conclude that a heuristic or qualitative criterion for predicting the breakdown of low-order models may be based on comparison of the mode shape of the excited mode with the initial nonlinear shape. When major deviations occur, low-order models are likely to fail.

5. Acknowledgments

This work was partially supported by 1997 Italian M.U.R.S.T. funds (60 %) and by the National Science Foundation under Grant No. CMS-9423774.

6. References

1. Nayfeh, A. H., Nayfeh, S. A., and Pakdemirli, M. (1995) On the discretization of weakly nonlinear spatially continuous systems, in W. Kliemann and N. Sri Namachchivaya (eds.), *Nonlinear Dynamics and Stochastic Mechanics*, 175–200.
2. Pakdemirli, M. and Boyaci, H. (1995) Comparison of direct-perturbation methods with discretization-perturbation methods for non-linear vibrations, *Journal of Sound and Vibration* **186**, 837–845.
3. Lacarbonara, W. (1999) Direct treatment and discretizations of non-linear spatially continuous systems, *Journal of Sound and Vibration* **221**, 849–866.
4. Lacarbonara, W., Nayfeh, A. H., and Kreider, W. (1998) Experimental validation of reduction methods for weakly nonlinear distributed-parameter systems:

Analysis of a buckled beam, *Nonlinear Dynamics* **17**, 95–117.

5. Rega, G., Lacarbonara, W., Nayfeh, A. H., and Chin, C-M (1999) Multiple resonances in suspended cables: direct versus reduced-order models, *International Journal of Non-Linear Mechanics* **34**, 901-924.
6. Nayfeh, A. H. (1998) Reduced-order models of weakly nonlinear spatially continuous systems, *Nonlinear Dynamics* **16**, 105–125.
7. Benedettini, F., Rega, G., and Alaggio, R. (1995) Non-linear oscillations of a four-degree-of-freedom model of a suspended cable under multiple internal resonance conditions, *Journal of Sound and Vibration* **182**, 775–798.
8. Irvine, H. M. and Caughey, T. K. (1974) The linear theory of free vibrations of a suspended cable, *Proceedings of the Royal Society London* **A341**, 299–315.
9. Nayfeh, A. H. (1973) *Perturbation Methods*, Wiley-Interscience, New York.
10. Nayfeh, A. H. (1981) *Introduction to Perturbation Techniques*, Wiley-Interscience, New York.
11. Rega, G., Alaggio, R., and Benedettini, F. (1997) Experimental investigation of the nonlinear response of a hanging cable. Part I: Local analysis, *Nonlinear Dynamics* **14**, 89–117.

ASYMPTOTIC METHOD FOR INVESTIGATION OF m - FREQUENCY OSCILLATION SYSTEMS

A.M. SAMOILENKO
Institute of Mathematics
National Academy of Sciences of Ukraine
3 Tereshchenkovskaya Str.
252601 Kiev 4, Ukraine

In $\mathbb{R}^{2n}$, consider a system of equations of the form

$$\frac{dx}{dt} = Ax + \varepsilon X(x), \tag{0.1}$$

where A is a constant matrix, $X \in \mathbb{K}[x]$, $\mathbb{K}[x]$ is the polynomial ring over $\mathbb{R}, x \in r^{2n}$, and ε is a small positive parameter. Assume that, for $\varepsilon = 0$, a general solution of system (0.1) is quasiperiodic with frequency basis $\omega = (\omega_1, \ldots, \omega_m)$, where $m \leq n$. Under these assumptions, system (0.1) reduces to the form

$$\frac{dx}{dt} = \lambda Hx + \varepsilon X(x), \tag{0.2}$$

where

$$\lambda H = diag\{\lambda_1 H_1, \ldots, \lambda_n H_n\}, \quad H_\nu^2 = -E, \quad \nu = \overline{1, n}, \tag{0.3}$$

E is the two-dimensional identity matrix, $\lambda_\nu > 0, x = (x_1^1, \ldots, x_n^n)$, $x_\nu^\nu \in \mathbb{R}^2, \nu = \overline{1.n}$.

We can apply to system (0.2) the general method of asymptotic integration of nonlinear systems given in [1, 2], as well as the Bogolyubov theory of quasiperiodic solutions [3].

1. Asymptotic Method

Consider the system of equations (0.2). For $\varepsilon = 0$, its fundamental matrix of solutions $e^{\lambda Ht}$ determines the m-parameter family of limit matrices as $|t| \to \infty$. The form of this family of matrices depends on the choice of the frequency basis ω. In the nonresonance case, we have $\lambda = \omega$ and the limit family is determined by the matrix

$$\Phi(\varphi) = e^{H\varphi} = diag\{sin\varphi_1 H_1 + cos\varphi_1 E, \ldots, sin\varphi_n H_n + cos\varphi_n E\}.$$

In the resonance case, we assume that

$$\lambda = K\omega, \tag{1.1}$$

where K is an integer-valued $n \times m$ matrix of rank m.

The limit family as $|t| \to \infty$ is now determined by the matrix $\Phi(K\Psi)$, where $\Psi = (\Psi_1, \ldots, \Psi_m) \in \mathcal{J}_m$.

N. Van Dao and E.J. Kreuzer (eds.),
IUTAM Symposium on Recent Developments in Non-linear Oscillations of Mechanical Systems, 247–256.

According to [2], we seek the asymptotic expansion of a solution $x = x(t, \varepsilon)$ of system (0.2) in the form of the series

$$x = y + \varepsilon u_1(y) + \cdots + \varepsilon^p u_p(y) + \ldots, \tag{1.2}$$

where $y = y_1^1, \ldots, y_n^n), y_\nu^\nu \in I\!R^2$, $y = y(t, \varepsilon)$ is a solution of the averaged equation

$$\frac{dy}{dt} = \lambda H y + \varepsilon Y_1(y) + \cdots + \varepsilon^p Y_p(y) + \ldots, \tag{1.3}$$

$u_\nu, \nu = 1, 2 \ldots,$ is a solution of the homological equation

$$\mathcal{L} u_\nu = X_\nu(y) - Y_\nu(y),\ \nu = 1, 2, \ldots,$$

under the condition

$$S u_\nu(y) = 0, \quad \nu = 1, 2, \ldots.$$

Here,

$$\mathcal{L} = \frac{\partial}{\partial y} \lambda H y - \lambda H$$

is the homological operator of the method, S is the averaging operator of the method, and

$$X_1(y) = X(y), \quad X_2(y) = \frac{\partial X(y)}{\partial y} u_1(y), \ldots,$$

$$X_\nu(y) = \frac{1}{\nu!} \frac{d^\nu}{d\varepsilon^\nu} \left[\varepsilon X(y + \varepsilon u_1(y) + \cdots + \varepsilon^{\nu-1} u_{\nu-1}(y))\right] |_{\varepsilon=0}, \tag{1.4}$$

. . . .

$$Y_\nu(y) = S X_\nu(y), \quad \nu = 1, 2, \ldots.$$

In view of the expressions for the limit values of the matrix $e^{\lambda H t}$ as $|t| \to \infty$, we transform the formulas [1] for S and $\mathcal{L}^{-1}$ to the form

$$SX(y) = \frac{1}{(2\pi)^m} \int_0^{2\pi} \cdots \int_0^{2\pi} \Phi(-K\Psi) X(\Phi(K\Psi) y) d\Psi_1 \ldots d\Psi_m, \tag{1.5}$$

$$\mathcal{L}^{-1}[X] = \sum_{k \neq 0} \frac{X_k^1(y)}{(k, \omega)},$$

where $(k, \omega) = k_1\omega_1 + \cdots + k_m\omega_m$, $k \in \boldsymbol{Z}^m$,

$$X_k^1(y) = -\frac{1}{(2\pi)^m} \int_0^{2\pi} \cdots \int_0^{2\pi} \Phi(-K\Psi) X(\Phi(K\Psi) y) sin(k, \Psi) d\Psi_1 \ldots d\Psi_m. \tag{1.6}$$

Relations (1.2)-(1.6) determine the asymptotic method of integration of system (0.2). The system in which

$$X(0) = 0, \quad \frac{\partial X(0)}{\partial x} = 0$$

is an important case of system (0.2). For this case, for $\varepsilon = 1$, the asymptotic method gives an expansion of the normal form [5] regularized by powers of ε for the system under consideration and the same expansion for the transformation leading to the normal form.

2. Splitting of the Averaged System of Equations

According to [2], the functions Y_ν belong to the kernel of the operator $\mathcal{L}$ and, therefore, satisfy the relation

$$Y_\nu(e^{\lambda Ht}y) = e^{\lambda Ht}Y_\nu(y), \quad \nu = 1, 2, \ldots. \tag{2.1}$$

In the limit as $|t| \to \infty$, relation (2.1) turns into the equality

$$Y_\nu(\Phi(K\Psi)y) = \Phi(K\Psi)Y_\nu(y), \quad \nu = 1, 2, \ldots, \tag{2.2}$$

which is valid for all $\Psi \in \mathcal{J}_m$ and $y \in I\!R^{2n}$.

Equality (2.2) allows one to split the averaged system of equations (1.3) by separating the system of equations with slow variables form it. For this purpose, instead of y, we introduce in (1.3) polar coordinates $(\varphi, h) \in \mathcal{J}_n \times I\!R^{+n}$ according to formulas

$$y = e^{H\varphi}Bh, \tag{2.3}$$

where B is a $2n \times n$ matrix defined by the following conditions

$$B^+HB = 0, \quad B^+B = E,$$

B^+ is the matrix pseudoinverse to B, and 0 and E are the zero and identity matrices of the corresponding dimension.

In the variables h, φ, equation (1.3) takes the form

$$\begin{aligned} \frac{dh}{dt} &- \sum_{\nu \geq 1} \varepsilon^\nu B^+ e^{-H\varphi} Y_\nu(e^{H\varphi}Bh), \\ \frac{d\varphi}{dt} &= \lambda - \sum_{\nu \geq 1} \varepsilon^\nu B^+ H e^{-H\varphi} Y_\nu(e^{H\varphi}Bh)/h, \end{aligned} \tag{2.4}$$

where the sign / denotes the componentwise division of a vector.

In the nonresonance case, according to (2.2), we have

$$e^{-H\varphi}Y_\nu(e^{H\varphi}Bh) = Y_\nu(Bh), \quad \nu = 1, 2, \ldots,$$

and system (2.4) takes the form of the split system of equations

$$\frac{dh}{dt} = \sum_{\nu \geq} \varepsilon^\nu B^+ Y_\nu(Bh), \quad \frac{d\varphi}{dt} = \lambda - \sum_{\nu \geq 1} \varepsilon^\nu B^+ H Y_\nu(Bh)/h.$$

In the case of resonance, one should continue the transformation of system (2.4) by using equality (1.1). For this purpose, we choose an integer-valued $n \times (n - m)$ matrix Q of rank $n - m$ from the condition of its orthogonality to the matrix K, i.e.,

$$K^T Q = 0,$$

where K^T is the transposed matrix. This can easily be done if, as Q, we take the matrix consisting of $n-m$ linearly independent columns of the matrix

$$d_1(E - KK^+), \tag{2.5}$$

where K^+ is the matrix pseudoinverse to K and d_1 is the least integer number that guarantees that matrix (2.5) is integer-valued.

We now change the variables by introducing the variables Ψ, θ instead of φ in (2.4) according to the formulas

$$\varphi = K\Psi + Q\theta, \quad \Psi = K^+\varphi, \quad \theta = Q^+\varphi,$$

where Q^+ is the matrix pseudoinverse to Q.

In the variables h, θ, Ψ, system (2.4) takes the form of the split system

$$\begin{aligned}
\frac{dh}{dt} &= \sum_{\nu \geq 1} \varepsilon^\nu B^+ \Phi(-Q\theta) Y_\nu(\Phi(Q\theta)Bh), \\
\frac{d\theta}{dt} &= \sum_{\nu \geq 1} \varepsilon^\nu Q^+ B^+ H \Phi(-Q\theta) Y_\nu(\Phi(Q\theta)Bh)/h, \\
\frac{d\psi}{dt} &= \omega - \sum_{\nu \geq 1} \varepsilon^\nu K^+ B^+ H \Phi(-Q\theta) Y_\nu(\Phi(Q\theta)Bh)/h.
\end{aligned} \tag{2.6}$$

It follows from (2.6) that under the change of variables (2.3) the hyperplanes $y_\nu = 0$ generate the poles $h_\nu = 0$ of system (2.6). Therefore, in the investigation of the trajectories of the averaged system of equations (1.3) starting from the hyperplanes $y_\nu = 0$, one should introduce polar coordinates only for y that do not belong to the indicated hyperplanes.

3. Analysis of Averaged Equations in the Nonresonace Case

In the nonresonance case, it is convenient to write the averaged equations (1.3) in the form of the following system with respect to two-dimensional parameters $y^j = (y_{2j-1}, y_{2j})$, $j = \overline{1,n}$:

$$\frac{dy^j}{dt} = \lambda_j H_j y^j + \sum_{\nu \geq 1} \varepsilon^\nu Y_\nu^j(y^1, \dots, y^n), \quad j = \overline{1,n}. \tag{3.1}$$

Equations (2.2) now take the form

$$\begin{aligned}
&Y_\nu^j(\Phi_1(\varphi_1)y^1, \dots, \Phi_j(\varphi_j)y^j, \dots, \Phi_n(\varphi_n)y^n) \\
&= \Phi_j(\varphi_j) Y_\nu^j(y^1, \dots, y^j, \dots, y^n), \quad j = \overline{1,n},
\end{aligned} \tag{3.2}$$

and determine the character of symmetry of the right-hand sides of system (3.1).

By setting $\varphi_j = \pi$ and $\varphi_\nu = 0$ for $\nu \neq j$ in (3.2), we get

$$Y_\nu^j(y^1, \dots, -y^j, \dots, y^n) = -Y_\nu^j(y^1, \dots, y^j, \dots, y^n) \tag{3.3}$$

for all $\nu = 1, 2, \dots$ and arbitrary $j = \overline{1,n}$.

By setting $\varphi_l = \pi$ and $\varphi_\nu = 0$ for $l \neq j$ and $\nu \neq l$ in (3.2), we obtain

$$Y_\nu^j(y^1, \dots, -y^l, \dots y^n) = Y_\nu^j(y^1, \dots, y^l, \dots, y^n) \tag{3.4}$$

for all $\nu = 1, 2, \dots$ and arbitrary $1 \leq l \leq n$, $l \neq j$, $j = \overline{1,n}$.

According to (3.3) and (3.4), the functions $Y_\nu^j(y^1, \dots, y^n)$ are odd with respect to the variable y^j and even with respect to the other variables y^l for $l \neq j, 1 \leq j \leq n$.

Differentiating (3.2) with respect to φ_j and setting $\varphi_1 = \varphi_2 = \dots\varphi_n = 0$, we get

$$-H_j \frac{\partial Y_\nu^j(y)}{\partial y^j} H_j y^j = Y_\nu^j(y), \quad j = \overline{1,n}, \quad \nu = 1, 2, \dots. \tag{3.5}$$

According to (3.5), the averaged equations (3.1) have the form

$$\frac{dy^j}{dt} = \lambda_j H_j y^j - \sum_{\nu \geq 1} \varepsilon^\nu H_j \frac{\partial Y_\nu^j(y)}{\partial y^j} H_j y^j, \quad j = \overline{1,n}. \tag{3.6}$$

Consequently, the hyperplanes

$$y^j = 0, \quad j = \overline{1,n},$$

and their intersections are invariant for the averaged equations.

The polar coordinates h_j, Ψ_j are introduced instead of y^j according to the formulas

$$y^j = (sin\varphi_j H_j + cos\varphi_j E) B_j h_j, \quad j = \overline{1,n},$$

and transform equations (3.6) to the form

$$\begin{aligned} \frac{dh_j}{dt} &= -\sum_{\nu \geq 1} \varepsilon^\nu B_j^+ H_j \frac{\partial Y_\nu^j(Bh)}{\partial y^j} H_j B_j h_j, \\ \frac{d\varphi_j}{dt} &= \lambda_j + \sum_{\nu \geq 1} \varepsilon^\nu B_j^+ \frac{\partial Y_\nu^j(Bh)}{\partial y^j} H_j B_j, \quad j = \overline{1,n}. \end{aligned} \tag{3.7}$$

Equalities (3.3)-(3.5) guarantee that the functions

$$\frac{\partial Y_\nu^j(Bh)}{\partial y^j} H_j B_j, \quad j = \overline{1,n}, \quad \nu = 1, 2, \dots,$$

are even with respect to any of the variables $h_1, \dots, h_n$. Then, taking into account that $Y_\nu^j(Bh)$ belongs to the polynomial ring $I\!K[h]$ for $j = \overline{1,n}$, $\nu = 1, 2, \dots$, we get

$$\frac{\partial Y_\nu^j(Bh)}{\partial y^j} H_j B_j = R_\nu^j(h_1^2, \dots, h_n^2)$$

for $j = \overline{1,n}$, $\nu = 1, 2, \dots$; here, $R_\nu^j(r_1, \dots, r_n)$ belongs to $I\!K[r]$.

In the variables

$$r_j = h_j^2, \quad j = \overline{1,n},$$

the system of equations (3.7) takes the form

$$\frac{dr_j}{dt} = -2\sum_{\nu\geq 1} \varepsilon^\nu B_j^+ H_j R_\nu^j(r) r_j, \quad \frac{d\varphi_j}{dt} = \lambda_j + \sum_{\nu\geq 1} \varepsilon^\nu B_j^+ R_\nu^j(r), \quad j = \overline{1,n},$$

or, in obvious notation,

$$\frac{dr}{dt} = r\sum_{\nu\geq 1} \varepsilon^\nu F_\nu(r), \quad \frac{d\varphi}{dt} = \lambda + \sum_{\nu\geq 1} \varepsilon^\nu G_\nu(r), \quad j = \overline{1,n}.$$

As follows from equations (3.7), their right-hand sides are also defined on the boundary of the domain $\mathbb{R}^{+n}$, i.e., system (3.7) can be used for the investigation of trajectories in the entire subspace

$$h_\nu \geq 0, \quad \nu = \overline{1,n}. \tag{3.8}$$

Moreover, the reasoning concerning the investigation of system (3.7) in domain (3.8) can be justified. Below, in particular, we present a result that generalizes [3]. To formulate it , we write the amplitude equations of the first approximation of system (3.7) in the form

$$\frac{dh}{dt} = \varepsilon F(h). \tag{3.9}$$

Let

$$h = h_0 \geq 0 : F(h_0) = 0 \tag{3.10}$$

be an equilibrium point of system (3.9) and let

$$H = \frac{\partial F(h_0)}{\partial h} \tag{3.11}$$

be the coefficient matrix of the variational equations of system (3.9) corresponding to the equilibrium point (3.10).

Let

$$u_1(y) = \sum_{k\neq 0} \frac{X_k(y)}{(k,\lambda)}$$

be the function from the asymptotic expansion (1.2) of solutions of system (0.2).

Theorem 1. *Assume that the right-hand side of system (0.2) is such that*

(i) for certain integer $2 \leq s \leq l$, the function $X(x)$ is l times continuously differentiable in the domain $D \subseteq \mathbb{R}^{2n}$ and $u_1(x)$ is s times continuously differentiable in D;

(ii) equation (3.9) has an equilibrium point (3.10) such that the torus

$$x = e^{H\varphi} B^+ h_0$$

belongs to the domain D and the eigenvalues of matrix (3.11) do not have zero real parts.

Then one can find $\varepsilon_0 > 0$ such that, for all $0 \leq \varepsilon \leq \varepsilon_0$, the system of equations (0.2) has an $(m-p)$-dimensional invariant torus

$$x = f(\Psi, \varepsilon),$$

where p is the number of zero coordinates of the vector h_0, $f \in C^{s-2}_{Lip}(\mathcal{J}_{m-p})$, $\Psi = (\varphi_{j_1}, \ldots, \varphi_{j_{n-p}})$, *and* $j_\nu, \nu = \overline{1, n-p}$, *are the indices of nonzero coordinates of the vector* h_0.

This torus satisfies the condition

$$\lim_{\varepsilon \to 0} \|f(\Psi, \varepsilon) - e^{H\varphi} h_0\|_{s-2} = 0$$

and is exponentially stable if the real parts of the eigenvalues of the matrix H are negative, exponentially unstable if they are positive, and exponentially dichotomous, otherwise.

Here, $C^{s-2}_{Lip}(\mathcal{J}_{m-p})$ is the space of $(s-2)$-differentiable functions on the torus $\mathcal{J}_{m-p}$ whose $(s-2)$ th derivatives satisfy the Lipschitz condition, $\|.\|_p = max_{0 \le |\rho| \le p} \|D^\rho.\|_0$, $D^\rho = \frac{\partial^{|\rho|}}{\partial \Psi_1^{\rho_1} \ldots \partial \Psi_{n-p}^{\rho_{n-p}}}$, $\|\rho\| = \sum_{\nu=1}^{n-p} \rho_\nu$, and $\|.\|_0 = max_{\Psi \in \mathcal{J}_{m-p}} \|.\|$.

In the case where the numbers λ satisfy the condition of "strong incommensurability"

$$|(k, \lambda)| \ge \frac{\mathcal{K}}{(1 + |k|)^d}, \quad k \in \mathbf{Z}^n, \quad k \ne 0,$$

according to [4], the assumption on u_1 is true for any l and s satisfying the inequality

$$l - d - \frac{n}{2} > s.$$

Theorem 1 can be extended to the case of analytic right-hand sides of system (0.1). In this case, it takes the form of the corresponding statement from[5].

4. Specific Features of the resonance Case

We say that system (0.2) is a resonance system if the frequency basis ω has the number of frequencies less than n. The basic relation for the characterization of the resonance case is relation (1.1), which connects the frequencies λ of own oscillations of the unperturbed system of equations with the frequency basis ω.

It follows from (1.1) that

$$\omega = K^+ \lambda, \tag{4.1}$$

where K^+ is the matrix pseudoinverse to K. Taking relations (1.1) and (4.1) into account, we say that the matrix K is the *determining matrix for the frequency basis.*

Let us clarify the role of the matrix K in the characterization of resonances.

A vector $k \in \mathbf{Z}^n, k \ne 0$, is called a *resonance* vector if

$$(k, \lambda) = 0.$$

Let Q be the matrix formed by all resonance vectors. Then

$$Q + \{Ker K^T \cap \mathbf{Z}^n\} \setminus \{0\}. \tag{4.2}$$

Relation (4.2) follows from the equality.

$$(k,\lambda) = (k, K\omega) = (K^T k, \omega)$$

and the fact that the frequencies ω are incommensurable

According to (4.2), we have

$$rank\ Q = n - m$$

and the quantity $(n-m)$ can be called the *resonance rank* of system (0.2). With the use of this notion, we can introduce the hierarchy of resonances, regarding that the greater the resonance rank, the more complicated the resonance.

According to (4.2), the averaging operator S for systems (0.2) with matrices K and K_1 of the same rank, which determine the frequency basis of system (0.2), is the same if

$$Ker K^T = Ker K_1^T.$$

This is possible only if

$$K_1 = KR, \quad R = K^+K, \quad det R \neq 0. \tag{4.3}$$

Thus, the passage from one frequency basis to another in system (0.2) does not change the averaging operator, as well as it does not change under the passage form λ to λ_1, whenever the matrices K and K_1 determining the frequency bases of λ and λ_1 satisfy relations (4.3).

Denote by Q_1 the subset of Q defined by the condition

$$|k| = \sum_{\nu=1}^{n} |k_\nu| \le l.$$

Let $X(x)$ be a polynomial of degree N. Then $\Phi(-\varphi)X(\Phi(\varphi)x)$ is a trigonometric polynomial with respect to φ, which contains the harmonics $e^{i(k,\varphi)}$ with $|k| \le N+1$. Therefore, the functions

$$\Phi(-\varphi)X(\Phi(\varphi)y) = \sum_{k\in Z^n} X_k(x)e^{i(k,\varphi)}$$

satisfy the condition

$$X_k(x) = 0, \quad |k| > N+1,$$

and the series $Y_1(y)$, which determines the averaged value of the function $X(x)$, is truncated to the sum

$$Y_1(y) = X_0(y) + \sum_{k\in Q_{N+1}} X_k(y).$$

Considering finite approximations of the asymptotic method of integration of system (0.2) for different λ, we must compare the sets Q_l for such l. In particular, if the matrix K is such that

$$Q_{N+1} = \phi, \tag{4.4}$$

then, in the first approximation, we have

$$Y_1(y) = X_0(y).$$

In this case, the equations of the first approximation for resonance coincide with equations of the first approximation in the nonresonance case. In this case, we split the averaged equations of the first approximation by introducing the amplitude-phase coordinates h, φ. Furthermore, the correspondence between the equilibrium points of the amplitude equations and invariant tori of system (0.2) described in Theorem 1 is preserved.

If condition (4.4) is not satisfied, we split the averaged equations according to the general scheme of the method. By using the general theory of perturbation of invariant tori, we can easily establish the correspondence between "rough" quasistatic equilibrium points

$$h = h_0, \quad \theta = \theta_0, \quad h_0 > 0,$$

of the split equations of the first approximation and m - dimensional invariant tori of the original system of equations (0.2) for resonance (1.1).

Among systems considered, there exist systems that split in the course of passing from the Euclidean to polar coordinates. The asymptotic integration of such system is significantly simplified immediately on passing to the polar coordinates.

Let us give an example of such a system. For this purpose, consider the following system of equations with cubic nonlinearity:

$$\frac{dx_\nu}{dt} = \left[(\lambda_\nu + \varepsilon g_\nu(x))H_\nu + \varepsilon\alpha_\nu + \beta_\nu(S_\nu x_\nu, x_\nu)E\right] x_\nu, \quad \nu = \overline{1,n}, \tag{4.5}$$

where g_ν is polynomial of the second degree in x,

$$S = diag\{S_1, \dots, S_n\},$$

$S_\nu, \nu = \overline{1,n}$ are symmetric matrices satisfying the conditions

$$S_\nu H_\nu + H_\nu^T S_\nu = 0, \quad H_\nu^T S_\nu H_\nu = S_\nu,$$

and α_ν and β_ν are constant parameters. The introduction of the polar coordinates

$$x_\nu = (H_\nu sin\varphi_\nu + Ecos\varphi_\nu)B_\nu H_\nu, \quad \nu = \overline{1,n},$$

splits system (4.5) and reduces it to the form

$$\frac{dr}{dt} = 2\varepsilon rF(\alpha + \beta sr), \quad \frac{d\varphi}{dt} = \lambda + \varepsilon g(\Phi(\varphi)Bh). \tag{4.6}$$

Here, the multiplication of vectors is understood in the coodinatewise sense,

$$\alpha = (\alpha_1, \ldots, \alpha_n), \quad \beta = (\beta_1, \ldots, \beta_n), \quad s = (s_1, \ldots, s_n), \quad s_\nu = (S_\nu B_\nu, B_\nu).$$

If the parameters α and β are chosen so that

$$\alpha\beta s < 0,$$

then the amplitude equation (4.6) has the following equilibrium points in the region $r \geq 0$:

$$er_0 = -e\alpha/\beta s,$$

where e is a vector whose coordinates may be equal only to 0 or 1. This implies that the original system of equations (4.5) has C_n^p p-dimensional tori of the form

$$x = e^{H\varphi}Beh_0, \quad h_0^2 = r_0$$

for any $1 \leq p < n$. Consequently, the system has at least 2^n invariant tori (including the point $x = 0$).

References

1. Yu. A. Mitropol'skii and A.M. Samoilenko, "*On the problem of asymptotic expansions in nonlinear mechanics,*" Ukr. Mat. Zh., 31, No. 1,42-53 (1979).

2. Yu. A. Mitropol'skii and A.M. Samoilenko, *General Problems of the Theory of Asymptotic Integration of Systems in Nonlinear Mechanics* [in Russian], Preprint No. 87.41, Institute of Mathematics, Ukrainian Academy of Sciences, Kiev (1987).

3. N.N. Bogolyubov, Yu. A. Mitropol'skii and A.M. Samoilenko *Method of Accelerated Convergence in Nonlinear Mechanics* [in Russian], Naukova Dumka, Kiev (1969).

4. A.D. Bryuno, *Local Method for Nonlinear Analysis of Differential Equations* [in Russian], Nauka, Moscow (1997).

5. A. M. Samoilenko, *Elements of the Mathematical Theory of Multifrequency Oscillations* [in Russian], Nauka, Moscow 1987.

NONLINEAR OSCILLATIONS OF VEHICLES IN CONVOY

WERNER SCHIEHLEN and AXEL FRITZ
Institute B of Mechanics
University of Stuttgart
Pfaffenwaldring 9
70550 Stuttgart, Germany

1. Introduction

Highway automation requires intelligent vehicle control in longitudinal direction. In this paper linear and nonlinear controllers are presented for a dual task: that each vehicle in a convoy operates in a stable fashion, and that the platoon is string stable in longitudinal direction. For this analysis a convoy of five passenger cars is used.

The paper is organized as follows: In section 2 vehicle and actuator models describing the longitudinal dynamics are introduced for simulation and for controller design, respectively. In section 3 the formulated control objectives are verified for the different controllers, depending on whether communication between the vehicles is available or not. Finally, in section 4 simulation results are discussed.

2. Vehicle Models

The modeling of the longitudinal dynamics of vehicles is discussed in detail in Mitschke [1] and Popp/Schiehlen [2]. The overall motion shows oscillations with respect to the nominal motion due to transitions of the traveling speed.

2.1. SIMULATION MODEL

The simulation model for the longitudinal dynamics is based on the method of multibody systems [3]. In the following the index i indicates the state of the i-th vehicle in the convoy, $i = 1(1)5$. The model of one car consists of the car body, four wheels, a differential and a shift gear, see Figure 1. The generalized coordinates are the longitudinal displacement x_i of the car body, the angular displacements of the four wheels, $\phi_{fli}, \phi_{fri}, \phi_{rli}$ and ϕ_{rri}, and the rotations at the differential and shift gear, ϕ_{dli}, ϕ_{dri} and ϕ_{gi}. The applied forces are the longitudinal tire forces $f_{tfl}, f_{tfr}, f_{trl}$ and f_{trr} which will be calculated from Pacejka's tire formula, the gravity forces and the nonlinear forces resulting from air, rolling and climbing drag, f_l, f_r and f_c. The power train consists of two drive shafts with the stiffness c_d, a differential gear, a cardan shaft with the stiffness c_c and a shift gear. The engine torque t_{ei} depends nonlinearly on the engine speed and the throttle angle α_{thi}, respectively. The combustion process in the engine is regarded as a first order system with a time constant

N. Van Dao and E.J. Kreuzer (eds.),
***IUTAM Symposium on Recent Developments in Non-linear Oscillations of Mechanical Systems*, 257–268.**

T_1. Braking is also implemented. The braking torque t_{bi} is a function of the hydraulic pressure p_{bi}.

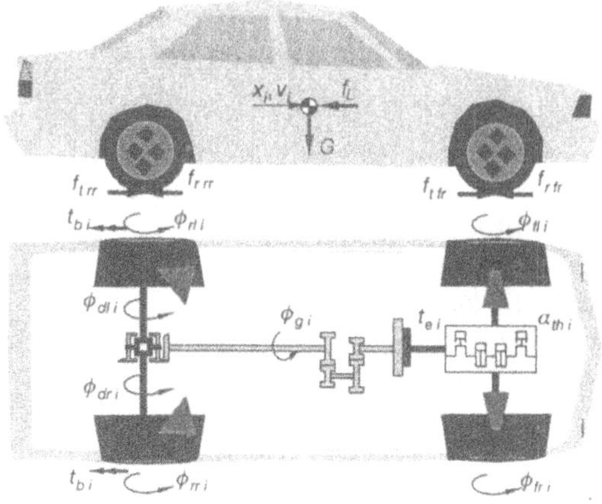

Figure 1. Simulation model of vehicle i

As a result one gets a set of strongly nonlinear equations of motion for the longitudinal dynamics of vehicle i which can be written in state space form as

$$\dot{\boldsymbol{x}}_{si} = \boldsymbol{f}_{si}(\boldsymbol{x}_{si},\ \boldsymbol{u}_{si}) \ . \tag{1}$$

2.2. ACTUATOR MODEL

The simulation model of each of the following cars in the convoy is extended by a nonlinear model of a controlled servo-motor used as an actuator transferring the throttle position claimed by the adaptive cruise controller. The eigendynamics of the actuator is represented by a cascade control loop shown in Figure 2.

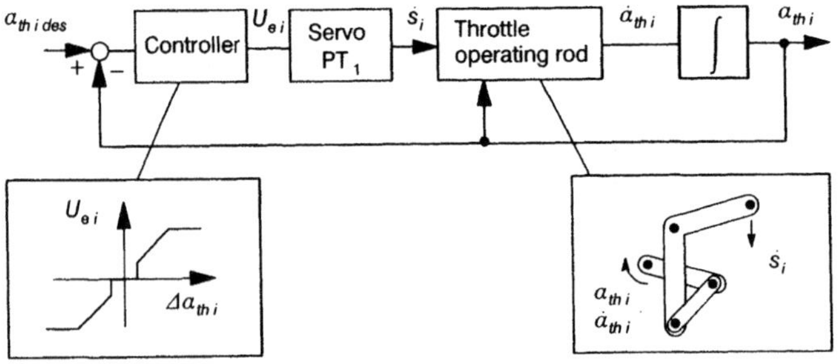

Figure 2. Actuator model of vehicle i

The controller characteristic converts the control error $\Delta\alpha_{thi}$ given as the difference between the desired throttle angle α_{thides} and the actual throttle angle α_{thi} into the input voltage U_{ei} of the servo motor. The servo-motor drives the Bowden wire with the velocity

$\dot{s}_i$ resulting in a first order delay PT_1. Then, the angular velocity of the throttle $\dot{\alpha}_{thi}$ follows from the kinematics of the throttle operating rod

$$\dot{\alpha}_{thi} = k_4^\star(k_3^\star\alpha_{thi}^3 + k_2^\star\alpha_{thi}^2 + k_1^\star\alpha_{thi} + k_0^\star)\dot{s}_i \quad , \tag{2}$$

where $k_i^\star, i = 0(1)3$ is found from the throttle operating rod characteristic. The nonlinear function $k_4^\star$ is determined by measurements. Finally, the angular velocity of the throttle is integrated to get α_{thi}.

2.3. DESIGN MODEL

Since the vehicle model is too complex to be used for controller design the model is partially linearized. The engine torque t_{ei} is considered as a linear function of engine speed and throttle position. Neglecting the slip and the elasticity of the power train, the engine speed is proportional to the vehicle speed $\dot{x}_i$. Then, the engine torque reads as

$$t_{ei} = M_0 + M_1\dot{x}_i + M_2\alpha_{thi} \ , \tag{3}$$

where $M_i, i = 0(1)2$ are constants.

Since the delay of the actuator dynamics is comparable to the dynamics of the combustion process the actuator has to be included in the controller design. However, the actuator controller and the throttle operating rod characteristic (2) are approximated by a tanh–function resulting in

$$\dot{\alpha}_{thi} = k_1 \tanh\left[k_2(\Delta\alpha_{thi} + k_3)\right] + k_4 \quad , \tag{4}$$

where $k_i, i = 1(1)4$ is found by identification.

With all these simplifications the equations of motion for the design model read as

$$\begin{bmatrix} 1 & 0 & 0 & 0 \\ 0 & r^2M + J & 0 & 0 \\ 0 & 0 & T_1 & 0 \\ 0 & 0 & 0 & 1 \end{bmatrix} \underbrace{\begin{bmatrix} \dot{x}_i \\ \dot{v}_i \\ \dot{t}_{exi} \\ \dot{\alpha}_{thi} \end{bmatrix}}_{\dot{\boldsymbol{x}}_i} = \underbrace{\begin{bmatrix} v_i \\ rK_M t_{exi} - t_{bi} - r^2(f_r + f_l + f_c) \\ M_0 + M_1 v_i + M_2\alpha_{thi} - t_{exi} \\ k_1 \tanh\left[k_2(\alpha_{thides} - \alpha_{thi} + k_3)\right] + k_4 \end{bmatrix}}_{\boldsymbol{f}(\boldsymbol{x}_i, \boldsymbol{u}_i)} . \tag{5}$$

In these equations M denotes the mass of the car body and the four wheels, J is the moment of inertia of the power train reduced to the rear wheels, r is the wheel radius and K_M is a coefficient characterizing the power train transmission. The state vector $\boldsymbol{x}_i$ contains the displacement x_i, the velocity v_i, the delayed engine torque t_{exi} and the throttle angle α_{thi} while the input vector $\boldsymbol{u}_i$ consists of the desired throttle angle α_{thides} and the braking torque t_{bi}.

3. Adaptive Cruise Control

In this section, first the measured quantities and control objectives are formulated. Then, a linear and a nonlinear controller design is discussed regarding the control objectives.

3.1. CONTROL LOOP PROPERTIES

The following measured quantities are available: the on board sensors measure the velocity v_i, the acceleration a_i and the throttle angle α_{thi} of each vehicle in the convoy as well as the range R_i between two cars dedicated by a laser. The range rate $\dot{R}_i$ is obtained by differentiating and filtering the measured distance data. Data of the leading vehicle like acceleration a_{i-1} is supposed to be optional available over communication.

The control error of the cruise control is the difference between the desired headway R_{ides} and the distance R_i calculated by

$$R_i = (x_{i-1} - l_{hi-1}) - (x_i + l_{vi}) \; , \tag{6}$$

see Figure 3, where l_{hi-1} is the distance between the center of gravity and the rear bumper of the $(i-1)$-th vehicle and l_{vi} the distance between the center of gravity and the front bumper of the ith vehicle.

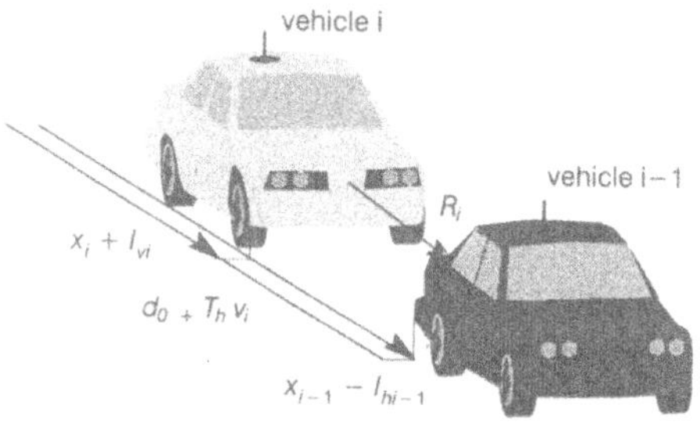

Figure 3. Range between vehicle $i-1$ and i

The desired headway R_{ides} can be expressed in the form

$$R_{ides} = d_0 + T_h v_i \; , \tag{7}$$

where d_0 is a constant distance and T_h the headway time. This formulation is called headway control strategy, see Swaroop *et al.* [4]. Then, the control error e_i reads as

$$\begin{aligned} e_i &= R_{ides} - R_i \\ &= d_0 + T_h v_i - \Big((x_{i-1} - l_{hi-1}) - (x_i + l_{vi}) \Big) \; . \end{aligned} \tag{8}$$

The following platoon objectives are supposed for the cruise control of a convoy of vehicles, see Eyre *et al.* [5].

- The steady state spacing errors of all vehicles in the platoon should be zero. This includes that the closed loop system of each vehicle should be asymptotically stable. In the following that is called individual vehicle stability.
- The phase margin guaranteeing vehicle stability should be as high as possible.
- The maximum absolute spacing R_i of the i-th vehicle should be less compared to that of the $(i-1)$-th vehicle. Therefore, the gain of the spacing transfer function $G_i(s)$ should be less than one,

$$|G_i(s)| = \left| \frac{R_i(s)}{R_{i-1}(s)} \right| < 1 \; . \tag{9}$$

This requirement is called L_2 string stability.

- String stability without overshoot is given, if equation (9) is true and the impulse response of the inverse Laplace transformation of $G_i(s)$ remains positive. In the following the inverse Laplace transformation of $G_i(s)$ is called $g_i(t)$.

In the next two paragraphs these objectives are verified for a linear and a nonlinear controller design considering only acceleration maneuvers. The control design for braking can be done analogous.

3.2. LINEAR CONTROL

The control loop for the linear controller design consists of a linear PD_2 controller and a linearized vehicle model, see Figure 4.

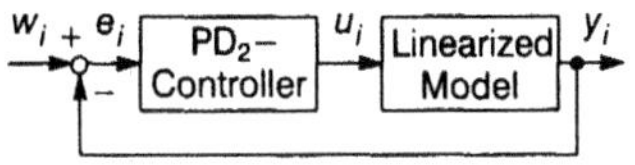

Figure 4. Control loop for linear control

The around the operating point $(v_0, \alpha_{thi0}, u_0)$ linearized state space equations of the vehicle model with actuator are

$$\dot{\boldsymbol{x}}_i = \underbrace{\begin{bmatrix} 0 & 1 & 0 & 0 \\ 0 & -2\frac{K_l v_0 r^2}{r^2 M+J} & \frac{r K_M}{r^2 M+J} & 0 \\ 0 & \frac{M_1}{T_1} & -\frac{1}{T_1} & \frac{M_2}{T_1} \\ 0 & 0 & 0 & -\frac{k_1 k_2}{\cosh^2(u_0^\star)} \end{bmatrix}}_{\boldsymbol{A}_i(v_0,\alpha_{thi0},u_0)} \boldsymbol{x}_i + \underbrace{\begin{bmatrix} 0 \\ 0 \\ 0 \\ \frac{k_1 k_2}{\cosh^2(u_0^\star)} \end{bmatrix}}_{\boldsymbol{b}_i(\alpha_{thi0},u_0)} u_i \ , \tag{10}$$

where $u_0^\star = k_2(u_0 - \alpha_{thi0} k_3)$. The output vector y_i is given by the measurements as

$$y_i = [-1 \ \ -T_h \ 0 \ 0]\, \boldsymbol{x}_i = \boldsymbol{c}_i^T \boldsymbol{x}_i \quad . \tag{11}$$

The input output behavior of the plant can be formulated as

$$G_{Pi}(s) = \frac{y_i}{u_i} = \boldsymbol{c}_i^T (\boldsymbol{I}s - \boldsymbol{A}_i)^{-1} \boldsymbol{b}_i \ , \tag{12}$$

and the transfer function for the chosen linear PD_2-controller is

$$G_{Ci}(s) = \frac{u_i}{e_i} = P(1 + T_{d1}s + T_{d2}s^2) \quad . \tag{13}$$

Since the relative acceleration between two vehicles is required for the calculation of the second derivative of e_i, the control parameter T_{d2} is chosen zero if no communication is available.

3.2.1. *Control Objectives of Linear Control*

With $e_i = w_i - y_i$, see Figure 4, and the equations (12) and (13) one gets

$$e_i(s) = \frac{1}{1 + G_{Pi}(s) G_{Ci}(s)} w_i(s) \quad . \tag{14}$$

Using the final value theorem for (14) with $w_i(s) = w_0/s$ one obtains

$$\lim_{t->\infty} e_i(t) = \lim_{s->0} s e_i(s) = \frac{w_0}{1 + G_{Pi}(0) G_{Ci}(0)} = 0 \ . \tag{15}$$

Therefore, individual vehicle stability with a zero steady state error is achieved independent of the operating point. Depending on the parameters chosen, the degree of stability may be small.

For the following analysis $T_h = 1s$ and two operating points ($v_0 = 10m/s, gear = 2, \alpha_{thi0} = u_0$ and $v_0 = 30m/s, gear = 4, \alpha_{thi0} = u_0$) are assumed. The Figure 5 shows the Bode plots of the linear controller for the operating point ($v_0 = 30m/s, gear = 4, \alpha_{thi0} = u_0$).

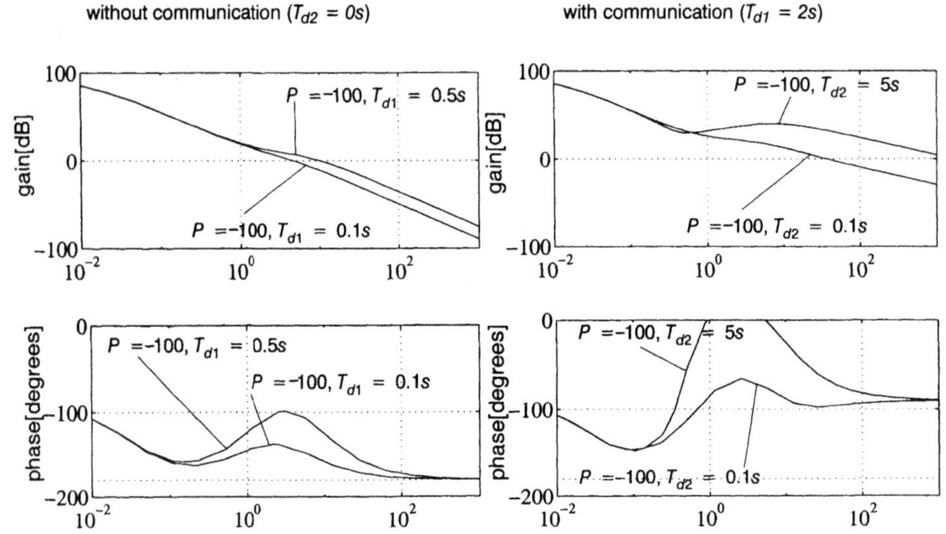

Figure 5. Bode plots of linear controller without (left) and with (right) communication for operating point ($v_0 = 30m/s, gear = 4$)

These plots show the individual vehicle stability of each car in the convoy. Further, one recognizes that without communication the phase margin becomes greater if T_{d1} is larger. A comparison of both Bode plots shows that the phase margin is clearly higher if communication is available.

Examing string stability, the gain of the transfer function $G_i(s)$ has to be less than one. For the linear controller the transfer function is given by

$$G_i(s) = \frac{-T_{d2}s^2 - T_{d1}s - 1}{c_4 s^4 + c_3 s^3 + c_2 s^2 + c_1 s - 1} \tag{16}$$

with

$$\begin{aligned} c_4 &= \frac{1}{Pt_4} \ , & c_3 &= \frac{t_3}{Pt_4} - T_{d2}T_h \ , \\ c_2 &= \frac{t_2}{Pt_4} - T_{d2} - T_{d1}T_h \ \text{and} & c_1 &= \frac{t_1}{Pt_4} - T_{d1} - T_h \ , \end{aligned} \tag{17}$$

where $t_k,\ k = 1(1)4$ is depending on the elements of $\boldsymbol{A}_i$. Setting equation (16) in (9) leads to

$$(T_{d2}T_h t_4 P - t_3)^2 + 2t_4 P(T_{d2} + T_{d1}T_h) - 2t_2 > 0\,, \qquad (18a)$$

$$(t_1 - t_4 P T_h)^2 + 2t_4 P(t_2 - t_1 T_{d1}) > 0\,, \qquad (18b)$$

$$(t_2 - T_{d1}t_4 P T_h)^2 - 2t_3 t_1 - 2Pt_4(1 - T_{d2}(T_h t_1 - t_2 - T_h^2 t_4 P) - t_3(T_h + T_{d1})) > 0\,. \qquad (18c)$$

Regarding the limit case equal zero, Figure 6 shows the inequalities depending on the parameter P, T_{d1} (without communication) and P, T_{d2} (with communication $T_{d1} = 2s$), respectively.

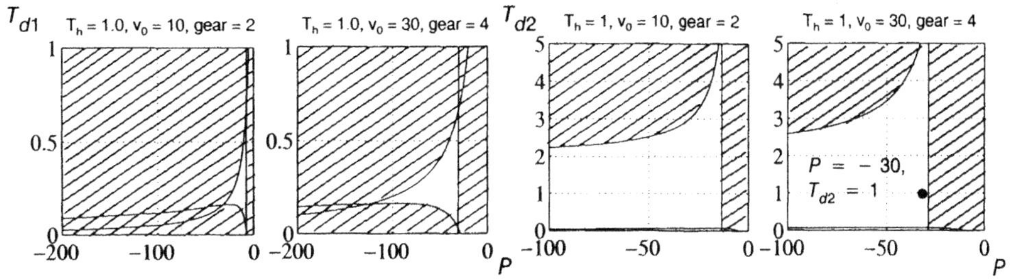

Figure 6. String stable areas without (left) and with (right) communication

The hatched regions mark string instability of the linear controller. It is important to mention that if no communication is available the stability area moves with the operating point while with communication only the size of the string stable area changes. Therefore, it is possible to find a parameter set so that string stability is independent of the linearization point, e.g. $P = -30$ and $T_{d2} = 1s$.

Examing string stability without overshoot the impulse responses of $g_i(t)$ for the given operating points without communication are displayed in Figure 7.

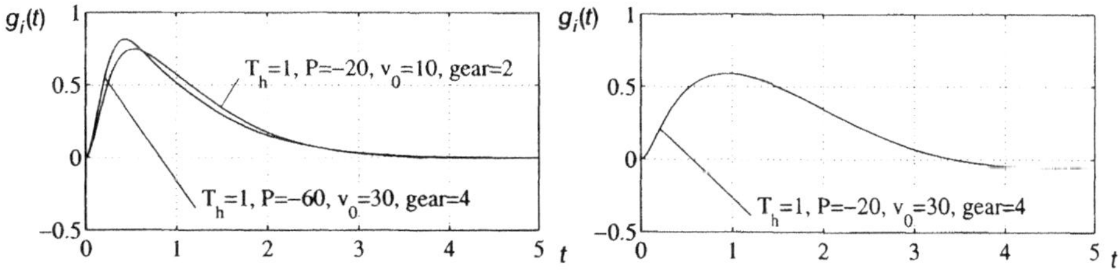

Figure 7. Impulse responses of $g_i(t)$ without communication

In the left picture string stable gains are used for the corresponding operation points ($P = -20$ for $v_0 = 10m/s, gear = 2$ and $P = -60$ for $v_0 = 30m/s$, gear=4). Then, the convoy is string stable without overshoot. Regarding $P = -20$ for $v_0 = 30m/s, gear = 4$ one recognizes that string stability without overshoot is no longer valid, see Figure 7 right. Therefore, string stability without overshoot depends on the linearization point if no communication is available.

The Figure 8 shows the same plots if communication is available for $P = -30, T_{d1} = 2s$ and $T_{d2} = 1s$. It demonstrates that string stability without overshoot is now independent of speed and gear.

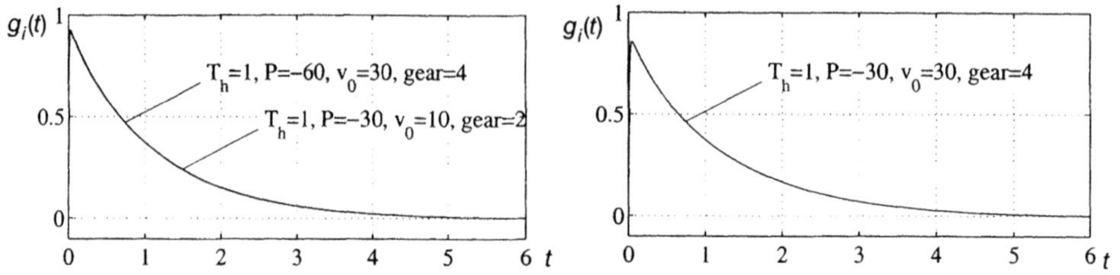

Figure 8. Impulse responses of $g_i(t)$ with communication

To improve the convoy dynamics and to overcome the dependency on the operating point or communication, respectively, a nonlinear controller will be now considered.

3.3. NONLINEAR CONTROL

In the following a nonlinear control concept is introduced based on the exact state linearization, see Isidori [6].

3.3.1. *Exact State Linearization and Tracking Control*

An important property of the nonlinear vehicle model given in (5) is its flatness in the sense of Fliess [7] subject to the output

$$y_i = x_i \quad . \tag{19}$$

Now, consider a transformation like $z_1 = y_i, z_2 = \dot{y}_i, \ldots, z_4 = y_i^{(3)}$ which leads to

$$\begin{bmatrix} \dot{z}_1 \\ \dot{z}_2 \\ \dot{z}_3 \\ \dot{z}_4 \end{bmatrix} = \begin{bmatrix} z_2 \\ z_3 \\ z_4 \\ \pi(\boldsymbol{z}, \alpha_{thides}) \end{bmatrix} \quad . \tag{20}$$

The function π is given in Fritz/Schiehlen [8]. To compensate these nonlinearities a new input w_i

$$w_i = \pi(\boldsymbol{z}, \alpha_{thides}) \tag{21}$$

is introduced. As a result of the compensation, one obtains a linear system in Brunovsky canonical form, see [6].

Because the vehicle model with actuator is flat, it is possible to design a linearized control law by solving (21) for α_{thides}

$$\alpha_{thides} = \pi^{\star}(\boldsymbol{z}, w_i) \tag{22}$$

which yields

$$y_i^{(4)} = w_i \quad . \tag{23}$$

The state transformation from the system given by equations (5) into Brunovsky canonical form is also given in Fritz/Schiehlen [8].

The next step is minimizing the starting control error $e_i = y_i - y_{ides}$ and changing the output y_i along the suitable desired trajectory. This can be solved by tracking the trajectory of the flat output in equation (19) along the desired output y_{ides} which is designed on the basis of equation (8)

$$y_{ides} = x_{i-1} - d_0 - l_{hi-1} - l_{vi} - T_h v_i \ . \tag{24}$$

The differential equation of the tracking error is given by

$$\underbrace{y_i^{(3)} - y_{ides}^{(3)}}_{e_i^{(3)} = x_i^{(3)} - x_{i-1}^{(3)} + T_h x_i^{(4)}} + \sum_{k=0}^{2} \ell_k \left(y_i^{(k)} - y_{ides}^{(k)} \right) = 0 \ . \tag{25}$$

The coefficients $\ell_k, k = 0, 1, 2$ provide stabilization of the control error. By solving (25) for $y_i^{(4)} = x_i^{(4)}$, it follows from (23)

$$w_i = \frac{1}{T_h} \left[x_{i-1}^{(3)} - x_i^{(3)} - \sum_{k=0}^{2} \ell_k \left(y_i^{(k)} - y_{ides}^{(k)} \right) \right] \ . \tag{26}$$

Using (26) the static feedback law in equation (22) results in

$$\alpha_{thides} = \pi^\star \left(z, \frac{1}{T_h} [x_{i-1}^{(3)} - x_i^{(3)} - \sum_{k=0}^{2} \ell_k (y_i^{(k)} - y_{ides}^{(k)})] \right) \ . \tag{27}$$

In the following appropriate coefficients $\ell_k, k = 0, 1, 2$ are chosen to meet the control objectives.

3.3.2. *Control Objectives of Nonlinear Control*

The eigenvalues of the error dynamics are assigned with a negative real part by chosing

$$\ell_k > 0 \quad \text{for} \quad k = 0, 1, 2 \quad \text{and} \tag{28a}$$

$$\ell_0 - \ell_1 \ell_2 < 0 \ . \tag{28b}$$

This guarantees individual vehicle stability and an asymptotically disappearing control error.

String stability can be examined regarding the spacing transfer function

$$G_i(s) = \frac{\kappa \ell_2 s^2 + \ell_1 s + \ell_0}{T_h s^4 + (\ell_2 T_h + 1) s^3 + (\ell_1 T_h + \ell_2) s^2 + (\ell_0 T_h + \ell_1) s + \ell_0} \ . \tag{29}$$

If communication is available κ is one, else zero. Evaluating equation (9) with (29) leads to

$$T_h^2 \ell_0^2 - 2(1 - \kappa) \ell_0 \ell_2 > 0 \ , \tag{30a}$$

$$T_h^2 (\ell_1^2 - 2\ell_0 \ell_2) - 2\ell_1 + (1 - \kappa) \ell_2^2 > 0 \text{ and} \tag{30b}$$

$$T_h^2 (\ell_2^2 - 2\ell_1) + 1 > 0 \ . \tag{30c}$$

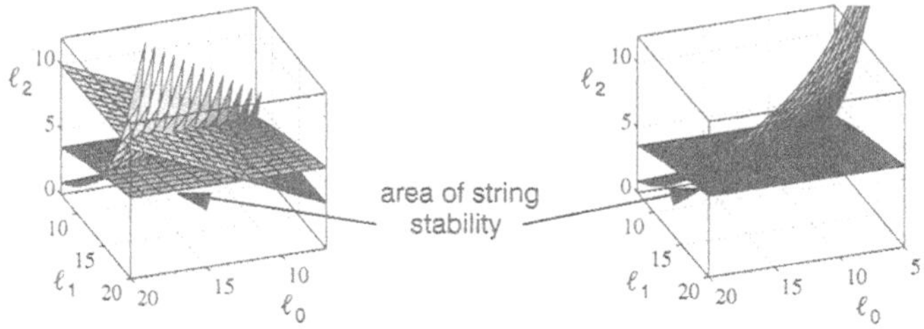

Figure 9. Areas of string stability without (left) and with (right) communication

These inequalities are illustrated for $T_h = 1s$ and various coefficients ℓ_0, ℓ_1 with and without communication in Figure 9. The string stable area fulfilling all three inequalities is marked by an arrow. Since inequality (30a) disappears in the case of communication, string stability is more easily achieved.

Considering string stability without overshoot the impulse response of the inverse Laplace transform of $G_i(s)$ has to be regarded. This function is displayed with and without communication in Figure 10 for $T_h = 1.0$, $\ell_0 = 8.0$, $\ell_1 = 8.5$ and $\ell_3 = 4.1$. Since both trajectories are always positive, string stability without overshoot is independent of communication.

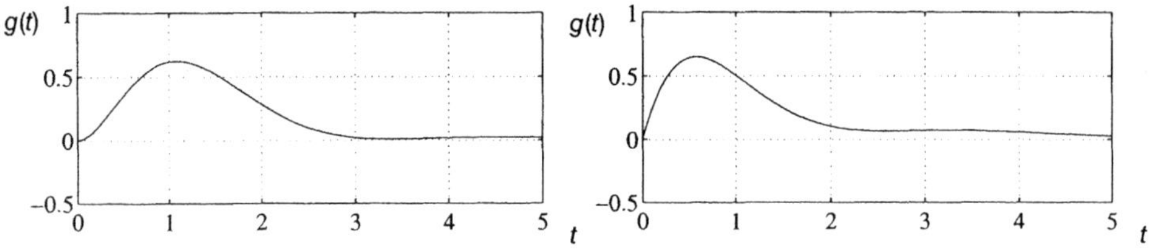

Figure 10. Impulse response of the inverse Laplace transform of G_i without (left) and with (right) communication

4. Simulation Results

For the following simulations a step disturbance in the throttle position of the leading vehicle from $\alpha_{th1} = 11°$ to $\alpha_{th1} = 15°$ is used. The convoy exists of five vehicles traveling for $t = 0s$ with a constant velocity of $v_0 = 10m/s$. The desired range is calculated with the constant distance $d_0^\star = 16m$ and the headway time $T_h = 1s$.

In Figure 11 the oscillations of the headway errors e_i, $i = 2(1)5$ of the linear PD_2-controller with $P = -20, T_{d1} = 0.32s$ and the nonlinear controller with $\ell_0 = 8.0, \ell_1 = 8.5, \ell_2 = 4.1$ are compared. Communication is not available. The graphs show that both controllers are string stable while the performance of the nonlinear controller is much better.

In the following the value for v_0 is changed from $10m/s$ to $30m/s$. All other controller parameters will not be changed. Since the nonlinear controller is independent of speed and gear, string stability is still given. On the other hand the linear controller generates larger errors for the following vehicle than for the leader vehicle, see Figure 12. The reason

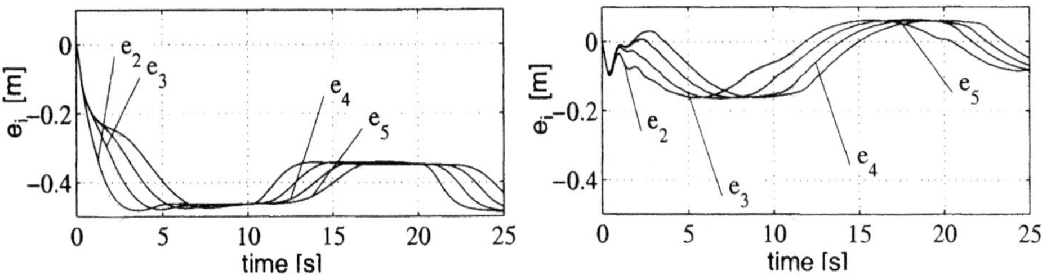

Figure 11. Tracking errors of linear (left) and nonlinear controller (right) without communication, $v_0 = 10m/s$

therefore is that the string stable parameters of the linear controller without communication depend on speed and gear as shown in section 3.2. Using a higher gain, e.g. $P = -60$ enables string stability again.

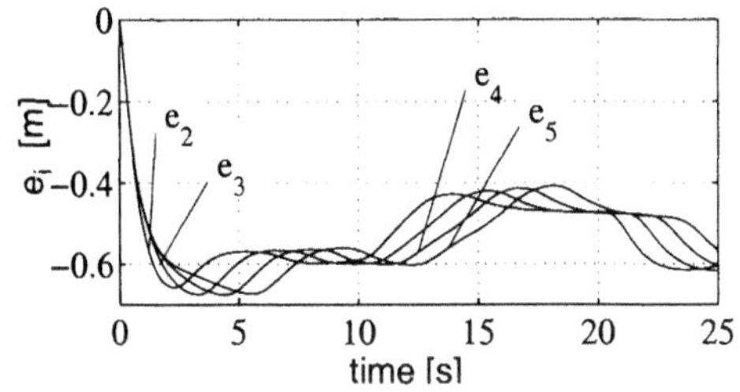

Figure 12. Tracking errors of linear controller without communication, $v_0 = 30m/s$

In the simulations displayed in Figure 13 communication is regarded for $v_0 = 10m/s$. For the linear controller the parameter $P = -30, T_{d1} = 2s, T_{d2} = 1s$ are used while the parameters of the nonlinear controller are maintained.

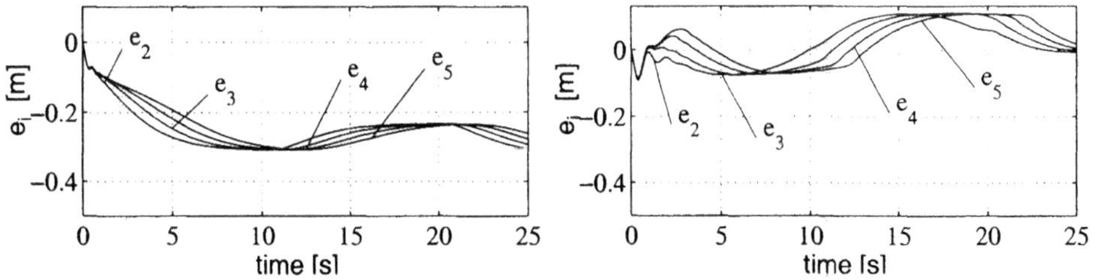

Figure 13. Tracking errors of linear (left) and nonlinear controller (right) with communication, $v_0 = 10m/s$

The tracking errors of both controllers are string stable. A comparison of these simulation results with Figure 11 show that the communication improves the dynamical behavior.

In the last simulation v_0 is set to $30m/s$. The simulation results in Figure 14 demonstrate that the string stability of the linear controller is independent of speed and gear if communication between the vehicles is used.

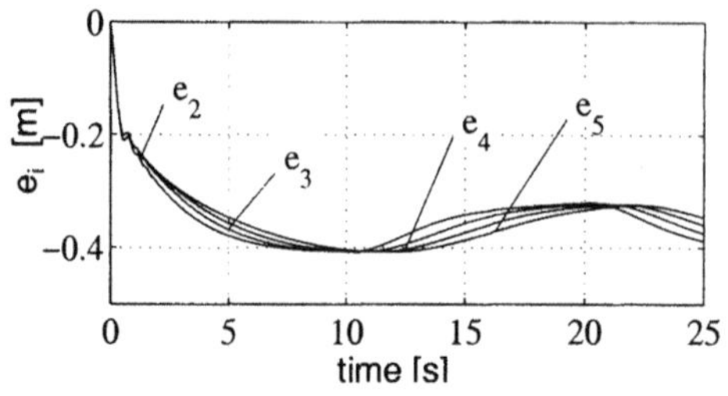

Figure 14. Tracking errors of linear controller with communication, $v_0 = 30m/s$

5. Conclusion

In this paper, a linear and a nonlinear cruise controller are developed for a convoy of vehicles. The control objectives are individual vehicle stability and string stability without overshoot.

Due to the linearization of the model the linear controller are to be designed for just one operating point given by speed and gear. Nevertheless, the individual vehicle stability of the linear controller is independent on speed and gear. On the other hand, the string stability depends strongly on speed and gear, if no communication is used. If communication is available, string stability is independent of the operating point, too.

In the case of the nonlinear controller individual vehicle stability and string stability without overshoot are guaranteed independent of the operating point.

A comparison of the both controller schemes shows that in all cases the oscillatory performance of the nonlinear controller is better than the one of the linear controller. Finally it has to be mentioned that communication improves the dynamical behavior of both controllers.

References

1. Mitschke, M. *Dynamik der Kraftfahrzeuge*, Band A, Springer-Verlag, 1982.
2. Popp, K.; Schiehlen, W. *Fahrzeugdynamik*, Teubner Verlag, 1983.
3. Schiehlen, W. *Multibody System Dynamics: Roots and Perspectives*, Multibody System Dynamics, Vol.1, No.2, 1997, 149–188.
4. Swaroop; Hedrick, J. K.; Chien, C. C. Ioannou, P. *A Comparison of Spacing and Headway Control Laws for Automatically Controlled Vehicles.* Vehicle System Dynamics, Vol. 23, (1994), 597–625.
5. Eyre, J.; Yanakiev, D.; Kanellakopoulos, I. *A Simplified Framework for String Stability Analysis of Automated Vehicles.* Vehicle System Dynamics, Vol. 30, (1998), 375–405.
6. Isidori, A. *Nonlinear Control Systems*, Springer-Verlag, 1989.
7. Fliess, M. et al. *On differentially flat nonlinear systems*, Nonlinear Control Systems Design, Oxford, 1992, 408-412.
8. Fritz, A.; Schiehlen, W. *Automatic Cruise Control of a Mechatronically Steered Vehicle Convoy*, AVEC 98, 1998, 729–734.

INSTABILITY PHENOMENA OF FOLDABLE STRUCTURES WITH UNILATERAL CONTACT

M. SCHULZ
Robert Bosch GmbH, Corporate Research and Development, FV/SLT2,
P.O. Box 30 02 40, 70442 Stuttgart, Germany

1. Introduction

In [1] a new antenna design has been proposed. A physical antenna model revealed, unexpectedly, that contacts develop between panels of adjacent wings of the reflector during deployment. Due to the contacts, additional constraints are imposed on the motion, reducing the mobility to zero and thus, requiring some deformation for deployment. This paper adresses foldable structures of this kind. They can be modelled in terms of a multibody system, the mobility of which is reduced to zero due to the occurrence of unilateral contact.

Introducing generalized coordinates φ and a control parameter p we can simulate a quasi-static deployment or retraction by determining the static equilibrium φ^* of the system for varying values of the control parameter p. In the literature, this procedure is referred to as tracing the equilibrium path of a system. For the satellite antenna, the driven hinge angle serves as control parameter.

The paper is laid out as follows. Section 2 presents the first order equilibrium equations of a multibody system with unilateral constraints. The equilibrium path can be traced, for example, by integrating these equations. Instability phenomena are discussed. In section 3, the equilibrium path of the satellite antenna is computed. The antenna serves as an example system with a corner limit point, a kind of limit point where the equilibrium path is non-smooth. Section 4 describes the effect of unilateral contact and related instability phenomena on the dynamical behaviour of the antenna.

2. Equilibrium Paths of Unilaterally Constrained Multibody Systems

We consider a conservative multibody system that is described by a set of n generalized position coordinates

$$\varphi = [\varphi_1 \quad \varphi_2 \quad \cdots \quad \varphi_n]^T . \tag{1}$$

and take into account m unilateral, frictionless contacts

N. Van Dao and E.J. Kreuzer (eds.),
IUTAM Symposium on Recent Developments in Non-linear Oscillations of Mechanical Systems, 269–277.

$$g_j(\varphi) \geq 0 \ , \ j \in I \ , \ I = \{1,2,\ldots,m\} \ . \tag{2}$$

The functions g_j may denote distances between two parts of the multibody system. The set of generalized coordinates φ is assumed to be a minimal set of coordinates when all unilateral contacts are inactive, i.e. $g_j > 0 \ \ \forall j \in I$. The potential energy U of the system is assumed to be a function of the generalized coordinates and a single control parameter p. We require that the constraint functions as well as the energy function are C^2-continuous. With the Lagrangian function

$$L(\varphi, p) = U - \sum_{j \in I} \lambda_j g_j \tag{3}$$

the equations of equilibrium are the well-known Kuhn-Tucker conditions which read

$$\nabla_\varphi L(\varphi^*, \lambda^*, p) = 0 \ , \tag{4 a}$$

$$g_j(\varphi^*) \geq 0 \ , \ \lambda_j^* \geq 0 \ , \ \lambda_j^* g_j^* = 0 \ , \ j \in I \tag{4 b}$$

when evaluated at a point of equilibrium φ^*, λ^*, p. Here, we assume that the gradient vectors ∇g_j^* of all active constraints $j \in I^*$ are linearly independent (regularity assumption) where

$$I^* = \left\{ j \in I \,\middle|\, g_j(\varphi^*) = 0 \right\} , \tag{5}$$

Excluding bifurcational behaviour for the moment, we can express the equilibrium path with the aid of a parameter s:

$$\varphi^* = \varphi^*(s) \ , \ \lambda^* = \lambda^*(s) \ , \ p = p(s) \ . \tag{6}$$

The equilibrium path in the φ-p-space is in general non-smooth, but exhibits kinks where unilateral constraints become active or inactive. Whereas for systems with bilateral constraints the equilibrium path would be smooth and the path derivatives would be continuous, for unilateral constraints we are only allowed to presuppose the existence of left-hand side and right-hand side derivatives which need not coincide.

We require that a point of equilibrium $\varphi^*(s = p = 0)$ is given and that a single path goes through this point. The equilibrium path can then be traced for example by integrating the first order equilibrium equations, i.e. the first order path derivatives of the equations of equilibrium. They have been derived in [2] without using advanced mathematical concepts such as B-differentiability [3] and read

$$H_+^* \dot{x}^* + T_+^{*T} (\nabla_\varphi U'^*) \dot{p} - T_+^{*T} E_0^* \dot{\lambda}_0^* = 0 \ , \tag{7 a}$$

$$\dot{g}_0^* \geq 0 \ , \ \dot{\lambda}_0^* \geq 0 \ , \ \dot{g}_0^{*T} \dot{\lambda}_0^* = 0 \ . \tag{7 b}$$

Here and in the following, the dot indicates a right-hand side derivative with respect to s and the prime denotes a partial derivative with respect to p. With the index set of strongly active constraints

$$I_+^* = \left\{ j \in I \mid g_j^* = 0, \lambda_j^* > 0 \right\} \tag{8}$$

and the index set $I^* \setminus I_+^*$ of weakly active constraints the corresponding constraint matrices

$$E_+^* = [\nabla g_j^*] \ , \ j \in I_+^* \ ; \ E_0^* = [\nabla g_j^*] \ , \ j \in I^* \setminus I_+^* \tag{9}$$

and column vectors of right-hand side path derivatives of Lagrange multipliers

$$\dot{\lambda}_+^* = \{\dot{\lambda}_j^*\} \ , \ j \in I_+^* \ ; \ \dot{\lambda}_0^* = \{\dot{\lambda}_j^*\} \ , \ j \in I^* \setminus I_+^* \tag{10}$$

have been defined. The column vectors of the matrix T_+^* span the nullspace $\aleph(E_+^{*T})$ of the matrix E_+^{*T}. The path derivatives of the generalized coordinates satisfy

$$E_+^{*T} \dot{\varphi}^* = 0 \ , \tag{11}$$

such that we can set

$$\dot{\varphi}^* = T_+^* \dot{x}^* \tag{12}$$

with $\dot{x}^* \in R^{n-m_+}$ and m_+, the number of strongly active constraints. The reduced Hessian is defined by

$$H_+^* = T_+^{*T} (\nabla_\varphi^2 L^*) T_+^* \ ; \tag{13}$$

and the column vector

$$\dot{g}_0^* = \{\dot{g}_j^*\} \ , \ j \in I^* \setminus I_+^* \tag{14}$$

can be computed according to

$$\dot{g}_0^* = E_0^{*T} T_+^* \dot{x}^* \ . \tag{15}$$

The inequalities (7 b) are to be interpreted in the sense that all elements of the column vectors are greater than or equal to zero.

It is not guaranteed that the reduced Hessian H_+^* is positive definite; and we cannot expect Eq. (7) to have a unique solution. The path derivatives are not unique when the mobility of the system for a fixed value of p is greater than zero. An initially zero mobility, for example, can become greater than zero when advancing on the equilibrium path and an initially active unilateral constraint becomes inactive. We will now distinguish different cases depending on the features of the reduced Hessian and of a certain other matrix, and discuss the solution of the first order equilibrium equations.

2.1 REDUCED HESSIAN IS REGULAR

We can solve Eq. (7 a) for $\dot{x}^*$ and substitute the solution into Eq. (16), yielding together with Eq. (7 b) a Linear Complementarity Problem (LCP)

$$\dot{g}_0^* = G^* \dot{\lambda}_0^* + \dot{p}\,\hat{g}^* \,, \tag{16 a}$$

$$\dot{g}_0^* \geq 0 \,,\; \dot{\lambda}_0^* \geq 0 \,,\; \dot{g}_0^{*T} \dot{\lambda}_0^* = 0 \tag{16 b}$$

with

$$G^* = E_0^{*T} T_+^* H_+^{*-1} T_+^{*T} E_0^* \,, \tag{17 a}$$

$$\hat{g}^* = -E_0^{*T} T_+^* H_+^{*-1} T_+^{*T} \nabla_\varphi U'^* \,. \tag{17 b}$$

After solving the LCP, we can compute the right-hand side path derivatives of φ^* and λ_+^* according to

$$\dot{\varphi}^* = T_+^* H_+^{*-1} T_+^{*T} \left(E_0^* \dot{\lambda}_0^* - (\nabla_\varphi U'^*) \dot{p} \right) , \tag{18 a}$$

$$\dot{\lambda}_+^* = \left(E_+^{*T} E_+^* \right)^{-1} E_+^{*T} \left((\nabla_\varphi^2 L^*) \dot{\varphi}^* + (\nabla_\varphi U'^*) \dot{p} - E_0^* \dot{\lambda}_0^* \right) . \tag{18 b}$$

2.1.1. *Reduced Hessian is Positive Definite*

In this case, the matrix G^* is also positive definite due to the regularity assumption. Therefore, the LCP (16) has a unique solution for $\dot{p} < 0$ as well as for $\dot{p} > 0$. The two solutions correspond to two opposite directions which can be defined on the path, see Fig. 1.

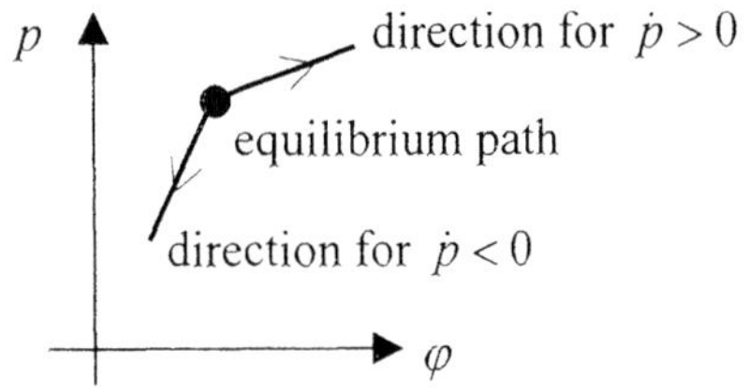

Figure 1. Solution of the LCP for positive definite reduced Hessian

2.1.2. *Reduced Hessian is Not Positive Definite*

In this case, the uniqueness of the solution of the LCP is not guaranteed. Because there are matrices that are not positive definite but regular, *bifurcational* behaviour can occur although the reduced Hessian is not singular. This is different from systems with bilateral constraints.

If H_+^* is for example negative definite G^* is also negative definite. The matrix G^* can even be negative definite in case of an indefinite reduced Hessian. We want to restrict the discussion to a point where only one constraint is weakly active and $G^* < 0$, $\hat{g}^* \neq 0$. This case might be rather common. The LCP, then, has two solutions for $\dot{p}\,\hat{g}^* > 0$ and no solution for $\dot{p}\,\hat{g}^* < 0$. Thus, if there are two solutions for $\dot{p} < 0$ ($\dot{p} > 0$), there is no solution for $\dot{p} > 0$ ($\dot{p} < 0$). The equilibrium path has reached a maximum (minimum) in the φ-p-space. We denote such a point a *corner limit point*. In contrast to a *common limit point* the condition $\dot{p} = 0$ does not necessarily hold at a corner limit point (Fig. 2). The reduced Hessian is singular at a common limit point whereas at a corner limit point this matrix can be regular but is not positive definite.

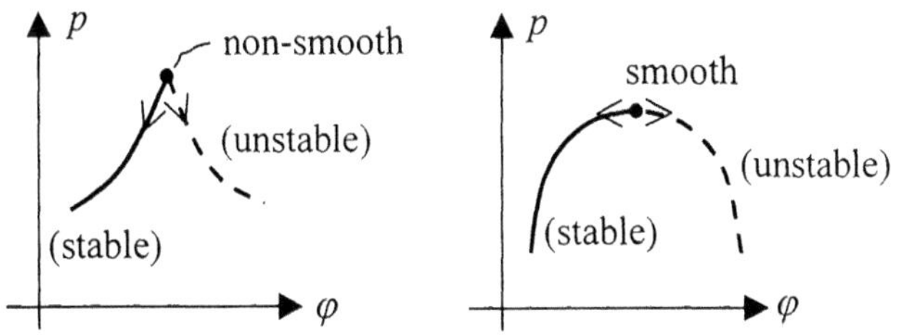

a) Corner Limit Point   b) Common Limit Point

Figure 2. Corner Limit Point

In [2] the change of stability at a corner limit point has been discussed. It was found that the behaviour at a common and at a corner limit point is the same. When advancing on the equilibrium path and passing a corner limit point where a single constraint transforms from active into inactive, the dimension of the reduced Hessian increases by one. With the dimension also the number of negative eigenvalues and thus the degree of instability changes by one.

2.2 REDUCED HESSIAN IS SINGULAR

The singularities that can occur if the index set $I^* \setminus I_+^*$ is empty are of the same nature as for a system with bilateral constraints. The situation appears to be more complex if the index set $I^* \setminus I_+^*$ is non empty. This case has been discussed in [2].

3. Satellite Antenna

Fig. 3 shows a rigid-panel deployable antenna recently developed at Cambridge University [1]. The antenna consists of six wings which are divided into five panels, respectively, and form a symmetric paraboloidal reflector. Each wing is connected to the next by a bar. The connections between adjacent panels, between the first panel of each wing and a central hub, and between the bars and the wings are made by revolute joints. The antenna is deployed by simultaneously driving the six joints between the connecting bar and the last panel of each wing. This produces a complex motion in which the wings first unwrap in a six-fold symmetric fashion, and then rotate about the hub.

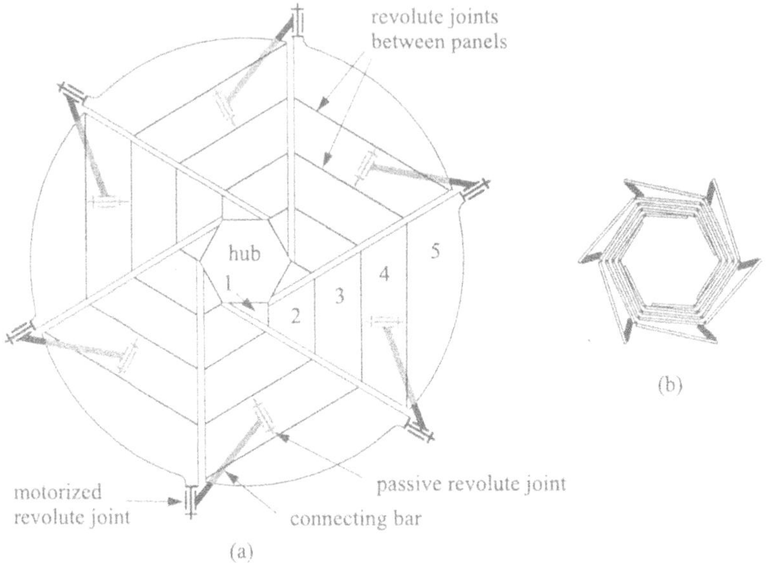

Figure 3. Rigid-panel deployable antenna
(a) Fully deployed; (b) Fully folded, the panels are shown flat for clarity

For a simulation we assume that the reflector is not paraboloidal but flat and that panel 5 is triangular. The topology of the antenna, however, is retained. Hub, panels and joints are assumed to be rigid. The elasticity of the connecting bar is taken into account by two identical extensional springs of stiffness k which are attached to the points $S1$ or $S3$ of the connecting bar and to points $S2$ or $S4$ of panel 4, see Fig. 4. Moreover, we restrict ourselves to a six-fold symmetric deployment and retraction such that for a certain value p of the motorized angle, a state of equilibrium can be given by the angles $\varphi_2,\ldots,\varphi_5$ between adjacent panels of one wing and by the angle φ_1 between panel 1 and the hexagonal hub. The angles are $p=\varphi_i=0$, $i=1,2,\ldots,5$, when the antenna is fully folded and $\varphi_1=90^\circ$, $\varphi_2=\varphi_3=\varphi_4=\varphi_5=60^\circ$ and $p=180^\circ$ when fully deployed. In both configurations, the global coordinates of the points $S1$ and $S2$ as well as $S3$ and $S4$ are the same and the springs are undeformed. The potential energy of the system is equal to the strain energy of the two springs and can be expressed as a function of the angles φ_i and the driven angle p.

We take into account the unilateral constraints of Tab. 1; they are of the following kind:

(1) Point C_i of wing 2 can touch a panel of wing 1, see Fig. 4.

(2) The hinge angles are constrained by brackets.

In case (2), the mathematical constraints are simple lower or upper bounds. For constraints of type (1), the distance between point C_i and the corresponding panel serves as a constraint function. Note that the constraint functions depend only on the generalized coordinates.

TABLE 1. Unilateral constraints

nr. i_G	constraint	nr. i_G	constraint
1	*C*1 - panel 2	8	*C*5 - panel 2
2	*C*1 - panel 3	9	$\varphi_5=60^\circ$ fully opened
3	*C*2 - panel 3	10	$\varphi_4=60^\circ$ fully opened
4	*C*2 - panel 4	11	$\varphi_3=60^\circ$ fully opened
5	*C*3 - panel 4	12	$\varphi_2=60^\circ$ fully opened
6	*C*3 - panel 5	13	$\varphi_1=0$ fully closed
7	*C*4 - panel 3		

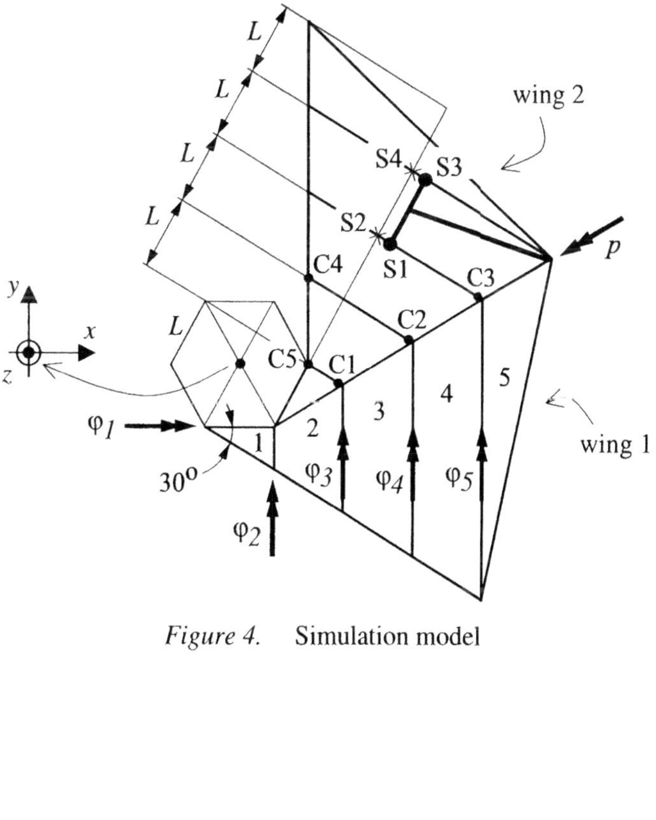

Figure 4. Simulation model

The equilibrium path has been traced by simultaneously solving the LCP and integrating Eqs. (18). As a starting point, the fully folded antenna has been chosen and the control parameter p has been used as a variable of integration. As the integration proceeds, the constraint functions of the inactive constraints and the Lagrange multipliers of the active constraints must be checked for roots. In this way the index sets of the weakly and strongly active constraints and of the inactive constraints can be updated.

The equilibrium path exhibits a corner limit point $L1$ and a common limit point $L2$, see Fig. 5. At point $L1$, constraints 2 and 10 are strongly active and constraint 8 is weakly active. In order to determine the path derivatives at point $L1$ the following LCP must be solved:

$$\dot{g}_0^* = -0.0015\dot{\lambda}_0^* - 0.0269\dot{p} \ , \qquad (19\text{ a})$$

$$\dot{g}_0^* \geq 0 \ , \ \dot{\lambda}_0^* \geq 0 \ , \ \dot{g}_0^{*T}\dot{\lambda}_0^* = 0 \ . \qquad (19\text{ b})$$

Since the factors of $\dot{\lambda}_0^*$ and $\dot{p}$ are negative, the LCP has two solutions for $\dot{p} < 0$ and no solution for $\dot{p} > 0$. The stability of the equilibrium path changes at $L1$ as mentioned above. The path is unstable between $L1$ and $L2$, as indicated by a thin line, and otherwise stable, as indicated by a bold line.

At p=-2.591586..., close to the common limit point $L2$, the integrator stops because the reduced Hessian of the Lagrangian is nearly singular. In the neigbourhood of this point, we switch the variable of integration to φ_1.

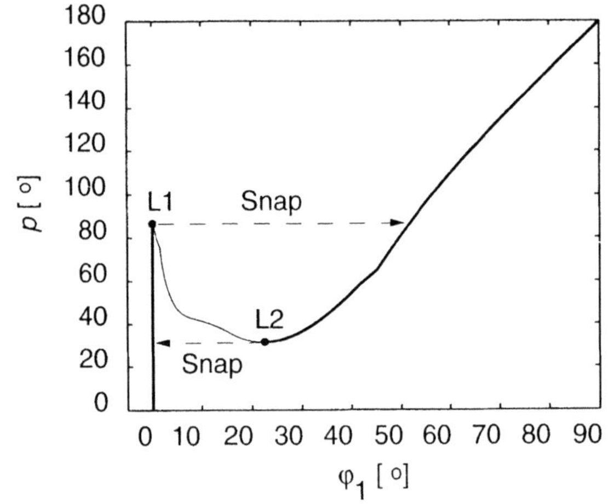

Figure 5. Equilibrium path of the satellite antenna

When quasi-statically increasing p and thus deploying the antenna, the system follows the stable equilibrium path. At the corner limit point $L1$, the antenna dynamically snaps to another stable part of the equilibrium path as indicated by the arrow in Fig. 5. The snap involves strong vibrations and a sudden movement of the antenna where the wings rotate as rigid bodies around the hub. This behaviour, has actually been observed during the deployment of a physical model of the antenna.

4. Conclusions

The first order equilibrium equations of a unilaterally constrained multibody system and possible solutions have been summarized. A rigid-panel deployable antenna served as an example for a deployable structure for which the mobility is reduced to zero due to the occurrence of unilateral contact. The equilibrium path that has been traced by integrating the first order equilibrium equations exhibits a corner limit point and a common limit point. A corner limit point is a point where the equilibrium path is non-smooth and reaches a minimum or maximum in the φ-p-space. Such a point is peculiar to systems with unilateral constraints. The equilibrium path becomes unstable at the corner limit point, leading to a dynamical snap after an initially quasi-static antenna deployment. This dynamical snap involves a sudden movement of the antenna and possibly strong vibrations. Thus, we have shown that a quasi-static deployment and retraction in general might not be possible if the occurrence of unilateral contact leads to instability phenomena of this kind.

Acknowledgement

This project has been carried out in the framework of the Training and Mobility of Researchers Programme in the Deployable Structures Laboratory of Cambridge University Engineering Department (England). Financial support from the European Commission and the invitation by Dr. S. Pellegrino to work in his laboratory are gratefully acknowledged.

References

[1] S.D. Guest and S. Pellegrino, "A new concept for solid surface deployable antennas", *Acta Astronautica*, **38**, 103-113, (1996).

[2] M. Schulz and S. Pellegrino, "Equilibrium Paths of Multibody Systems with Unilateral Contact – Part I: Theory", accepted for publication in *The Royal Society, Proceedings: Mathematical, Physical and Engineering Sciences*.

[3] G. Björkman, "Path following and critical points for contact problems", *Computational Mechanics*, **10**, 231-246 (1992).

SIMILARITIES IN THE YAW AND ROLL DYNAMICS OF SHIPS IN EXTREME ASTERN SEAS

K.J. Spyrou
Ship Stability Research Centre, University of Strathclyde,
48 North Portland Street, Glasgow, G1 1XM, UK.
also currently at:
Ship Design Laboratory, National Technical University of Athens
9 Iroon Polytechniou, Athens 15773, Greece.

1. Introduction

In large following waves there exist some interesting analogies between the dynamic behaviour of a steered ship in the yaw and roll directions. It is well known, for example, that capsize may occur due to fluctuations of the roll righting-arm [1]. Similar fluctuations in the stiffness term may take place also in yaw, originating from the combined effect of rudder control with the wave induced yaw moment. This may give rise to course instability which will be realized as deviation from the desired heading and broaching [2].

Consider a ship traveling in long following sinusoidal waves. In order to avoid coupling complications let us assume further that, due to high natural frequencies in heave and in pitch compared to the encounter frequency, the ship can maintain a state of quasi-static equilibrium on the vertical plane. If the waves are relatively steep, the geometry of the submerged part of the hull will vary noticeably, on the basis of the ship's position on the wave. This will be reflected in roll's righting-arm. Reduced or even negative roll restoring may arise when the middle of the ship is near to a wave crest, due to substantial "loss of waterline" (typically such a trend is more pronounced when there is low freeboard at midship combined with strong flare at the ends [3]). If roll restoring remains negative for sufficient time, so that heel finds the time to develop unopposed well beyond the "vanishing angle", then capsize due to the so-called pure-loss of stability mechanism will be realised [4]. In this case the magnitude of roll damping affects little the survivability of the ship. Capsize can occur of course also in a typical parametric resonance fashion and here damping will be much more important

N. Van Dao and E.J. Kreuzer (eds.),
IUTAM Symposium on Recent Developments in Non-linear Oscillations of Mechanical Systems, 279–290.

[5]. Practically, the variation of restoring must be however quite intensive, so that large amplitude roll can build-up within a small number of wave cycles.

The onset of yaw instability, in a similar wave environment, is a slightly more complex process because yaw is always coupled with sway. Furthermore, a control law for the rudder must be considered. Unlike with roll, if there is no active control in yaw, no restoring force exists in still water; but this may be created by the movement of the rudder which tends to bring the ship back on the correct course. If waves of length equal to the ship length or longer meet the ship from behind, they will create a moment that will be dependent on the angle between the direction of wave propagation and the ship's heading. This wave yaw moment works as a positive restoring component when the ship passes from crests (stabilizing effect). The opposite will be realized however in the vicinity of wave troughs where the waves will tend to orientate the ship perpendicular to the direction of their propagation. The relative magnitudes of the rudder's and the wave's moment will determine whether restoring becomes, in the region of the wave trough, negative. But even if it remains positive, a parametric mechanism with the potential to destabilize the horizontal-plane motion of the ship will have been set in place. The commonality of the underlying dynamics of yaw and roll is prevalent.

Our first objective in this paper is to identify the correspondence between yaw and roll parameters from the perspective of these Mathieu-type phenomena. Furthermore, we shall introduce an approach for assessing the effect of surge motion. As is well known, when the waves are large, the nonlinearity of surge cannot be neglected [6]. A manifestation of this nonlinearity, is a virtual rescaling of time as the ship is spending longer time on the crests than on the troughs of the wave. In spite of the significance of this mechanism for the safety-critical motions, there has been no systematic analysis earlier on it.

2. Equations of motion for yaw and for roll

Consider the linear differential equations of sway and yaw [7], with the addition of wave excitation terms at their right-hand side:

$$\text{Sway:} \quad \left(m' - Y'_{\dot{v}}\right)\dot{v}' - Y'_v v' + \left(m'x'_G - Y'_{\dot{r}}\right)\dot{r}' + \left(m' - Y'_r\right)r' = Y'_\delta \delta + Y'_{(wave)} \tag{1}$$

$$\text{Yaw:} \quad \left(m'x'_G - N'_{\dot{v}}\right)\dot{v}' - N'_v v' + \left(I'_z - N'_{\dot{r}}\right)\dot{r}' + \left(m'x'_G - N'_r\right)r' = N'_\delta \delta + N'_{(wave)} \tag{2}$$

In the above v', r' are respectively sway velocity and yaw angular velocity, δ is the rudder angle, m' is ship mass and x_G is the longitudinal position of the centre of gravity; $Y'_{\dot{v}}$, $Y'_{\dot{r}}$, $N'_{\dot{v}}$, $N'_{\dot{r}}$ are acceleration coefficients (added masses/moments of inertia) and Y'_v, Y'_r, N'_v, N'_r, Y_δ are velocity coefficients (hydrodynamic damping terms). The wave's sway force and yaw moment are respectively, $Y'_{(wave)}$, $N'_{(wave)}$. The prime indicates nondimensionalised quantity and the overdot differentiation over time.

At first instance we shall assume that the yaw and sway velocities are restrained from building-up to high values (thus they remain small; as a matter of fact the resulting damping forces may be considered linear) through use of appropriate rudder control. Additionally, ship behaviour is examined at "some distance" from the region of surf-riding, so that, for this first part of the paper, it is not unrealistic to assume that surge velocity is constant.

We express the wave terms $Y'_{(wave)}$, $N'_{(wave)}$ in respect with the frequency of encounter (rather than as functions of absolute wave frequency and position). Also we neglect some phase difference (relatively to the wave) which might exist in these two types of excitation:

$$Y'_{(wave)} = Y'_w \, \psi \sin(\omega'_e t') \tag{3}$$

$$N'_{(wave)} = N'_w \, \psi \cos(\omega'_e t') \tag{4}$$

The following notation is applied: Y'_w, N'_w are wave force/moment coefficients; ψ is the ship's heading relatively to the wave ($\psi = 0$ when the sea is exactly following – generally, ψ is assumed small).

Consider further rudder control with a linear law based on two gains, k_1 and k_2: k_1 multiplies the instantaneous heading deviation from the desired course ψ_r, while k_2 multiplies yaw's angular velocity:

$$\delta = -k_1(\psi - \psi_r) - k'_2 r' \tag{5}$$

Substituting (3), (4) and (5) in (1) and (2), uncoupling yaw from sway and using well known expressions for system gain and time constants, K', T'_1, T'_2, T'_3 [7], a differential equation of heading angle with the following structure is obtained:

$$\psi'^{(3)} + b\ddot{\psi}' + p[1 + f\cos(\omega'_e t' - \sigma)]\dot{\psi}' + q^2[1 - h\cos(\omega'_e t' - \rho)]\psi = j \tag{6}$$

The above third-order differential equation has time-dependent coefficients in two places. As is well known however, if T'_1 is much greater than T'_2 and T'_3, we can use the so-called simplified yaw response model of Nomoto [8]. In that case the order of equation (6) is reduced by one:

$$T'\ddot{\psi}' + \dot{\psi}' = K'\delta + A'\psi\cos(\omega'_e t' - a) \tag{7}$$

K', T' are respectively system gain and time constants, ψ is relative heading angle (assumed small), δ is rudder angle, A' is wave excitation amplitude, ω_e' is the encounter frequency and a is a phase angle.

By coupling (7) with the autopilot equation (5) and dropping for simplicity the phase angle a, we obtain after some rearrangement:

$$\ddot{\psi}' + \gamma\,\dot{\psi}' + {\omega'_{0(yaw)}}^2 [1 - h\cos(\omega_e' t')]\psi = j \tag{8}$$

In the above $\omega'_{0(yaw)} = \sqrt{k_1 K'/T'}$, $\gamma = (1 + k_2' K')/T'$ (damping), $h = A'/k_1 K'$ (amplitude of parametric variation of restoring), $j = k_1 K' \psi_r / T'$. It is easily recognized that (8) is Mathieu's equation with the addition however of bias-like external static forcing term, j.

For stability, positive T' is required as $1/T'$ is the inverse of the damping of the unsteered vessel. However, large positive T' implies slow convergence towards the corresponding steady rate-of-turn which is determined by the value of the static gain K'. A trend exists for large T' to appear in conjunction with large K' which gives a nearly straight-line *spiral curve*. The effect of active control on damping is represented by the quantity $k_2' K'/T'$. It depends thus on the yaw rate ("differential") gain term in the autopilot. If $T' < 0$, suitable choice of k_2' can turn the damping of the system positive since k_2' multiplies the positive quantity K'/T', thereby yielding stability for the steered ship in calm sea. The wave effects are lumped into the restoring and independent-periodic-forcing terms since the quantities K' and T' were assumed to be, at first approximation, unaffected by the wave. If the amplitude of wave excitation A' exceeds $k_1 K'$, then on the basis of (11) negative yaw restoring will arise around the trough. Should the duration of operation under negative restoring be long enough, undesired turning motion will be initiated ("broaching"). From a dynamics perspective there is complete equivalence with a capsize event of the so-called "pure-loss" type. It can be avoided if the proportional gain k_1 is chosen to be always greater than A'/K' even for the most extreme wave environment where the ship will operate (it should be a matter of further investigation to what extend this is technically feasible). A notable difference between the manifestation of this instability in roll and in yaw is that in roll it arises near the crest of the wave, whereas in yaw the ship becomes vulnerable near a trough.

We may rewrite (8) on the basis of heading error $\psi_1 = \psi - \psi_r$ and then apply the transformation $\tau = \omega'_{0(yaw)} t'$:

$$\frac{d^2\psi_1'}{d\tau^2}+2\zeta\frac{d\psi_1'}{d\tau}+[1-h\cos(\Omega\tau)]\psi=f\cos\Omega\tau \tag{9}$$

where $\Omega=\omega_e'/\omega_{0(yaw)}'$ and $h=A'/(k_1\,K')$. Also, $f=A'\,\psi_r/(T'\,\omega_{0(yaw)}'^{\,2})$ which means that for $\psi_r=0$ the external forcing term of (9) will be zero. The damping ratio is given by the expression: $2\zeta=(1+k_2K')/\sqrt{k_1K'/T'}$ (the presence of k_1 inside ζ should be noted).

It is obvious from (9) that parametric instability of yaw may also arise, very much like that of roll. To establish the analogy we remind that the generic linearised in x equation of roll for a following sea is:

$$\frac{d^2\varphi'}{d\tau^2}+2\zeta\frac{d\varphi'}{d\tau}+[1-h\cos(\Omega\tau)]\varphi'=0 \tag{10}$$

φ' is the normalised roll angle, $\varphi'=\varphi/\varphi_v$, with φ the true roll angle and φ_v the angle of vanishing stability. Although for the damping ratio, scaled time, amplitude of forcing and frequency ratio we have used the same symbols as in roll, the expressions from which we derive their values will be of course different in the yaw case. The damping ratio will be: $2\zeta=B\,\omega_{0(roll)}/(M\,g\,(GM))$ where B is the dimensional linear damping coefficient, M is ship mass and (GM) is the metacentric height. Also, $\Omega=\omega_e'/\omega_{0(roll)}'$ and $\tau=\omega_{0(roll)}t$, with roll's natural frequency given by $\omega_{0(roll)}=\sqrt{Mg\,(GM)/(I+\Delta I)}$. With substitution of $\omega_{0(roll)}$ in ζ we may obtain further: $2\zeta=B/\sqrt{(I+\Delta I)\,M\,g\,(GM)}$. Also, the amplitude of the parametric is $h=\delta(GM)/(GM)$ where $\delta(GM)$ is the difference in the values of metacentric height at the crest and in still water. This is a common assumption which may be sufficient for the preliminary character of this study but of course it results in a highly idealised formulation because the average (GM) has no reason to be identical with the still water (GM). In addition, the variation from trough to crest may not be sinusoidal.

2.1. CONDITIONS AT EXACT RESONANCE

For overtaking waves the frequency of encounter will be positive and for the case where no damping exists the condition of exact resonance will be: $\omega_e/\omega_0=2/\eta$, $\eta=1,2,3,\ldots$ (ω_0 may be the frequency of encounter of yaw or of roll). Thus with increasing η the vertices will tend to accumulate nearer to the zero frequency of encounter.

The expression of the encounter frequency for a following sea is $\omega_e = (2\pi/\lambda)(c-U)$. In yaw, time is commonly nondimensionalised on the basis of U/L (noted the resulting time-dependence). Therefore, the expression of the nondimensional frequency of encounter in yaw is: $\omega'_e = 2\pi L(c-U)/(\lambda U)$. With the substitutions $\omega'_e = 2\omega'_{0(yaw)}/\eta$ and $c/U = Fn_{wave}/Fn$ (Fn_{wave} is the Froude number corresponding to wave celerity) the parametric equation of the vertices of the corresponding <u>undamped</u> system is $Fn = Fn_{wave}/(1+\lambda\omega'_{0(yaw)}/\eta\pi L)$,. Given that $Fn_{wave} = \sqrt{\lambda/(2\pi L)}$ we may write further:

$$Fn = \sqrt{\lambda/(2\pi L)}\Big/\left[1+\omega'_{0(yaw)}\lambda/(n\pi L)\right] \tag{11a}$$

Consider further the domain of variation for the yaw natural frequency, which, as was found earlier, is expressed as: $\omega'_{0(yaw)} = \sqrt{k_1 K'/T'}$. It is known that for conventional ships, the ratio K'/T' usually takes values within the range $[0.3-1.4]$ (see for example [9]). It is derived that $\omega'_{0(yaw)}$ should lie in the range $[0.55\sqrt{k_1}$ - $1.18\sqrt{k_1}]$. With a proportional gain k_1 between 1.0 and 2.0, $\omega'_{0(yaw)}$ should then be between 0.55 and 1.67. In Fig. 1a is shown how the critical Fn would vary as function of the wave length-to-ship-length ratio λ/L, for three different values of $\omega'_{0(yaw)}$: 0.5, 1.0 and 1.5.

We shall consider now roll motion: The natural frequency is nondimensionalised on the basis of ship length and acceleration of gravity, $\omega'_0 = \omega_0\sqrt{L/g}$ (thus it is not speed dependent, at least in an explicit sense). The difference in the nondimensionalisation method between yaw and roll results in different parametric expressions of the critical Froude number:

$$Fn = \frac{\sqrt{\dfrac{\lambda}{L}}}{\sqrt{2\pi}} - \frac{\omega'_{0(roll)}\left(\dfrac{\lambda}{L}\right)}{\pi n} \tag{11b}$$

It is well known that container vessels are sometimes susceptible to parametric resonance, one of the reasons being that their operational speed falls near to the region of principal resonance which is the most dangerous. Extensive model tests have been carried out recently in Japan in order to identify the critical conditions for capsize. The occurrence of parametric instability was one of the investigated scenarios [10]. For an examined containership the measured natural frequency was $\omega'_{0(roll)} = 0.566$. In Fig. 1b are shown, for a range of roll natural frequencies, the critical Froude numbers for the first few resonances (as for the similar equation for yaw, damping is not included).

We should note that, unlike the parametric instability of roll which is well verified experimentally, for yaw little has been attempted so far on the experimental front.

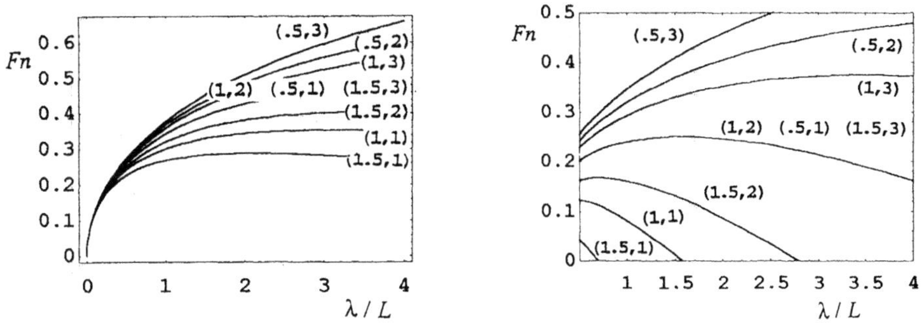

(a) for yaw (b) for roll

Fig. 1: Critical Froude numbers for stability of idealised undamped system. We varied the pair (ω'_e, η) where ω'_e is the corresponding nondimensional natural frequency, and η is the order of the resonance.

2.2. MAGNITUDE AND EFFECT OF DAMPING

Even when a ship is equipped with bilge keels and fins, the damping ratio is usually quite low and very rarely goes above a value of, say, 0.2 (it should be noted however that ζ depends not only on the hydrodynamic characteristics of the hull but also on factors such as the metacentric height and the moment of inertia).For yaw on the other hand, the damping ratio depends strongly on the autopilot's gains. Common values are known to be in the range $0.8 < \zeta < 1.0$ [11]. Here lies therefore a very significant difference between the roll and yaw equations: The damping ratio of yaw is normally very large. This practically means that in order to be placed in a resonance region, the loss of yaw restoring at the trough should be very considerable. Usually the requirement implied by this is the existence of very steep waves. It is very interesting, and perhaps relevant, that about 40 years ago, during model experiments of broaching, it had been observed that as the encounter frequency departs from the zero value, the required wave steepness for broaching shows a very considerable increase [12].

While expressions of the stability boundary are not so difficult to find for relatively low damping, see for example [13], the same may not be said for the very large ζ appearing in the yaw equation. As a first indication of the effect of ζ on the critical h we may use the expression of Gunderson, Rigas & VasnVleck (1974) which is applicable for relatively large damping values:

$$h = \left(1-\zeta^2\right)\tanh\left(\pi 2\zeta\sqrt{\frac{\omega_0^2}{\omega_e^2}}\right) \tag{12}$$

For ζ as low as 0.3 the required h, according to (12), is 0.67 and 0.87 respectively for the principal and the fundamental resonance. One should bear in mind however that these values reflect long-term behaviour. For the build-up of significant motion within a small number of wave cycles (in a practical context this is most relevant) considerably higher values of h are required.

2.3. NONLINEARITY ·

In the roll equation nonlinearity exists in the restoring term (strong) and in damping (mild). Their effect is now relatively well understood, see for example [10, 13-16].

In yaw, nonlinearity is possible to appear in the damping term if the yaw velocity is allowed to become large (for example when the autopilot gain values are low). This relates with the S-shaped curve ("spiral curve") connecting the steady rate-of-turn with the angle of the rudder in still water for directionally unstable ships. It is common to take account of this nonlinearity through a cubic term of yaw velocity. After coupling with the autopilot equation we obtain the following nonlinear version of equation (8) which is left for future consideration:

$$\ddot{\psi}' + \gamma\,\dot{\psi}' - a\,\dot{\psi}'^3 + \omega_{0(yaw)}'^{\,2}[1 - h\cos(\omega_e' t')]\psi = j \tag{13}$$

3. Static and dynamic loss of stability

When roll stability in a following sea is examined, it is customary in naval architecture to distinguish between two mechanisms of capsize: (a) *Pure-loss of stability*, where the ship departs from the state of upright equilibrium due to negative restoring on a wave crest. Then, heel increases monotonically until the ship is overturned. In this mode the magnitude of damping plays little role. (b) *Parametric instability*, which is the classical Mathieu-type mechanism where the build-up is

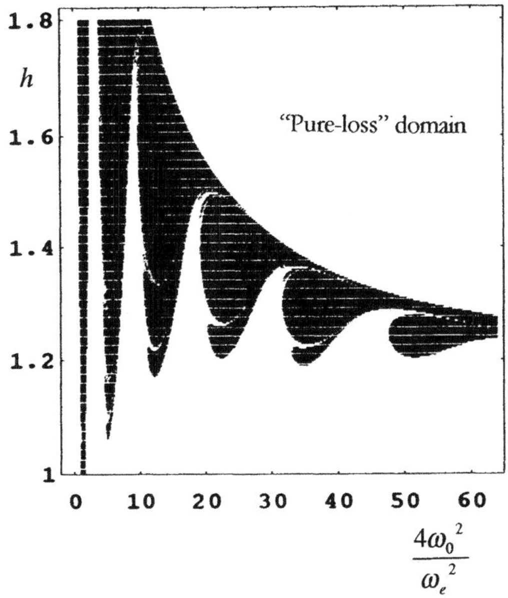

Fig. 2: A 'unifying' view of the domains of "pure loss" and of parametric instability for cubic-type restoring. "Pure loss" occupies the upper and left part of the parametric instability domain.

oscillatory and the magnitude of damping is very important. In Fig. 2 are shown the domains of pure loss and of parametric instability for a ship with a generic cubic-type restoring curve and with linear damping.

Instabilities of a similar nature are possible in yaw as well, resulting in broaching behaviour (sudden turn and deviation from the desired course). Especially the instability usually termed as *broaching due to surf-riding*, happening at Froude numbers near to the wave celerity, may be paralleled with the pure-loss mechanism[2].

A parametric-type mechanism of broaching also exists, which is more likely to happen at lower Froude numbers [12]. For this mechanism the discussion given in sub-section (2.2) is most relevant. Higher wave steepness is required for the occurrence of this instability due to the dominant effect of the large damping factor.

4. The effect of surge

We shall consider now the effect of surge motion for pure-loss and for parametric instability. An implicit assumption in our analysis so far, and also underpinning all earlier studies on ship parametric instability, has been that the forward speed may be assumed as constant. Such an assumption is not however always consistent with the wider context of the analysis. For dangerous dynamic behaviour of roll to arise, steep and long waves are required. Waves of this kind will incur also significant nonlinear effects on surge. The characteristic of large-amplitude surging is that it is asymmetric and the ship stays longer near the crests than near the troughs. This effect is imported into the yaw and roll dynamics through the restoring terms of the corresponding equations.

Consider the roll motion first: The nonlinearity of surge is detrimental for stability because around the crest (where the ship stays longer) restoring capability is reduced. For yaw on the other hand, the effect is opposite. Yaw stability is not worsened because the passage of the ship from the trough is quicker. The danger arises in steeper waves and especially during the process of capture in surf-riding, Fig. 3.

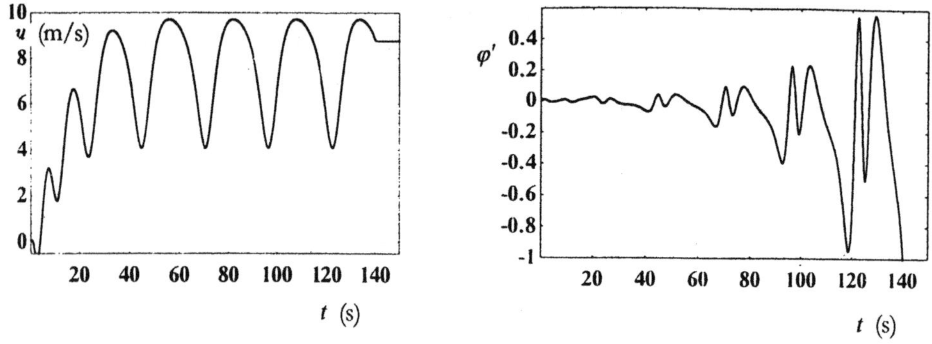

Fig. 3: Parametric instability and capsize due to large amplitude surging (very near to the boundary of surf-riding). Time-domain plots of surge velocity (left) and roll angle (right). The ship was initially with zero velocity at a crest.

The three main forces acting in the surge direction are the resistance, the wave and the propulsion force. As has been shown in [17] these forces result in the following differential equation for the surge motion:

$$\begin{aligned}(m - X_{\dot{u}})\frac{d^2x}{dt^2} + \{[3a_3c^2 + 2(a_2 - b_1)c + a_1] - b_2 n\}\frac{dx}{dt} + \\ +[3a_3c + (a_2 - b_1)](\frac{dx}{dt})^2 + a_3(\frac{dx}{dt})^3 + f\sin(kx) = \\ = b_1c^2 + b_2cn + b_3n^2 - (a_1c + a_2c^2 + a_3c^3)\end{aligned} \tag{14}$$

$-X_{\dot{u}}$ is the surge added mass, c is the wave celerity, n is the propeller's rate of rotation, x is the position of the ship on the wave measured from a moving system fixed on a wave trough; a_1, a_2, a_3 are the coefficients of the resistance polynomial. Likewise, b_1, b_2, b_3 are the thrust-related coefficients. The velocity u of the ship for an observer fixed on the earth is given from the relation $u = c - dx/dt$.

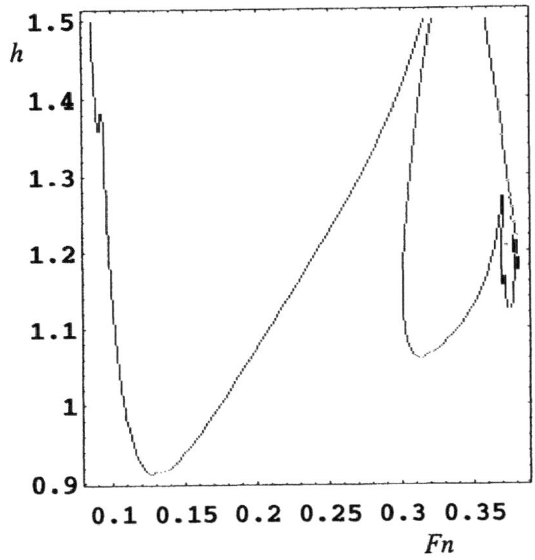

Fig. 4: Boundary lines of capsize due to parametric type instability when the nonlinear surge is taken into account. The upper-right boundary separates capsize from surf-riding. The other boundaries are interfaces with domains of ordinary periodic motion in surge.

Equation (14) is to be solved simultaneously with the equation of yaw or of roll, depending on whether the capsize or the broaching problem is considered. We have identified how the transition curves are modified when roll is coupled with surge. This coupling arises due to the existence of x in the restoring term of the roll equation:

$$\ddot{\varphi}' + 2\mu\dot{\varphi}' + \omega_0^2\left[1 - h\cos(kx)\right]\varphi' = 0 \tag{15}$$

On the basis of the above equation we have found for what combinations of Fn and h the normalised roll angle φ' exceeded the value of 1.0 (from an initial perturbation 0.01 and with zero initial velocity) within a specified time ($t = 200$ s). The calculations were based on a ship with $\omega_{0(roll)} = 0.84$ ($\omega'_{0(roll)} = 1.577$) and $\mu = 0.0585$. Of course, having exceeded the value 1.0 does not necessarily mean capsize, since we may still lie inside the safe basin. But for a practical analysis this is a good basis for comparisons. Fig. 4 provides clear evidence that surge motion has a profound effect on the "capsize" domains. Further investigations are currently underway on this matter. For the considered ship, the principal resonance could not be realised in following waves because a negative Froude number is required for this (the ship should be backing rather than going forward). The lower part of the fundamental is the only place where there is some commonality with the conventional ('damped') Strutt diagram. The upper part of the fundamental has become considerably wider. The next resonance occupies an enlarged domain; but the two after this seem to degenerate. This may relate with the emergence of the surf-riding domain where the behaviour of the ship is stationary.

There, the ship will travel with the speed of the wave, having its middle located near to a trough.

5. References

1. Grim, O. (1952) Rollschwingungen, Stabilität und Sicherheit im Seegang, *Schiffstechnik*, **1**, 1, 10-21.
2. Spyrou, K.J. (1996) Dynamic instability in quartering waves: the behaviour of a ship during broaching. *Journal of Ship Research*, **40**, 1, 46-59.
3. Paulling, J. R. A comparison of stability characteristics of ships and offshore structures. *Proceedings*, 2nd International Conference on Stability of Ships and Ocean Vehicles, STAB '82. Published by the Society of naval Architects of Japan, Tokyo, October 1982, 581-587.
4. Paulling, J.R. (1961) The transverse stability of a ship in a longitudinal seaway, *Journal of Ship Research*, **4**, 37-49.
5. Kerwin, J.E. (1955) Notes on rolling in longitudinal waves, *International Shipbuilding Progress*, **2**, 16, 597-614.
6. Kan, M. (1990) Surging of large amplitude and surf-riding of ships in following seas. *Selected Papers in Naval Architecture and Ocean Engineering*, The Society of Naval Architects of Japan, **28**.
7. Clarke, D., Gendling, P. & Hine, G. (1983) Application of manoeuvring criteria in hull design using linear theory, *Trans.* RINA, **125**, 45-68.
8. Nomoto, K., Taguchi, K., Honda, K. & Hirano, S. (1957) On the steering qualities of ships. *International Shipbuilding Progress*, **4**, 35, July, 354-370.
9. Barr, R., Miller, E.R., Ankudinov, V. & Lee, F.C. (1981) Technical basis for manoeuvring performance standards. Technical Report 8103-3, Hydronautics, Inc., submitted by the United States to the International Maritime Organization (IMO).
10. Hamamoto, M., Umeda, N., Matsuda, A. & Sera, W. (1995) Analyses of low cycle resonance of a ship in astern seas, *Journal of the Society of Naval Architects of Japan*, **177**, 197-206.
11. Fossen, T.I. (1994) *Guidance and Control of Ocean Vehicles*, John Wiley and Sons, Chichester, UK.
12. Spyrou, K.J. (1997) Dynamic instability in quartering seas-Part III: Nonlinear effects on periodic motions. *Journal of Ship Research*, **41**, 3, 210-223.
13. Spyrou, K.J. Designing against parametric instability in following seas, accepted in *Ocean Engineering*, 1999, 42 pages.
14. Zavodney, L.D., Nayfeh, A.H. & Sanchez (1990) Bifurcations and chaos in parametrically excited single-degree-of-freedom systems. *Nonlinear Dynamics*, **1**, 1, 1-21.
15. Blocki, W. (1980) Ship safety in connection with parametric resonance of the roll, *International Shipbuilding Progress*, **27**, 36-53.
16. Soliman, M.S. & Thompson, J.M.T. (1992) Indeterminate sub-critical bifurcations in parametric resonance, *Proceedings of the Royal Society of London*, A, **438**, 511-518.
17. Spyrou, K.J. (1997) On the nonlinear dynamics of broaching-to. *Proceedings*, International Conference on Design for Abnormal Waves, The Royal Institution of Naval Architects, Glasgow, October, 12 pages.

NONLINEAR THREE-DIMENSIONAL OSCILLATIONS OF FLUID CONVEYING VISCOELASTIC TUBES WITH AN ADDITIONAL MASS

B. ALBRECHT, A. STEINDL AND H. TROGER

Vienna University of Technology

A–1040 Vienna, Austria

Abstract. The loss of stability of the trivial downhanging equilibrium position of a slender circular tube conveying incompressible fluid flow is studied. The tube is clamped at its upper end and free at its lower end. In addition, the tube is carrying a concentrated mass. Increasing the flow rate quasistatically the loss of stability of the downhanging equilibrium position is studied. For this system experimental results are available in (Copeland, Moon, 1992) for the mass located at the end of the tube. The main objective of this paper is to compare the experimental results of (Copeland, Moon, 1992) with theoretical results obtained by means of Equivariant Bifurcation Theory which is used for the non-linear analysis of the $\mathbf{O}(2)$ (rotationally) - symmetric system.

1. Introduction

Thanks to the book (Golubitsky, Stewart, Schaeffer, 1988), Equivariant Bifurcation theory has now become a widespread method to treat non-linear stability problems of solutions of symmetric systems. Basically there are two model problems in fluid dynamics, namely the convection in a layer (Benard problem) and the convection in the gap between two concentric cylinders rotating at different angular velocities (Taylor Coutte problem) (Craik, 1985). For both problems extensive

N. Van Dao and E.J. Kreuzer (eds.),
IUTAM Symposium on Recent Developments in Non-linear Oscillations of Mechanical Systems, 291–300.

theoretical analyses have been partly compared with experimental findings. However, both problems as well in the mathematical as in the experimental analysis pose considerable difficulties. On the other hand, there exists a third model problem, well studied in engineering, possessing basically the same great variety of solutions, but both in the mathematical analysis and in the experimental realisation, it seems to be less complicated than the two mentioned problems from fluid dynamics: The vertically downhanging visco-elastic fluid conveying tube performing three-dimensional motions. Again this is a problem having $\mathbf{O}(2)$- symmetry. The recently published book (Paidoussis, 1998) gives an excellent overview of this problem.

Loss of stability of equilibrium positions of fluid conveying tubes is a problem which has been extensively treated in engineering ((Paidoussis, 1998) for a long time. Surprisingly, almost no papers in mathematical journals appeared with the exception of (Bajaj and Sethna, 1984). Later in a series of papers (Steindl, 1992), (Steindl, Troger, 1992), (Steindl, Troger, 1995a), (Steindl, Troger, 1995b) the three-dimensional motion of a downhanging tube constrained by an elastic point support is treated. Varying the stiffness of the support as second parameter, besides the flow rate as bifurcation parameter, all Codimendion Two bifurcations classified in (Golubitsky, Stewart, Schaeffer, 1988) could be found and physically interpreted. The nice feature of the tube problem is that almost everybody has an intuitive feeling for the qualitative interpretation of the various different solutions.

The tube with an elastic support was a nice model theoretically but, as it turned out later, when we tried to make experiments to check some of our theoretical results only few of the many different motions could be observed in our experiments. This was due to the fact that tubes which are commercially available are never perfectly straight and, hence, due to geometric imperfections only few of the theoretically found solutions could be observed in the experiment.

On the other hand, in (Copeland, Moon, 1992) careful experiments of three-dimensional tube motions are reported with a different tube system. There a mass is attached to the end of the tube. This mass had the pleasant effect in the experiment that even if the tube without mass was not perfectly straight the added

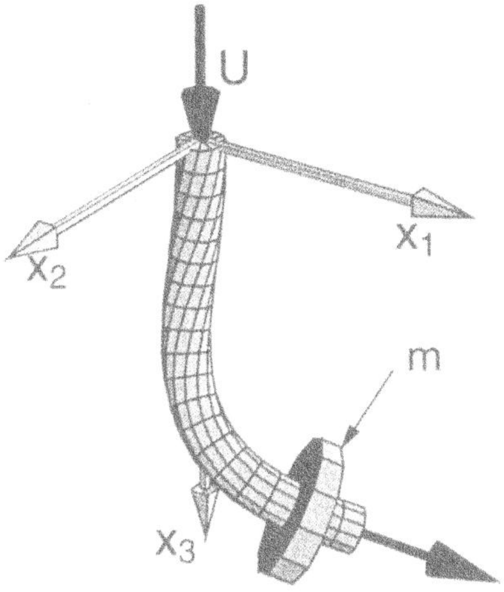

Figure 1. Mechanical model of the fluid carrying tube with an additional mass m

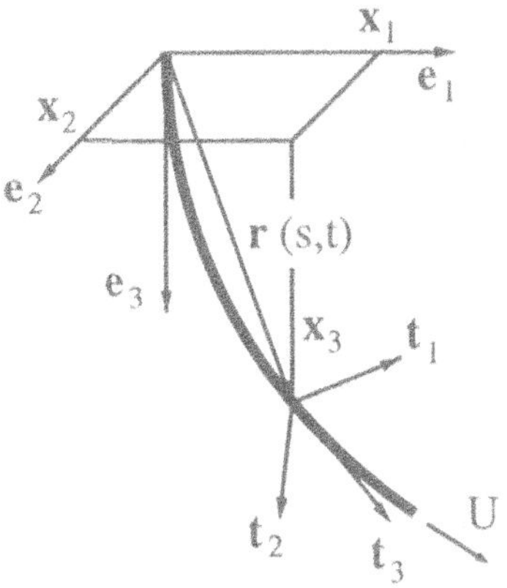

Figure 2. Fixed spatial frame $\{e_i\}$, and moving orthogonal frame $\{t_i\}$.

mass streched the tube and hence decreased or even extinguished small geometric imperfections.

We thought it worthwhile to study the tube system with a point mass (Fig. 1) to see how well we could reproduce theoretically both qualitatively and quantitatively the solutions found experimentally in (Copeland, Moon, 1992). We note that also in (Paidoussis, Semler, 1998) the same problem is treated experimentally and theoretically, however, restricted to planar motions of the tube.

In our theoretical model we assume the mass to be a point mass and that the position (ξ) of the mass along the tube could still be an additional parameter. For a fixed position (ξ) of the point mass we have a two parameter bifurcation problem with the flow rate U and the value m of the mass as the two parameters. Moreover, the considered system is $\mathbf{O}(2)$ - symmetric.

We have to point out one important difference between this work and our previous work concerning the elastically constrained tube. In the latter case we had two generic types of loss of stability varying the stiffness of the support. Namely loss of stability at a zero root (divergence bifurcation) for a stiff spring and at a pair of purely imaginary roots (Hopf bifurcation) for a soft spring. However, for the problem of the tube with the added mass, for example, attaching the mass at the end of the tube and varying its value, always loss of stability due to an imaginary pair of eigenvalues occured. Hence for this problem the coupling between divergence and flutter will not occur and the variety of solutions will be much less than reported for example in (Steindl, Troger, 1995a) or (Steindl, Troger, 1995b).

2. Mechanical model and equations of motion

We take the tube equations derived in (Steindl, Troger, 1996) which we only have to supplement by the condition following from the additional point mass. Hence we have the following relations:

$$\begin{aligned} \boldsymbol{r}' &= \mathbf{B}\boldsymbol{e}_3 \\ \mathbf{B}' &= \mathbf{B}\widehat{\boldsymbol{\Omega}} \\ \boldsymbol{T}' &= \boldsymbol{T}\times\boldsymbol{\Omega}+\boldsymbol{F}\times\boldsymbol{e}_3 \\ \boldsymbol{F}' &= \boldsymbol{F}\times\boldsymbol{\Omega}+\mathbf{B}^T(-\gamma\boldsymbol{e}_3+\ddot{\boldsymbol{r}}+\delta\dot{\boldsymbol{r}}+2\sqrt{\beta}\varrho\dot{\boldsymbol{r}}'+\varrho\boldsymbol{r}'') \end{aligned} \tag{1}$$

subject to the boundary conditions

$$\boldsymbol{r}(0)=\boldsymbol{0}\,,\qquad \mathbf{B}(0)=\mathbf{E}\,,\qquad \boldsymbol{F}(1)=\boldsymbol{0}\,,\qquad \boldsymbol{T}(1)=\boldsymbol{0}\,, \tag{2}$$

and the intermediate condition at the location (ξ) of the point mass

$$\boldsymbol{F}(\xi+)-\boldsymbol{F}(\xi-)=\mathbf{B}^T(\xi)(-\gamma_2\boldsymbol{e}_3+\Gamma\ddot{\boldsymbol{r}}(\xi)). \tag{3}$$

All variables and parameters in (1), (2) and (3) are dimensionless. They are defined from the physical quantities by

$$\tilde{s}=\frac{s}{\ell},\quad \tilde{\boldsymbol{r}}=\frac{\boldsymbol{r}}{\ell},\quad \tilde{t}=\nu t,\quad \nu=\sqrt{\frac{EJ}{m_T+m_F}},\quad \tilde{\boldsymbol{\Omega}}=\ell\boldsymbol{\Omega},\quad \tilde{\boldsymbol{T}}=\frac{\ell\boldsymbol{T}}{EJ},$$

$$\tilde{\boldsymbol{F}}=\frac{\ell^2\boldsymbol{F}}{EJ},\quad \tilde{\boldsymbol{q}}=\frac{\ell^3\boldsymbol{q}}{EJ},\quad \tilde{\delta}=\delta\nu,\quad \beta=\frac{m_F}{m_F+m_T},\quad \gamma_2=\frac{mg\ell^2}{EJ},\quad \gamma_3=\frac{GJ_T}{EJ};$$

$$\varrho=\sqrt{\frac{m_F}{EJ}}\ell U,\quad \tilde{\alpha}_i=\alpha_i\nu,\quad \gamma=\frac{\ell^3 g(m_T+m_F)}{EJ},\quad \Gamma=\frac{m}{(m_T+m_F)\ell}.$$

However, the tilde on the dimensionless quantities is dropped in (1), (2) and (3). According to Fig. 2 the deformation of the tube is described by the radius vector

$\boldsymbol{r}(s,t)$, which represents the displacement of the point s of the axis of the tube and by the rotation matrix $\mathbf{B}(s,t) \in \mathbf{SO}(3)$, which defines the orientation of the moving frame $\{\boldsymbol{t}_i\}$ with respect to the frame $\{\boldsymbol{e}_i\}$ which is fixed in space. $\boldsymbol{t}_1$ and $\boldsymbol{t}_2$ span the cross-section which is assumed to be orthogonal to the axis (no shear deformation) but may be twisted. The main parameters are ϱ, proportional to the flow rate and Γ to the point mass. ℓ is the length of the tube and $0 < \xi \leq 1$ describes the position of the mass along the tube. The resultant force and moment in the cross-section are denoted by $\boldsymbol{F}$ and $\boldsymbol{T}$. There is only a relationship between the components of $\boldsymbol{T}$ and the components of $\boldsymbol{\Omega}$ which represent curvature and twist, since the tube is assumed to be inextensible. For a slender tube we still can assume that large displacement of the tube will result in small strains and hence the simple linear Kelvin-Voigt law of viscoelasticity

$$T_1 = (\Omega_1 + \alpha_1\dot{\Omega}_1), \quad T_2 = (\Omega_2 + \alpha_2\dot{\Omega}_2), \quad T_3 = \gamma_3(\Omega_3 + \alpha_3\dot{\Omega}_3) \tag{4}$$

is stipulated. $EJ_1 = EJ_2 = EJ$ and GJ_T are bending stiffness and torsional rigidity, respectively. The material damping coefficients are $\alpha_1 = \alpha_2$ and α_3.

Our numerical investigations are based on numerical data taken from (Copeland, Moon, 1992) and concerning the internal damping from (Sugiyama et. al, 1985) where it is shown that the α_i are very small. Nevertheless it turns out that material damping has a major influence both on the qualitative and quantitative behavior of the tube at loss of stability of the straight configuration.

The equivariance conditions of the $\mathbf{O}(2)$-symmetric system are derived in (Steindl, Troger, 1995a).

3. Stability boundary in parameter space

In order to calculate the stability boundary (Troger, Steindl, 1991) in the (Γ, ϱ) - parameter plane, we perform the linearization of the equations of motion about the trivial downhanging equilibrium position

$$\boldsymbol{r} = s\boldsymbol{e}_3\,, \qquad \mathbf{B} = \mathbf{E}\,, \qquad \boldsymbol{T} \equiv \mathbf{0}\,, \qquad \boldsymbol{F} = [H(\xi - s)\,\Gamma + (1 - s)]\gamma\boldsymbol{e}_3, \tag{5}$$

which is a solution to (1) for all values of ϱ and Γ. The state (5) is invariant under all transformations in $\mathbf{O}(2) \times S^1$ (rotations and reflections about the vertical axis and arbitrary time shifts). The function H in (5) denotes the Heaviside unit step function.

The linearization explained in detail in (Steindl, Troger, 1995a) results in two identical bending equations in the two dependent variables x_1 and x_2 in the form

$$\ddot{x}_i + \delta\dot{x}_i + x_i^{IV} + \alpha_i\dot{x}_i^{IV} + 2\sqrt{\beta}\varrho\dot{x}_i' + \varrho^2 x_i'' + \gamma x_i' - F_3 x_i'' = 0 \qquad (6)$$

for $i = 1, 2$. For F_3 the proper component from (5) must be substituted. Further we have the equation $\gamma_3(\chi'' + \alpha_3\dot{\chi}'') = 0$ for the twisting angle χ. Similiarly to (6) also the boundary and jump conditions decouple

$$x_i(0) = 0, \quad x_i'(0) = 0, \quad \chi(0) = 0 \qquad (7)$$

$$\begin{aligned} x_i''(1) + \alpha_1\dot{x}_i''(1) &= 0 \\ \gamma_3(\chi'(1) + \alpha_3\dot{\chi}'(1)) &= 0 \\ x_i'''(1) + \alpha_1\dot{x}_i'''(1) &= \begin{cases} \Gamma(\gamma x_i'(1) + \ddot{x}_i(1)) & \text{if } \xi = 1 \\ 0 & \text{if } \xi < 1. \end{cases} \end{aligned} \qquad (8)$$

In case that $\xi < 1$ the jump condition at ξ must be added

$$x_i'''(\xi_+) + \alpha_1\dot{x}_i'''(\xi_+) - (x_i'''(\xi_-) + \dot{x}_i'''(\xi_-)) = -\Gamma\Big(\gamma x_i'(\xi) + \ddot{x}_i(\xi)\Big) . \qquad (9)$$

For our numerical calculations we assume that the mass is located close to the end of the tube ($\xi = .95$ and $\xi = 1$). The trivial state (5), obviously, is asymptotically stable for zero flow rate $\varrho = 0$. Now we increase the flow rate ϱ quasistatically until the trivial downhanging state of the tube becomes unstable. Varying the value of the end mass we look for zero and purely imaginary eigenvalues at the critical flow rate. However, a zero eigenvalue was never found. Hence the loss of stability always occurs due to a Hopf bifurcation. The stability boundary in (Γ, ϱ) - parameter space is shown in Fig. 3 and Fig. 4. We were very curious to see whether an experimental result reported in the paper (Copeland, Moon, 1992), which is rather unexpected, could be also reproduced from the theoretical analysis, namely that increasing the amount of the end mass from zero, first, has a destabilizing effect, and only after the end mass is increased beyond a critical value a stabilizing effect for the downhanging tube configuration can be noticed. In fact this property is also supplied by the theoretical analysis. Further we point out two interesting facts

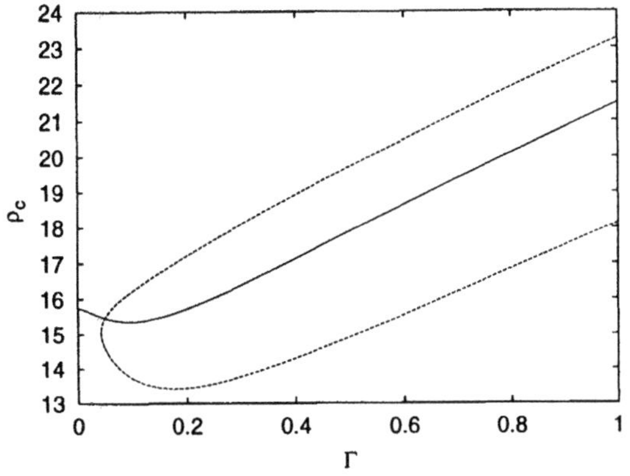

Figure 3. Stability boundaries in (Γ, ϱ)-space for $\xi = 0.95$ and $\alpha_1 = 0$. Two different curves are obtained, one of which doesn't intersect the $\Gamma = 0$ axis and would therefore be missed in simple continuation studies.

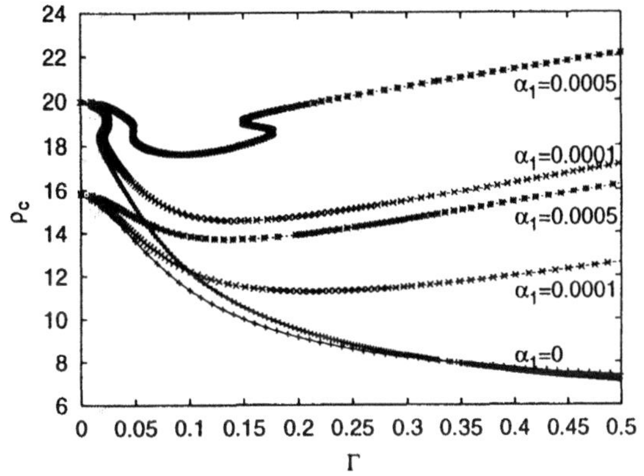

Figure 4. Stability boundaries in (Γ, ϱ)-space for $\xi = 1$ and different values of internal damping α_1. For $\alpha_1 = 0$ two boundaries corresponding to different mode numbers intersect at $\Gamma \approx 0.35$ leading to a codimension 2 bifurcation with Hopf/Hopf interaction.

to be concluded from Fig. 3 and Fig. 4. First, the strong qualitative influence which adding of slight internal damping has on the behaviour of the tube in comparison to an undamped tube model. And, second, that for the undamped model a coincident eigenvalue at $\Gamma \approx 0.35$ occurs (Fig. 4) which would result in a Hopf-Hopf interaction for the nonlinear system.

4. Center manifold reduction to bifurcation equations

The bifurcation equations, that is, the amplitude equations of the critical modes can be derived by Center Manifold theory (Troger, Steindl, 1991).

The key condition for the possibility of the reduction of the infinite dimensional system governed by partial differential equations (1) and the corresponding boundary (2) and intermediate (9) conditions to a low dimensional bifurcation system governed by ordinary differential equations is that besides the critical eigenvalues, which are located on the imaginary axis, all other eigenvalues must be strictly negative and a resolvent condition must be satisfied (Mielke, 1988).

For the calculation of the nonlinear terms in the bifurcation equations we have to note that though the bending and twisting motions decouple in the linearized case, there is a nonlinear coupling between bending and twisting in the nonlinear case, which must not be neglected (Steindl, Troger, 1995a).

Due to the symmetry properties each critical eigenvalue appears with multiplicity two and hence the bifurcation equations are a set of four first order ordinary differential equations in the generic case of a Hopf bifurcation. In the special case of a Hopf-Hopf interaction, mentioned in the caption of Fig. 4, one would obtain a system of eighth order (see for details in (Troger, Steindl, 1991)).

Here we only treat the generic case. The simplification of the equations obtained from center manifold theory due to symmetry and normal form theory is explained in detail in (Troger, Steindl, 1991) for the loss of stability at a purely imaginary pair which also applies here. Finally a set of equations of the form

$$\begin{aligned} \dot{r}_1 &= r_1(\lambda + c_1 r_1^2 + c_2 r_2^2) \\ \dot{r}_2 &= r_2(\lambda + c_2 r_1^2 + c_1 r_2^2) \end{aligned} \tag{10}$$

up to third order terms is obtained. The r_i are the amplitudes of wave motions. Stationary solutions of (10) are:

(1) $z_1 = z_2 = 0$: TS, downhanging tube
(2) $z_1 \neq 0, z_2 = 0$: TW, rotating tube with frequency ω
(3) $z_1 = z_2 \neq 0$: SW, planar oscillating tube with frequency ω
(4) $z_1 \neq 0, z_2 \neq 0$: MW, modulated rotating tube, motion on 2-torus.

Representing the solutions in the bifurcation diagram Fig. 5 we see how the two basic solutions for the Hopf bifurcation in a $\mathbf{O}(2)$-symmetric system, namely the travelling wave TW which is a rotating tube motion and the standing wave SW which is a planar tube oscillation, are partitioned in the c_1, c_2 plane. In addition their stability properties are given. We see that only two domains exist where stable planar standing oscillations (SW) or stable rotating solutions (TW) exist.

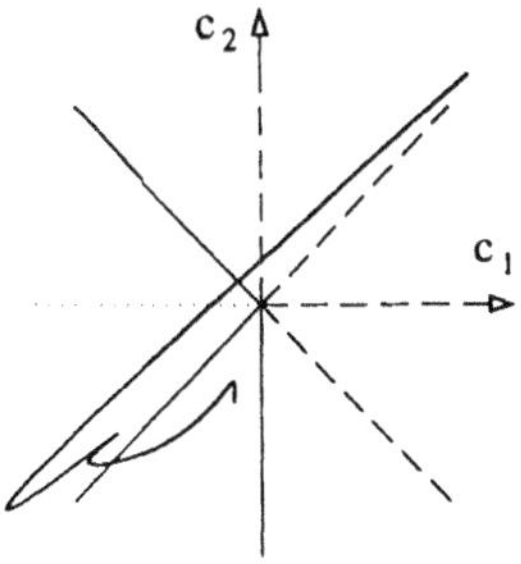

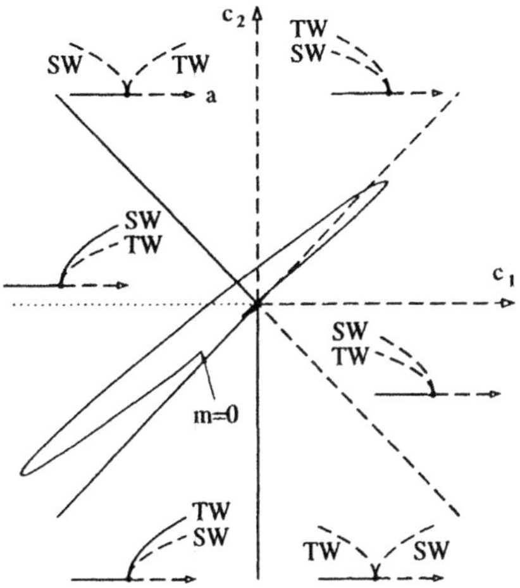

Figure 5. Bifurcation diagram for the O(2)-symmetric Hopf bifurcation.The amplitudes are shown in their dependence on the flow rate ϱ. (Full lines: stable solutions; broken lines: unstable ones). The curve in the diagram shows for increasing end mass which solution is attained. The sketch above is an enlargement about the origin

In fact the TW branch are two solutions, one rotating in clockwise direction and the other one in counter-clockwise direction. In Fig. 5 a path is show which shows which of these two solutions is present in the tube motion for increasing values of the mass. In those domains where no stable solutions are found from a third order analysis we would have to perform, first, a higher order analysis and, second, to look for the MW solution, that is the modulated wave or quasi-periodic motion.

5. Concluding remarks

The most important aspect of this analysis concerning its mathematical aspect is to show that theoretically obtained solutions are in good agreement with experimental findings reported in (Copeland, Moon, 1992). This shows that, first, our mechanical tube model obviously is quite accurate and, second, that Equivariant Bifurcation theory which replaces an infinite dimensional problem by a low dimensional system of amplitude equations is an accurate method of analysis. However, we have to mention that oviously the most important point in comparing theory with experiment is the accurate assigning of the amount of internal damping. This is a very influencial parameter both qualitatively and quantitatively.

Acknowledgement

This research project has been supported by the Austrian Science Foundation (FWF), under the project P13131-MAT.

References

Bajaj, A. K. and Sethna, P. R., 1984, *Flow Induced Bifurcations to Three-Dimensional Oscillatory Motions in Continuous Tubes,* SIAM J. Appl. Math., Vol. 44, pp. 270–286.

Copeland, G. S., Moon, F. C. 1992, *Chaotic Flow-Induced Vibration of a Flexible Tube with End Mass* AMD-vol. 152, Stability and Control of Pipes Conveying Fluid (eds. M. P. Paidoussis and N. S. Namachchivaya), pp. 63–77.

Craik A. D. D., 1985, *Wave interactions and fluid flows,* Cambridge University Press, Cambridge – London – New York.

Golubitsky, M., Stewart, I. and Schaeffer, D.,, 1985, 1988, *Singularities and Groups in Bifurcation Theory,* vol. 51 and 69 of *Appl. Math. Sciences,* Springer-Verlag, New York – Heidelberg – Berlin.

Mielke, A., 1988, *Reduction of quasilinear elliptic equations in cylindrical domains with applications,*. Mathematical Methods in the Applied Sciences, 10:51–66.

Paidoussis M. P., 1998 *Fluid - Structure Interactions, Slender Structures and Axial Flow,* Volum 1, Academic Press San Diego – London – New York.

Paidoussis, M. P. and Semler, C., 1998, *Nonlinear dynamics of a fluid-conveying cantilevered pipe with a small mass attached at the free end,* Int.J. of Non-Linear Mechanics, Vol. 33, pp. 15–32.

Steindl, A., 1992, *Hopf/Steady-State Mode Interaction for a Fluid Conveying Elastic Tube with* $\mathbf{D}_4$*-symmetric Support,* Int. Series of Num. Math., vol 104, Birkhäuser Verlag, p. 305-315.

Steindl, A. and Troger, H., 1992, *Nonlinear Three-Dimensional Oscillations of an Elastically Constrained Fluid Conveying Viscoelastic Tube with* $\mathbf{O}(2)$*-Symmetry,* AMD-vol. 152, Stability and Control of Pipes Conveying Fluid (eds. M. P. Paidoussis and N. S. Namachchivaya), pp. 47–62.

Steindl, A. and Troger, H., 1995a, *Nonlinear Three-Dimensional Oscillations of Elastically Constrained Fluid Conveying Viscoelastic Tubes with Perfect and Broken* $\mathbf{O}(2)$*-Symmetry,* Nonlinear Dynamics 7, 165-193.

Steindl, A. and Troger, H., 1995b, *One and Two-Parameter Bifurcations to Divergence and Flutter in the Three-Dimensional Motions of a Fluid Conveying Viscoelastic Tube with* $\mathbf{D}_4$*-Symmetry,* Nonlinear Dynamics 8, 161-178.

Steindl, A. and Troger, H., 1996, *Equations of motion of a fluid conveying tube* ZAMM, Vol. 76, pp. 555–558.

Sugiyama, Y., Tanaka, Y., Kishi, T. and Kawagoe, H., 1985, *Effect of a Spring Support on the Stability of Pipes Conveying Fluid,* J. Sound and Vibration, Vol. 100, pp. 257–270.

Troger, H. and Steindl, A., *Nonlinear Stability and Bifurcation Theory, An Introduction for Engineers and Applied Scientists,* Springer-Verlag, Wien – New York, 1991.

NONLINEAR DYNAMICS OF A RIGID ROTOR ON COMPLIANT JOURNAL BEARINGS

B.L. VAN DE VRANDE AND D.H. VAN CAMPEN
Department of Mechanical Engineering
Eindhoven University of Technology
P.O. Box 513, 5600 MB Eindhoven, The Netherlands

Abstract. This paper presents the influence of bearing liner compliance on the long-term dynamics of a symmetric rigid rotor on short journal bearings. Both balanced and unbalanced rotors are investigated for different values of the compliance, using (nondimensional) rotor speed as the bifurcation parameter. The results show a decrease in the onset speed of instability of the balanced rotor, and a considerable influence on the 1/2 subharmonic solutions in case of unbalance excitation.

1. Introduction

In the classical short journal bearing model, the bearing surface is assumed to be rigid, and the bearing reaction force can be expressed as an analytical function of the journal position and velocity. However, if the bearing is lined with an elastic material, deformations are no longer negligible, and the classical model fails. In this paper, a compliant bearing model is presented and applied to both balanced and unbalanced symmetric rigid rotors on short journal bearings.

Childs *et al.* (1977) introduced the journal bearing impedance method for rotordynamic applications, defining an analytical description of the bearing reaction force as a function of the journal position and velocity. A list of impedance definitions was published by Moes and Bosma (1981). Van de Vorst *et al.* (1996) investigated the long-term dynamics of a flexible rotor-bearing system with a journal bearing, using the impedance method. Higginson (1965) studied the static characteristics of a long bearing with a thin elastic liner. Nilsson (1978) determined dynamic coefficients of a finite length 180° partial arc journal bearing with a thin liner, lubricated by

N. Van Dao and E.J. Kreuzer (eds.),
IUTAM Symposium on Recent Developments in Non-linear Oscillations of Mechanical Systems, 301–309.

a fluid with a pressure dependent viscosity. Mao *et al.* (1983) and Zhang *et al.* (1986) calculated dynamic coefficients of a finite length compliant journal bearing, using the FEM to model bearing elasticity. A model of a dynamically loaded flexible short journal bearing was presented by Van der Tempel *et al.* (1985). With this model, pressure and film thickness distributions can be calculated if the load on the bearing is known as a function of time. Therefore, it is unsuitable for rotordynamic applications.

In contrast to bearings with a rigid surface, where the lubricant film thickness is a function of the position of the shaft, the film thickness of a compliant bearing is an unknown and has to be included as a DOF in the model of the rotor-bearing system. The time derivative of the film thickness follows from the Reynolds equation, and is added to the equations of motion of the rotor. The lubricant pressure is found from the film thickness, using the elastic model of the bearing liner, and integration of the pressure distribution yields the bearing reaction force.

AUTO 97 (Doedel *et al.*, 1998) is used to perform the calculations for different values of the bearing liner compliance, using (nondimensional) rotor speed as the bifurcation parameter. Time is discretized with 10 and 20 mesh intervals for harmonic and 1/2 subharmonic solutions, respectively, with 4 collocation points per interval. The results show that increasing the bearing liner compliance decreases the onset speed of instability of the balanced rotor. Also, a considerable influence of the compliance is found on the 1/2 subharmonic solutions in case of unbalance excitation.

2. Compliant Short Journal Bearing Model

Figure 1 shows the geometry of a plain journal bearing with an elastically deformed liner, where the dashed circle represents the undeformed situation. In the center of the undeformed bearing, the origin of the stationary x, y, z-system is located. The circumferential coordinate θ is measured from the positive y-axis. The journal with radius R is rigid and rotates with constant angular speed Ω. The position of its center is given by the eccentricity vector $\vec{e}$, with attitude angle γ. The radial clearance of the undeformed bearing is given by C, the bearing liner deformation is measured by U, and the lubricant film thickness by H.

To find H, the cosine rule of triangles is applied to Fig. 1:

$$(R+H)^2 = (R+C+U)^2 + e^2 - 2(R+C+U)e\cos(\pi - \theta + \gamma) \qquad (1)$$

Expanding this equation, dividing by R^2, and discarding 2nd order terms in H/R, C/R, e/R, and U/R yields:

$$H = C + e\cos(\theta - \gamma) + U = C + e_x \sin\theta - e_y \cos\theta + U \qquad (2)$$

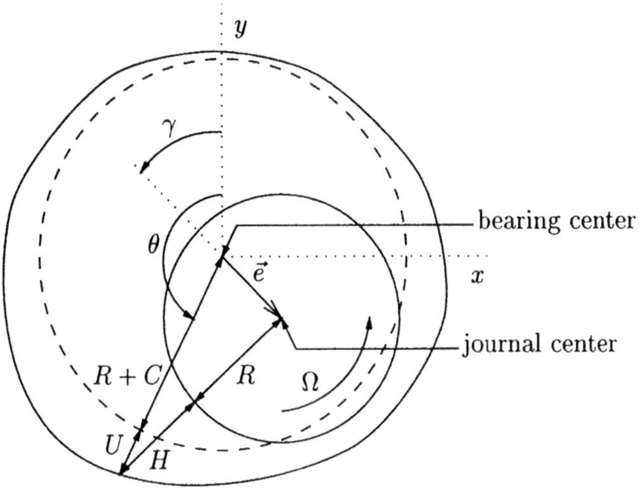

Figure 1. Plain journal bearing geometry

where e_x and e_y are the eccentricity components in the x- and y-directions, respectively. Introducing the nondimensional quantities $h = H/C$, $\epsilon = e/C$, and $u = U/C$, this equation is rewritten into:

$$h = 1 + \epsilon_x \sin\theta - \epsilon_y \cos\theta + u \tag{3}$$

The pressure p in the lubricant film is governed by the Reynolds equation for short journal bearings:

$$\frac{\partial}{\partial z}\left(H^3 \frac{\partial p}{\partial z}\right) = 6\mu\left(\Omega \frac{\partial H}{\partial \theta} + 2\dot{H}\right) \tag{4}$$

where μ is the lubricant viscosity, and an overdot denotes differentiation w.r.t. time t. Using the nondimensional quantities $\Omega^* = \Omega/\Omega_0$ (Ω_0 will be defined later), $p^* = (C/R)^2 p / 6(L/D)^2 \mu \Omega_0$ (L and D are the bearing length and journal diameter, respectively), $z^* = z/L$, and $\tau = \Omega_0 t$, this equation becomes:

$$\frac{\partial}{\partial z^*}\left(h^3 \frac{\partial p^*}{\partial z^*}\right) = 4\left(\Omega^* \frac{\partial h}{\partial \theta} + 2h'\right) \tag{5}$$

where a prime denotes differentiation w.r.t. τ. In lubrication literature, Ω is normally used to scale the equations; here however, Ω_0 is used for that purpose because Ω (actually Ω^*) will be used as bifurcation parameter.

Solving Eq. (5) yields for the (average) pressure:

$$p^* = -\frac{1}{3h^3}\left(\Omega^* \frac{\partial h}{\partial \theta} + 2h'\right) \tag{6}$$

Assuming plain strain and Poisson's ratio not close to 0.5, the deformation of a thin elastic liner on a rigid backing can be approximated by:

$$U = \frac{pd(1+\nu)(1-2\nu)}{E(1-\nu)} = \frac{p}{\lambda} \tag{7}$$

(Armstrong, 1986 and Hlaváček and Vokoun, 1993), where d, E, ν, and λ are the liner thickness, Young's modulus, Poisson's ratio, and stiffness, respectively. Introducing the nondimensional bearing liner compliance $B = 6(L/D)^2\mu\Omega_0/(C/R)^2\lambda C$, this equation is rewritten into:

$$u = Bp^* \tag{8}$$

For a rigid bearing surface, $B = 0$.

The time derivative of the film thickness is found by rearranging Eq. (6):

$$h' = -\tfrac{1}{2}\left(3h^3p^* + \Omega^*\frac{\partial h}{\partial \theta}\right) \tag{9}$$

Substituting Eq. (8) in (3), p^* can be expressed as a function of h, ϵ_x, and ϵ_y, so with the previous equation, h' can be expressed as a function of h, $\partial h/\partial\theta$, ϵ_x, and ϵ_y. By space discretization in the θ-direction to approximate $\partial h/\partial\theta$, a set of 1st order ODEs results for the discretized film thicknesses. Here, a 2nd order backward difference quotient is used to approximate $\partial h/\partial\theta$, discretizing with 50 mesh intervals, and using periodicity as the boundary condition.

The bearing reaction force components in the x- and y-directions are found by integration of the pressure distribution:

$$\begin{cases} F_x = LR\int p\sin\theta\,\mathrm{d}\theta \\ F_y = -LR\int p\cos\theta\,\mathrm{d}\theta \end{cases} \tag{10}$$

Introducing the nondimensional force $F^* = (C/R)^2F/(L/D)^2\mu\Omega_0 LD$, the previous equation is rewritten into:

$$\begin{cases} F_x^* = 3\int p^*\sin\theta\,\mathrm{d}\theta \\ F_y^* = -3\int p^*\cos\theta\,\mathrm{d}\theta \end{cases} \tag{11}$$

where again p^* can be expressed as a function of h, ϵ_x, and ϵ_y. The force components are included in the 2nd order equations of motion for ϵ_x and ϵ_y.

As an approximate cavitation model, negative pressures are set equal to zero in the evaluation of the integrals in Eq. (11), while in Eq. (9), negative values of p^* are allowed. This model introduces no discontinuities in the ODEs, and gives a good approximation of the full (discontinuous) cavitation model, in which constraints are used to prevent the discretized film thicknesses from causing negative pressures. Simpson's Rule is used to approximate the integrals.

The set of 1st order ODEs for the discretized film thicknesses is added to the 2nd order equations of motion of the rotor, to include the compliant short journal bearing in the model of the rotor-bearing system. The equations are coupled because of the fact that h' is a function of ϵ_x and ϵ_y, while ϵ_x'' and ϵ_y'' are functions of F_x^* and F_y^*, respectively, which are functions of h.

The nondimensional reaction force components in the x- and y-directions of a rigid-surface short journal bearing are given in appendix A.

3. Symmetric Rigid Rotor

Figure 2 shows a symmetric rigid rotor of mass m with unbalance a that is supported by two identical short journal bearings. A constant load F_0 is

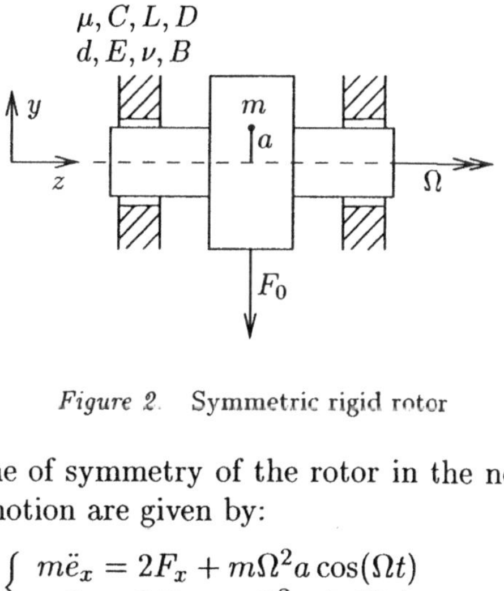

Figure 2. Symmetric rigid rotor

applied in the plane of symmetry of the rotor in the negative y-direction. The equations of motion are given by:

$$\begin{cases} m\ddot{e}_x = 2F_x + m\Omega^2 a\cos(\Omega t) \\ m\ddot{e}_y = 2F_y + m\Omega^2 a\sin(\Omega t) - F_0 \end{cases} \tag{12}$$

Defining now Ω_0 by $\sqrt{F_0/mC}$, this equation is rewritten into:

$$\begin{cases} \epsilon_x'' = 2F_x^*/F_0^* + \Omega^{*2}a^*\cos(\Omega^*\tau) \\ \epsilon_y'' = 2F_y^*/F_0^* + \Omega^{*2}a^*\sin(\Omega^*\tau) - 1 \end{cases} \tag{13}$$

where $a^* = a/C$. All calculations are done with $F_0^* = 1$.

3.1. BALANCED ROTOR

Figure 3 shows the bifurcation diagrams of the balanced rotor ($a^* = 0$) for different values of B, where $\max|\epsilon_y|$ is plotted against Ω^*. Stable and un-

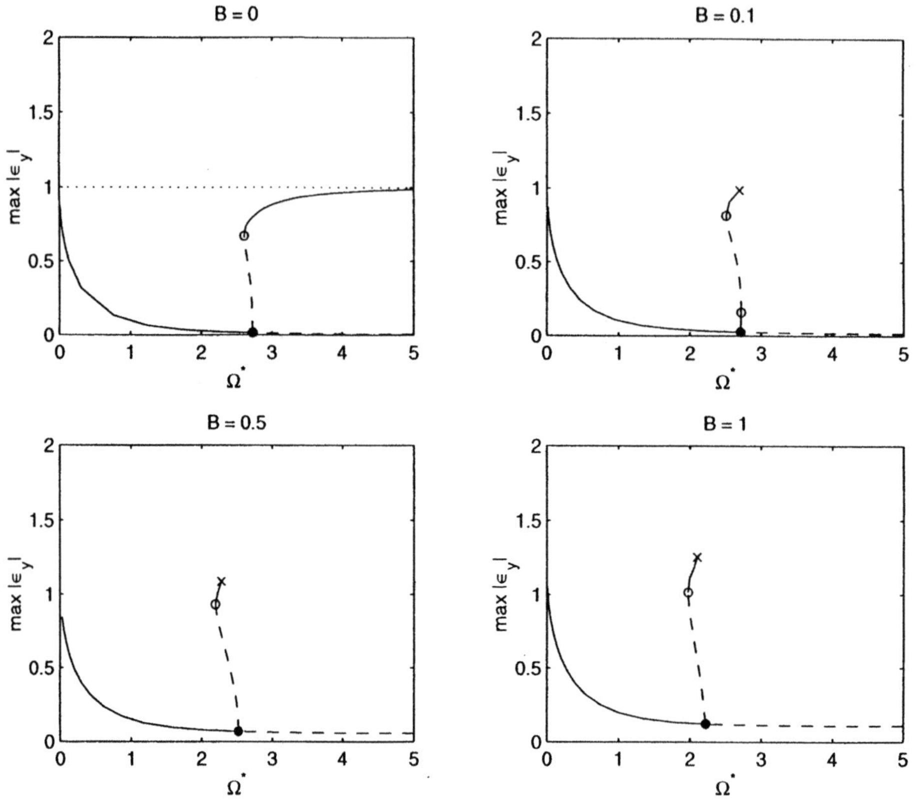

Figure 3. Bifurcation diagrams of balanced rotor

stable solution branches are depicted by solid and dashed lines, respectively; bifurcations and cyclic folds by dots and circles, respectively. The stationary solution branches become unstable via 1st Hopf bifurcations, given in Table 1. It can be seen from this table that the value of Ω^*, at which the

	Balanced rotor Ω^*		Unbalanced rotor ($a^* = 0.2$) Ω^*	
B	1st Hopf	cyclic fold	flip	cyclic fold
0	2.74	2.61	2.87, 5.76	≈ 20
0.1	2.71	2.50, 2.72	4.28, 5.96	6.87
0.5	2.52	2.19	3.29, 4.13	3.72, 4.14
1	2.22	1.97	2.35, 3.08	2.61, 3.09

TABLE 1. Bifurcation values

1st Hopf bifurcation occurs, (the onset speed of instability) decreases for increasing bearing liner compliance. The periodic solution branches emanating from the 1st Hopf bifurcations bend to the left, and become stable via cyclic folds, also given in Table 1. Therefore, stable periodic solutions with large values of $\max|\epsilon_y|$ exist for Ω^* less than the onset speed of instability. In contrast to the other values of B, for $B = 0.1$, the periodic solution branch starts stable and becomes unstable via a second cyclic fold.

At the points indicated in Fig. 3 by crosses, the calculation of the periodic solution branches is stopped, because of negative film thicknesses. Increasing the number of mesh intervals in the θ-direction to improve accuracy will probably solve this problem, but this is not investigated here.

3.2. UNBALANCED ROTOR

Figure 4 shows the bifurcation diagrams of an unbalanced rotor with $a^* = 0.2$ for different values of B. The harmonic solution branches possess two

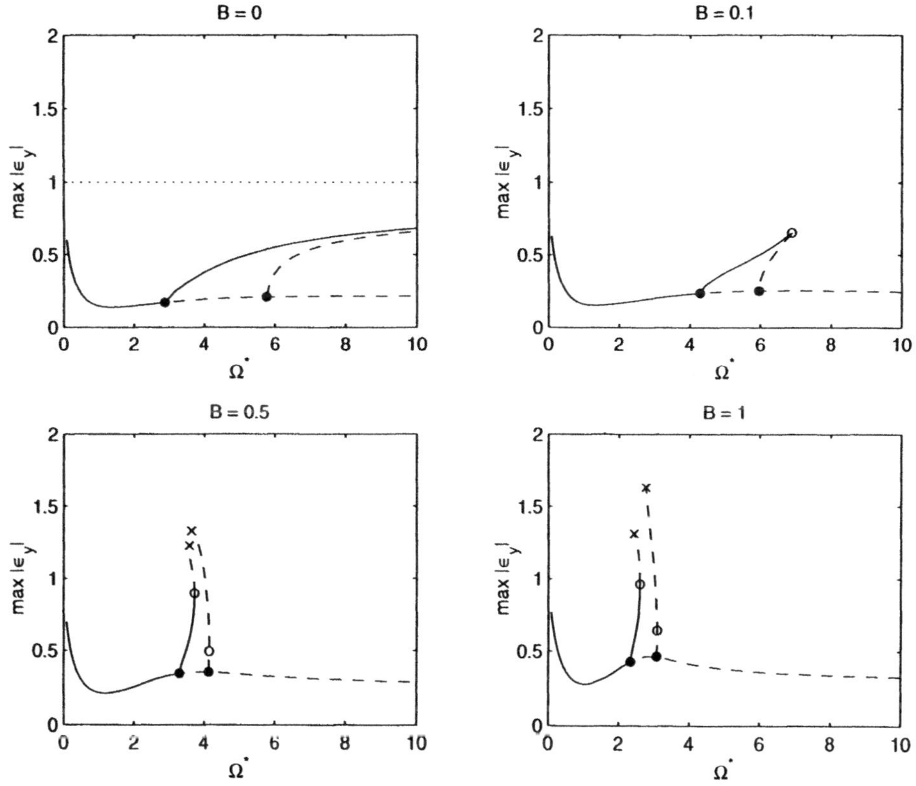

Figure 4. Bifurcation diagrams of unbalanced rotor ($a^* = 0.2$)

flip bifurcations, given in Table 1. It can be seen from this table that the value of Ω^*, at which the harmonic solution branch loses stability, first increases and then decreases below the value for $B = 0$, for increasing bearing liner compliance. At the second flip bifurcations, the harmonic solution branches stay unstable because one Floquet multiplier has a modulus greater than 1, while another multiplier crosses -1. The 1/2 subharmonic solution branches that emanate from these bifurcation are also unstable, and three unstable branches meet at these points. The cyclic folds in the 1/2 subharmonic solution branches emanating from the flip bifurcations are also given in Table 1. For $B = 0.5$ and 1, the 1/2 subharmonic solution branches stay unstable at the second cyclic folds because, again, one Floquet multiplier has a modulus greater than 1, while in these cases another multiplier crosses $+1$.

Again, at the points indicated by crosses, the calculations are stopped because of negative film thicknesses. For $B = 0$ and $\Omega^* > 20$, quasi-periodic behavior was found with $\max|\epsilon_y|$ close to 1. For $B > 0$, film thicknesses become negative in the regions where no stable solutions are found, if the system is integrated numerically.

4. Discussion and Conclusions

A model of a compliant short journal bearing is presented with approximate elasticity and cavitation models. The results show that if the bearing liner compliance is increased, the onset speed of instability of the balanced rigid rotor decreases. The compliance also has a considerable influence on the 1/2 subharmonic solution branch in case of unbalance excitation.

At some points during the calculations, film thicknesses become negative. Increasing the number of mesh intervals in the circumferential direction will probably solve this problem, but this is not investigated here.

In future research, an efficient continuation algorithm (see Lust and Roose, 1999) for PDEs (such as the Reynolds equation) will be used to save computation time.

References

Armstrong, C.G. (1986) An Analysis of the Stresses in a Thin Layer of Articular Cartilage in a Synovial Joint, *Engineering in Medicine* **15**, 55–61.

Childs, D., Moes, H., and Van Leeuwen, H. (1977) Journal Bearing Impedance Descriptions for Rotordynamic Applications, *Journal of Lubrication Technology* **99**, 198–210.

Doedel, E.J., Champneys, A.R., Fairgrieve, T.F., Kuznetsov, Y.A., Sandstede, B., and Wang, X. (1998) *AUTO 97: Continuation and Bifurcation Software for Ordinary Differential Equations*, Concordia University, Montreal, Canada.

Higginson, G.R. (1965) The Theoretical Effects of Elastic Deformation of the Bearing Liner on Journal Bearing Performance, *Proc. IMechE (Elastohydrodynamic Lubrication)* **180**, 31–38.

Hlaváček, M. and Vokoun, D. (1993) Lubrication of a Cylindrical Synovial Joint Considering Rolling Motion and Elastic Incompressible Cartilage, *Wear* **165**, 1–7.

Lust, K. and Roose, D. (1999) Computation and Bifurcation Analysis of Periodic Solutions of Large-Scale Systems, in *Large-Scale Dynamical Systems*, IMA Volumes in Mathematics and its Applications. Submitted. (see http://www.cs.kuleuven.ac.be/~kurt/)

Mao, Q., Han, D.-C., and Glienicke, J. (1983) Stabilitätseigenschaften von Gleitlagern bei Berücksichtigung der Lagerschalenelastizität (Stability Properties of Journal Bearings by Taking into Account the Bearing Shell Elasticity), *Konstruktion* **35**, 45–52, In German.

Moes, H. and Bosma, R. (1981) Mobility and Impedance Definitions for Plain Journal Bearings, *Journal of Tribology* **103**, 468–470.

Nilsson, L.R.K. (1979) The Influence of Bearing Flexibility on the Dynamic Performance of Radial Oil Film Bearings, in D. Dowson *et al.* (eds.), *Elastohydrodynamics and Related Topics (Proc. 5th Leeds-Lyon Symp. Trib.)*, 311–319.

Van der Tempel, L., Moes, H., and Bosma, R. (1985) Numerical Simulation of Dynamically Loaded Flexible Short Journal Bearings, *Journal of Tribology* **107**, 396–401.

Van de Vorst, E.L.B., Fey, R.H.B., De Kraker, A., and Van Campen, D.H. (1996) Steady-State Behaviour of Flexible Rotordynamic Systems with Oil Journal Bearings, *Nonlinear Dynamics* **11**, 295–313.

Zhang, Z., Mao, Q., and Xu, H. (1986) The Effect of Dynamic Deformation on Dynamic Properties and Stability of Cylindrical Journal Bearings, in D. Dowson *et al.* (eds.), *Fluid Film Lubrication—Osborne Reynolds Centenary (Proc. 13th Leeds-Lyon Symp. Trib.)*, 363–366.

A. Rigid-Surface Short Journal Bearing Force

The nondimensional reaction force components in the x- and y-directions of a rigid-surface short journal bearing are found from the impedances given by Moes and Bosma (1981):

$$\left\{ \begin{array}{l} F_x^* = -v_s(W_\xi^* \cos\zeta - W_\eta^* \sin\zeta) \\ F_y^* = -v_s(W_\xi^* \sin\zeta + W_\eta^* \cos\zeta) \end{array} \right. \tag{14}$$

where $v_s = \sqrt{v_{s,x}^2 + v_{s,y}^2}$, $\zeta = \arctan(v_{s,y}/v_{s,x})$,

$$\left\{ \begin{array}{l} v_{s,x} = \epsilon_x' + \frac{1}{2}\Omega^* \epsilon_y \\ v_{s,y} = \epsilon_y' - \frac{1}{2}\Omega^* \epsilon_x \end{array} \right. \tag{15}$$

and

$$W_\xi^* = \frac{6\{\frac{1}{3}(1 + 2\xi^2 - \eta^2)T + \xi\}}{(1-\epsilon^2)^2} \qquad W_\eta^* = \frac{6\eta(\xi T + 1 - \frac{1}{3}E)}{(1-\epsilon^2)^2} \tag{16}$$

where $T = 2\arctan\{\sqrt{1-\epsilon^2}/(\sqrt{1-\eta^2} - \xi)\}/\sqrt{1-\epsilon^2}$, $E = (1-\epsilon^2)/(1-\eta^2)$, $\xi = \epsilon\cos\alpha$, $\eta = \epsilon\sin\alpha$, $\alpha = \beta - \zeta$, and $\beta = \arctan(\epsilon_y/\epsilon_x)$.

NONLINEAR DYNAMIC BEHAVIOR OF SHAPE MEMORY ALLOY OSCILLATORS

F. VESTRONI and D. BERNARDINI

Università di Roma 'La Sapienza'
Dipartimento di Ingegneria Strutturale e Geotecnica
Via Eudossiana 18, 00184 Roma (Italy)

1. Introduction

Shape Memory Alloys (SMA) are successfully used in various fields of engineering. Recently they are receiving attention also for the realization of vibration reduction devices (Van Humbeeck, 1998, Clark *et al.*, 1995). This interest is mainly due to the fact that SMA exhibit, under cyclic loading, a peculiar hysteretic behavior that makes it possible to avoid, or recover, the residual strains at unloading. Such a behavior is quite different with respect to most common metals since the strains associated to the hysteresis loops are not due to irreversible crystallographic alterations, but rather, to the occurrence of thermoelastic phase transformations (Otsuka *et al.*, 1986). The resulting mechanical behavior is highly nonlinear and, in addition, under dynamical loads, relevant temperature variations are observed. Such thermal effects strongly affect the hysteresis loop and often become a relevant constraint in the design of applications. Despite SMA are often used under dynamical loads, their nonlinear dynamics has been only little studied. Some of the proposed studies address the problem making use of isothermal models (Feng *et al.*, 1996, Thomson *et al.*, 1995), while others adopt more refined models but without discussing dynamical aspects (Oberaigner *et al.*, 1995).

In this paper the forced nonlinear oscillations of a single degree of freedom oscillator with restoring force provided by a SMA device are studied taking into account the full thermomechanical behavior. The dynamics of the system turns out to be described by a four-dimensional differential equations system that is numerically solved to find the stationary solutions using an extended harmonic balance procedure. The dynamic characterization of the hysteretic system is then performed through the analysis of the frequency-response curves for different excitation amplitudes and different values of the material parameters. The comparison between isothermal and non-isothermal response highlights the influence of thermal aspects on the dynamics and the differences with respect to the classical hysteretic oscillators (Capecchi and Vestroni, 1990, Vestroni and Capecchi 1997).

N. Van Dao and E.J. Kreuzer (eds.),
IUTAM Symposium on Recent Developments in Non-linear Oscillations of Mechanical Systems, 311–320.

2. Shape Memory Alloys behavior

Most of the SMA applications are developed by taking advantage of the two effects of *pseudoelasticity* and *shape memory* (Van Humbeeck, 1998).

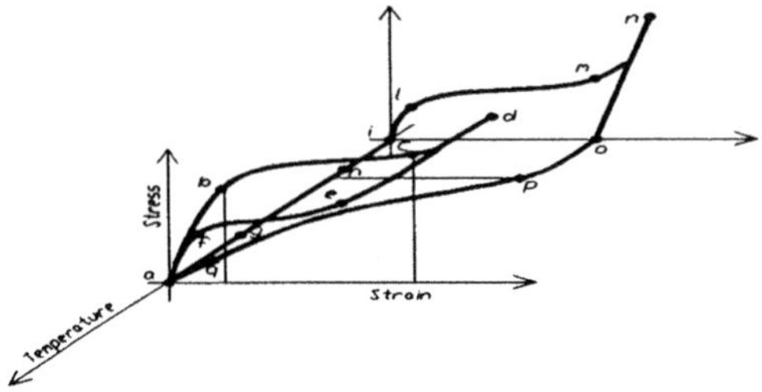

Figure 1. Typical thermomechanical uniaxial response of a SMA element.

The above two effects are due respectively to stress or thermally-induced martensitic transformations between two solid phases called Martensite (M) and Austenite (A). In Figure 1 the path *abcdefa* corresponds to *pseudoelastic* behavior that is characterized by the absence of residual strain at unloading. The path *ilmnoi* looks like a conventional plastic behavior but the residual strain can be recovered by heating the material, i.e. through the path *opqa* (*shape memory* effect).

In this work, oscillators with pseudoelastic restoring force are considered. In such a case, the forward transformation ($A \rightarrow M$) takes place during loading, while the reverse transformation ($M \rightarrow A$) during unloading.

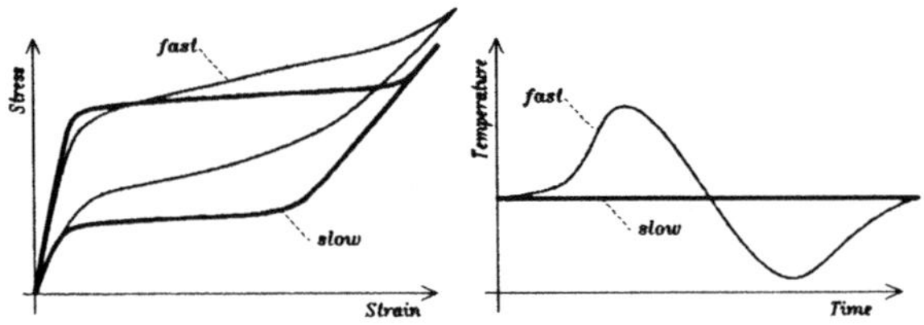

Figure 2. Influence of the loading rate on the mechanical and thermal response.

Under mechanical loads SMA show temperature variations increasing with the loading rate, so that, only in the limit case of very slow loads the material can be considered to behave isothermally as usually done in literature. When such temperature variations take place, the material response deviates from the isothermal conditions as qualitatively shown in Figure 2. Further details on the loading rate effect, supported by experimental results, can be found in (Bernardini and Van Humbeeck, 1999). It therefore appears that, in the study of the SMA dynamic behavior, the hypothesis of isothermal conditions is inappropriate, since the loading rate influences both the hardening and the hysteresis loop shape that, in turn, are important factors in determining the dynamic response.

3. Model for the material behavior

Several models exist to describe pseudoelasticity, see e.g. Huo and Muller, 1993. Here the interest is focused in capturing the non-isothermal behavior and the loading rate dependence of the mechanical response. To this end, a simple thermodynamic model where the phase transformation progress is taken into account by the evolution of the martensite fraction z is used to describe the force-displacement-temperature $(\widetilde{f}, \widetilde{x}, \widetilde{\vartheta})$ response of a SMA element in an environment at constant temperature $\widetilde{\vartheta}_A$. In view of successive adimensionalizations, the tilde is used to denote dimensional quantities. The formulation of the model is only briefly outlined, while details can be found in (Bernardini, 1998). For the development of the model it is useful to present it in a nondimensional form using $\widetilde{f}_r = 0, \widetilde{x}_r = 0, \widetilde{\vartheta} = \widetilde{\vartheta}_r$ as reference state. While for elastoplastic systems, forces and displacements are usually normalized with respect to their values at yielding, in the present case the counterpart of yielding, that is the forward transformation start, is not a constant since it depends on the environment temperature. Forces and displacements are then normalized with respect to the values, denoted by the subscript Ms, assumed at the forward transformation start at $\widetilde{\vartheta}_r$, while temperature is normalized with respect to $\widetilde{\vartheta}_A$

$$x = \frac{\widetilde{x}}{\widetilde{x}_{Ms}}, \qquad f = \frac{\widetilde{f}}{\widetilde{f}_{Ms}}, \qquad \vartheta = \frac{\widetilde{\vartheta} - \widetilde{\vartheta}_A}{\widetilde{\vartheta}_A} \tag{1}$$

The basic expression for the constitutive equation relating nondimensional restoring force and displacement is derived in such a way to give a symmetric behavior with respect to x

$$f = x - \operatorname{sign}(x)\, \lambda z \qquad \text{with } \lambda = \frac{\widetilde{\delta}}{\widetilde{x}_{Ms}} \tag{2}$$

where $\widetilde{\delta}$ is the maximum displacement due to phase transformation and $\widetilde{x}_{Ms}$ the value of the displacement at forward transformation start at reference temperature.

The activation of the phase transformation in SMA is governed by the quantity Π, called driving force

$$\Pi\,(x, \vartheta, z) = |x| + LJ\,(1 + \vartheta - \vartheta_r) - 1 + Pz \tag{3}$$

the above expression depends on the following nondimensional quantities

$$P = \mu - \lambda, \qquad \mu = \frac{2A\lambda}{E\widetilde{\delta}^2}, \qquad L = -\frac{\Delta\eta_0}{c}, \qquad J = \frac{c\widetilde{\vartheta}_A}{\widetilde{\delta}\sigma_{Ms}}, \qquad \vartheta_r = \frac{\widetilde{\vartheta}_r}{\widetilde{\vartheta}_A} \tag{4}$$

where E is the elastic modulus, $\Delta\eta_0$ is the entropy difference between the phases at the reference state, c is the specific heat, σ_{Ms} the forward transformation start stress at $\widetilde{\vartheta}_r$.

The activation of the phase transformations is assumed to take place when Π reaches a threshold value determined from experiments

$$\begin{array}{ll} \Pi \geq -BPz & \textit{forward} \text{ transformation} \\ \Pi \leq BP\,(1 - z) & \textit{reverse} \text{ transformation} \end{array} \tag{5}$$

where B is a further nondimensional material parameter governing the features of the subloops internal to the main hysteresis loop.

The prescription of an evolution equation for the martensite fraction z is a purely phenomenological task and several expressions are available in literature. For the sake of simplicity, a simple linear expression relating the phase fraction rate to the displacement and temperature rates is used

$$\dot{z} = \frac{1}{\lambda + H\mu} \left[(\text{sign}\,(x))\,\dot{x} + (LJ)\,\dot{\vartheta} \right] \tag{6}$$

where H is a nondimensional material parameter governing the hardening features of the phase transformation. In particular if $H = 0$ then an ideal pseudoelastic behavior is obtained, while values of $H > 1$ give rise to hardening.

Finally, equations (2) and (6) together with the activation conditions (5) for the driving force (3) furnish the complete set of thermomechanical constitutive equations of the SMA device.

4. Outline of the model behavior

In the following, the main features of the material behavior that can be reproduced by the model are presented.

In *isothermal* conditions, for complete loading-unloading cycles, the hysteresis loop shape is regulated by the two parameters λ, μ governing respectively the 'length' of the pseudoelastic plateaus and the ratio between the forward and reverse transformation forces, as shown in Figure 3.

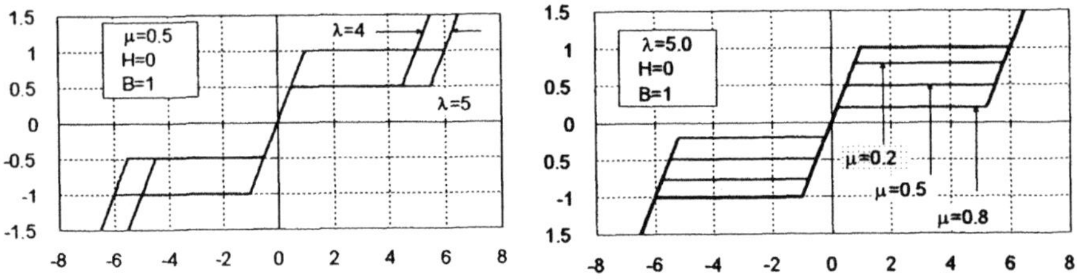

(a) (b)

Figure 3. Hysteretic force-displacement loops for different values of λ (a), μ (b)

If the loading history involves subcycles giving rise to incomplete phase transformations, then a feature typical of SMA is observed. Within the main outer hysteresis loop smaller, internal subloops take place. In this model the features of such subcycles are regulated by the parameter B. In particular, when $B = 0$ the model reduces to the classical Muller's model where phase transformations always activate on the diagonal line of the outer loop (Huo and Muller, 1993), while values of $B < 1$ lead to different activation modalities for internal subloops as it is qualitatively shown in Figure 4.

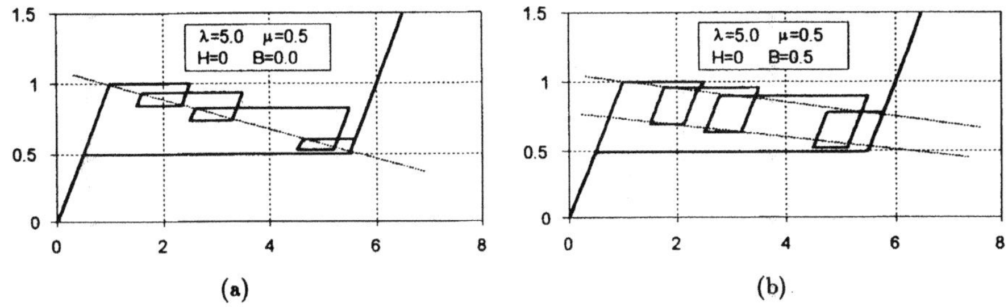

Figure 4. Hysteretic loops for B=0 (a) and B=0.5 (b)

When *non-isothermal* conditions are considered, two main differences arise in comparison with the previous cases: additional hardening and hysteresis loops shape modification. The influence of the thermal aspects is scaled by the parameter h_C which makes it possible to recover the two limit cases of isothermal ($h_C \to \infty$) and adiabatic ($h_C = 0$) conditions, while different values deliver intermediate conditions accounting for the loading rate dependence. The non-isothermal response is regulated by the two parameters L, J that have a combined effect as shown in Figure 5, where also the isothermal behavior is plotted for comparison. The variation of L, J clearly influences also the response in terms of temperature as shown in Figure 5.

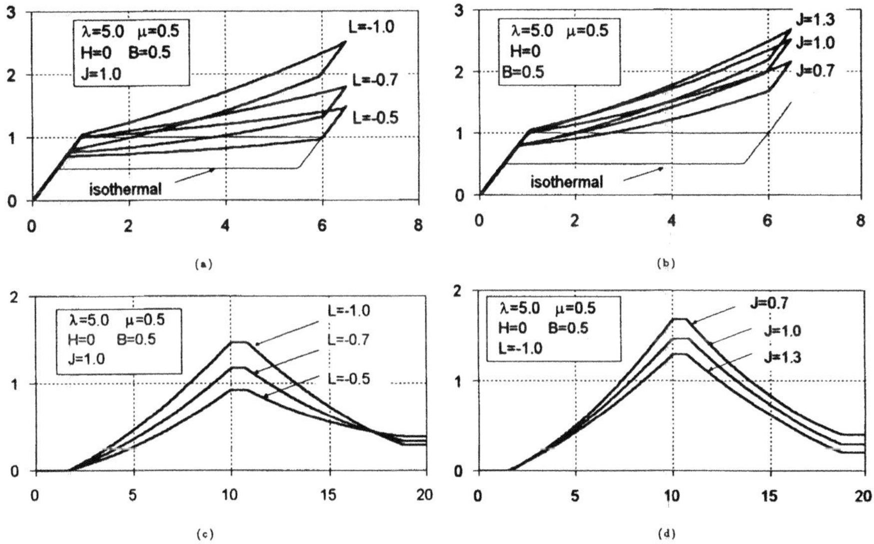

Figure 5. Non-isothermal force-displacement (a,b) and temperature-time (c,d) curves for different L,J.

5. Governing equations

The SMA oscillator is a mechanical system where a mass m can undergo a displacement $\widetilde{x}$ being counteract by a pseudoelastic restoring force $\widetilde{f}$ provided by a SMA device. The equations governing the dynamics of the system are given by two balance equations. Differently with respect to purely mechanical oscillators here the temperature have to be introduced within the state vector and the energy conservation equation have to be added to the system.

The equation of motion of the system under single harmonic excitation is given, in dimensional form, by

$$m\ddot{\tilde{x}} + c\dot{\tilde{x}} + \tilde{f} = \tilde{F}\cos\omega t \tag{7}$$

where $\tilde{f}$ is the restoring force, c the damping coefficient and $\tilde{F}, \omega$ the forcing amplitude and frequency. The energy conservation equation in dimensional form is

$$c\dot{\tilde{\vartheta}} + \left(\tilde{\Pi} - \Delta\eta_0\tilde{\vartheta}\right)\dot{z} = \dot{\tilde{Q}} \tag{8}$$

where $\dot{\tilde{Q}}$ is the thermal energy exchange rate with the environment, whose expression needs to be specified by phenomenological considerations. The linear expression $\dot{\tilde{Q}} = h_C\left(\tilde{\vartheta} - \tilde{\vartheta}_A\right)$ is used here to model convective heat exchange. After introducing the following nondimensional quantities

$$\omega_0^2 = \frac{k}{m}, \quad \tau = \omega_0 t, \quad \zeta = \frac{c}{2\omega_0 m}, \quad F = \frac{\tilde{F}}{\tilde{f}_{Ms}}, \quad \alpha = \frac{\omega}{\omega_0} \quad h_C = \frac{\tilde{h}_C}{\omega c} \tag{9}$$

with k being the initial elastic stiffness, and using the nondimensional temperature defined in (1), then the nondimensional balance equations turn out to be

$$\begin{aligned} \ddot{x}(\tau) + 2\zeta\dot{x}(\tau) + f(x(\tau), \vartheta(\tau)) &= F\cos\alpha\tau \\ \dot{\vartheta}(\tau) + \Gamma(x(\tau), \vartheta(\tau)) &= h_C\vartheta(\tau) \end{aligned} \tag{10}$$

where the quantity Γ is

$$\Gamma = \frac{\left(-\Delta\eta_0\tilde{\vartheta} + \tilde{\Pi}\right)\dot{z}}{c\tilde{\vartheta}_A} = \left[L(1+\vartheta) + \frac{\Pi}{J}\right]\dot{z} \tag{11}$$

The governing nondimensional equations can be written as a differential system in the state vector $[x, v, \vartheta, z]$

$$\begin{aligned} \dot{x} &= v \\ \dot{v} &= -x + \operatorname{sign}(x)\lambda z - 2\zeta v + F\cos\alpha\tau \\ \dot{z} - \frac{\operatorname{sign}(x)}{\lambda + H\mu}\dot{x} - \frac{LJ}{\lambda + H\mu}\dot{\vartheta} &= 0 \\ \dot{\vartheta} + \frac{LJ(1+\vartheta) + \Pi(x, \vartheta, z)}{J}\dot{z} &= h_C\vartheta \end{aligned} \tag{12}$$

6. Harmonic balance solution

Since the aim of the present work is the study of the stationary response, the solutions of the differential equations system (12) are searched by means of an extended harmonic balance method. In order to apply the method, it is useful to start from system (10) where f and Γ denote respectively the pseudoelastic restoring force and the heat associated to the phase transformation. Both these quantities are history-dependent through their dependence on z. The stationary response, given by the pair of displacement and temperature functions, is approximated by a trigonometric series truncated after N terms

$$\begin{aligned} x(\tau) &= \frac{a_0}{2} + \sum_{n=1}^{N} a_n \cos n\alpha\tau + b_n \sin n\alpha\tau \\ \vartheta(\tau) &= \frac{c_0}{2} + \sum_{n=1}^{N} c_n \cos n\alpha\tau + d_n \sin n\alpha\tau \end{aligned} \tag{13}$$

In a similar way the history-dependent quantities are approximated as

$$\begin{aligned} f(x(\tau), \vartheta(\tau)) &= \frac{C_0}{2} + \sum_{n=1}^{N} C_n \cos n\alpha\tau + S_n \sin n\alpha\tau \\ \Gamma(x(\tau), \vartheta(\tau)) &= \frac{B_0}{2} + \sum_{n=1}^{N} B_n \cos n\alpha\tau + R_n \sin n\alpha\tau \end{aligned} \tag{14}$$

by using the constitutive equations (2), (3), (6), (11) the coefficients C_n, S_n, B_n, R_n can be computed from the response coefficients a_n, b_n, c_n, d_n. Denoting $F_n = F$ for $n = 1$ and $F_n = 0$ for $n > 1$, the substitution of (13) and (14) into (10) yields

$$\begin{aligned} \frac{C_0}{2} + \sum_{n=1}^{N} \left(C_n - n^2\alpha^2 a_n + 2\zeta n\alpha b_n - F_n\right) \cos n\alpha\tau + \\ + \left(S_n - n^2\alpha^2 b_n - 2\zeta n\alpha a_n\right) \sin n\alpha\tau &= 0 \\ \frac{B_0 - h_c c_0}{2} + \sum_{n=1}^{N} \left(B_n + n\alpha d_n - h_c c_n\right) \cos n\alpha\tau + \\ + \left(R_n - n\alpha c_n - h_c d_n\right) \sin n\alpha\tau &= 0 \end{aligned} \tag{15}$$

The application of the Galerkin procedure delivers the following harmonic balance equations involving the coefficients of the series

$$\begin{cases} C_0 = 0 \\ B_0 - h_C c_0 = 0 \end{cases}$$
$$\begin{cases} C_n - F_n - n^2\alpha^2 a_n + 2\zeta n\alpha b_n = 0 \\ S_n - 2\zeta n\alpha a_n - n^2\alpha^2 b_n = 0 \\ B_n - h_C c_n + n\alpha d_n = 0 \\ R_n - h_C d_n - n\alpha c_n = 0 \qquad n = 1, .., N \end{cases} \tag{16}$$

Considering that the coefficients C_n, S_n, B_n, R_n of the approximation of the history-dependent quantities depend on the coefficients a_n, b_n, c_n, d_n of the displacement and temperature response, the system (16) is a system of $4N + 2$ nonlinear algebraic equations in the $4N + 2$ unknowns coefficients of the (x, ϑ) approximation.

7. Frequency-response curves

The main features of the stationary response of the SMA oscillator are discussed through the curves representing the maximum amplitudes of the displacement and temperature response versus forcing frequency at different forcing amplitudes and varying the parameters which govern the material behavior.

It is useful to illustrate first the *isothermal* response for forcing amplitudes ranging from $F = 0.1$ to $F = 0.4$ and the influence of the parameter μ that, as shown in Figure 3b, determines the area of the hysteresis loops in the force-displacement plane. Figure 6 compares the curves computed for different values of μ going from $\mu = 0.9$ that corresponds to a wider loop to $\mu = 0.3$ corresponding to a narrower loop.

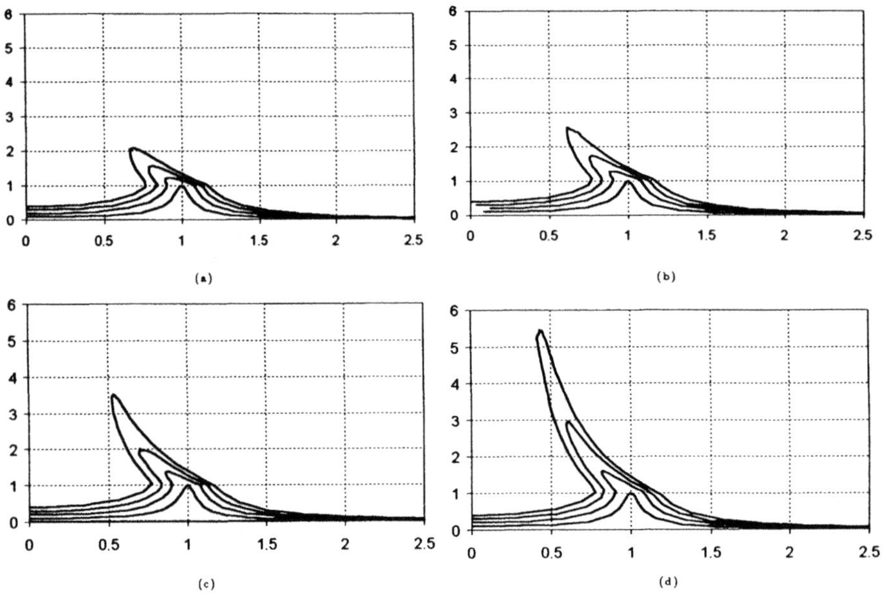

Figure 6. Isothermal frequency-response curves $\mu = 0.9$(a), 0.7(b), 0.5(c), 0.3(d)

From the above figures it is seen that when μ decreases, the maximum oscillation amplitudes increase considerably as a consequence of the reduction of the hysteresis loop area. Furthermore, jump phenomena with multiple coexisting solutions for the same frequency are present in all curves, as expected, considering the restoring force loop shape and the results obtained in (Capecchi and Vestroni, 1990).

In order to highlight the effect of the thermomechanical interactions on the dynamic response, the frequency response curves corresponding to the cases depicted in Figure 6 have been computed in *non-isothermal* conditions $J = 1.0, L = 1.0, h_C = 0.01$. In such a case two kind of curves are obtained, one relative to the maximum displacement amplitude (Figure 7) and another to the maximum temperature amplitude (Figure 8).

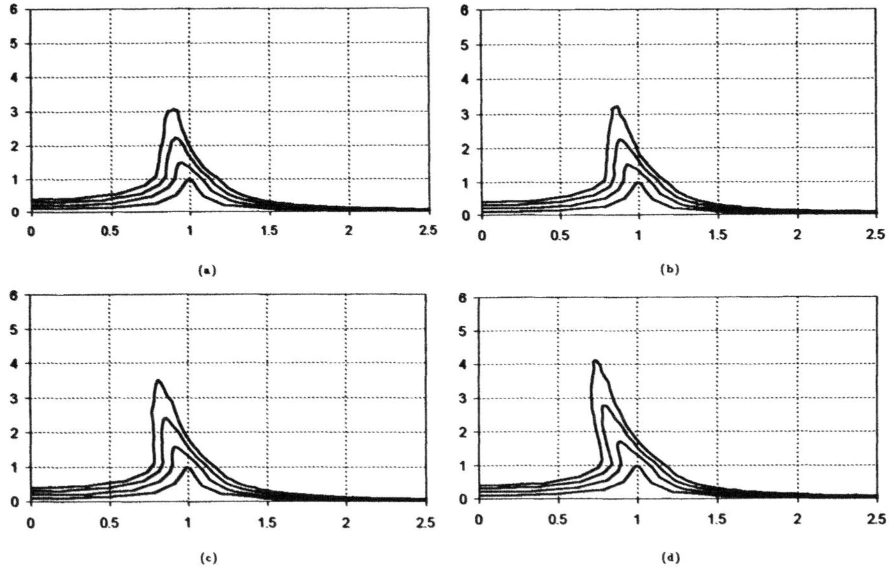

Figure 7. Non-isothermal frequency-response curves: $\mu = 0.9$(a), 0.7(b), 0.5(c), 0.3(d)

From the comparison of Figures 6 and 7 it turns out that, at least in the selected parameter range, the response modification due to the thermal aspects induces a relevant modification of the dynamic response. Moreover, for the same value of μ, the range of multiple solutions is notably reduced in non-isothermal conditions. This is related to the modifications of the loop shape. In particular the occurrence of the hardening produces a frequency increase with the oscillation amplitude that tends to destroy the jump phenomenon induced by the initial softening behavior.

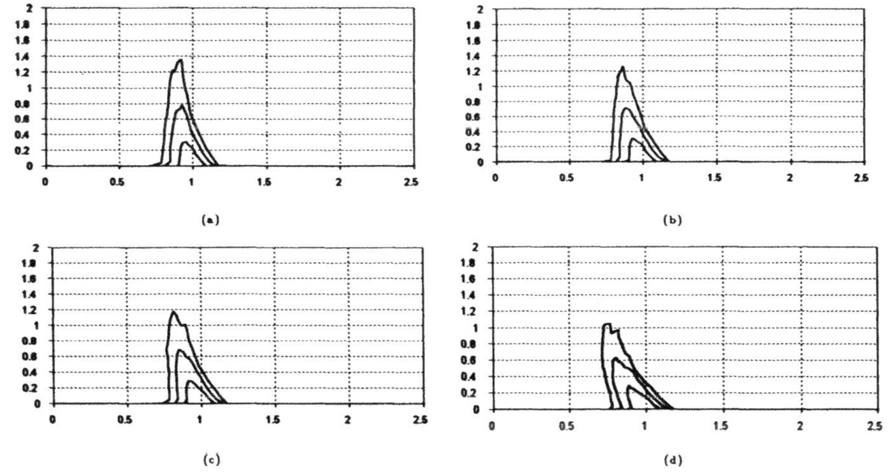

Figure 8. Frequency-temperatures curves corresponding to Figure 7.

It is to note that the oscillation of the temperature takes place only when the stationary response goes beyond the elastic region and phase transformation is activated and, as expected, the temperature curves are multivalued in the same range as the displacement ones.

8. Conclusions

In this paper the nonlinear dynamic response of a SMA oscillator has been studied by means of a simple thermodynamic model that makes it possible to take into account the full thermomechanical behavior. The novelty of the work is related to the consideration of the influence of the non-isothermal response, typically observed in real alloys, on the dynamic behavior. A first numerical investigation, based on an extended harmonic balance procedure, has been carried out and results have been presented through the frequency-response curves. For a selected set of values of the parameters L, J, h_C governing the thermomechanical material behavior, it has been found that the dynamic response of the system shows some peculiar characteristics with respect to the traditional hysteretic systems, already studied in literature. Notable differences have been observed passing from isothermal to non isothermal conditions. Further studies are actually ongoing with the aim to investigate the behavior of the oscillator at higher force amplitudes, where a richer dynamic response is observed.

9. References

Van Humbeeck J. (1998) Shape Memory Alloys Applications, in Proc. of *ICOMAT 98*, Bariloche, Argentina.

Clark P.W. , Aiken I.D. and Kelly J.M. (1995) Experimental and analytical studies of SMA dampers for structural control, in *Smart Struct. and Mat. 1995*, San Diego, CA, **2445**, 241-251.

Otsuka K. and Shimizu K. (1986) Pseudoelasticity and shape memory effects in alloys, *Int. Metals Rev.* , **31**, 93-114.

Feng Z.C. and Li D.Z. (1996) Dynamics of a mechanical system with a shape memory alloy bar, *J. of Intell. Mater. Syst. and Struct.*, **7**, 399-410.

Thomson P. , Balas G.J. and Leo P.H. (1995) Pseudoelastic SMA models and effects on passive structural damping, *Smart Struct. and Mat. 1995*, San Diego, CA, **2445**, 188-199.

Oberaigner E.R. , Tanaka K. and Fisher F.D. (1995) Investigation of the damping behavior of a vibrating SMA rod, in *Smart Struct. and Mat. 1995*, San Diego, CA, **2442**, 349-361.

Capecchi D. and Vestroni F. (1990) Periodic response of a class of hysteretic oscillators. *Int. J. of Nonlin. Mech.*, **25**, 309-317.

Vestroni F. and Capecchi D. (1997) Coupling and resonance phenomena in dynamical systems with hysteresis, in Proc. of *IUTAM Symp.on Nonlinear and Chaotic Dynamics in Mechanics.*, Cornell University, Ithaca, N.Y.

Bernardini D. and Van Humbeeck J. (1999) Temperature and loading rate effects in SMA wires, in preparation.

Huo Y. and Muller I. (1993) Nonequilibrium thermodynamics of pseudoelasticity, *Cont. Mech. and Thermod.*, **5**, 163-204.

Bernardini D. (1998) *Thermomechanics of Shape Memory Alloys: models for the macroscopic behavior*, PhD thesis, Università di Roma 'La Sapienza'.

NONLINEAR WAVES IN A FLUID–FILLED PLANAR DUCT WITH A FLEXIBLE WALL

J. WAUER
Institut für Technische Mechanik, Universität Karlsruhe,
Kaiserstraße 12, D–76128 Karlsruhe, Germany

Abstract. As a generalization considering small fluid–structural vibrations, the present paper examines the finite magnitude oscillatory motion of a compressible fluid layer bounded by two parallel walls where one of them is a flexibly supported membrane structure. The fluid is assumed to be inviscid and irrotational. A pertubation analysis is utilized to calculate not only the dynamic characteristics for small coupled oscillations but also the corrections due to the inherent nonlinearities of the vibroacoustic problem.

1. Introduction

Fluid–structural coupling is very important in many fields of engineering. For small oscillations where linear boundary value problems govern the vibrational behavior, several problems have been discussed during the last two decades in all details (see [1,3,4,7,8,10–12], for example) starting with a short note by Weidenhammer [16]. As typical one– and two–dimensional examples, an elastically supported rigid plate vibrating in contact with a layer of fluid and an elastically suspended rigid cylinder in a surrounding circular fluid–filled duct with rigid walls were analyzed.

The objective of the present contribution is to generalize these considerations to oscillations of finite magnitude so that a nonlinear description is required. Recalling a recent examination on finite vibrations of a flexibly supported rigid body immersed in an incompressible ideal liquid at rest by Seemann and Wauer [13], in the present contribution attention is focused to the corresponding problem of a fluid in contact with a structure which itself is flexible so that fluid–structural traveling waves appear. In detail, finite magnitude oscillatory motion of a compressible fluid in an infinitely long layer–shaped wave guide bounded by two walls are examined where one side is rigid and the other is flexible (in form of an elastic membrane on an elastic foundation). The fluid is assumed to be inviscid and irrotational and free transverse vibrations of the body are dealt with. The governing equations of motion are the fully nonlinear Euler equations together with the continuity equation, a state equation (here for an ideal isothermal gas), the (nonlinear) partial differen-

N. Van Dao and E.J. Kreuzer (eds.),
IUTAM Symposium on Recent Developments in Non-linear Oscillations of Mechanical Systems, 321–332.

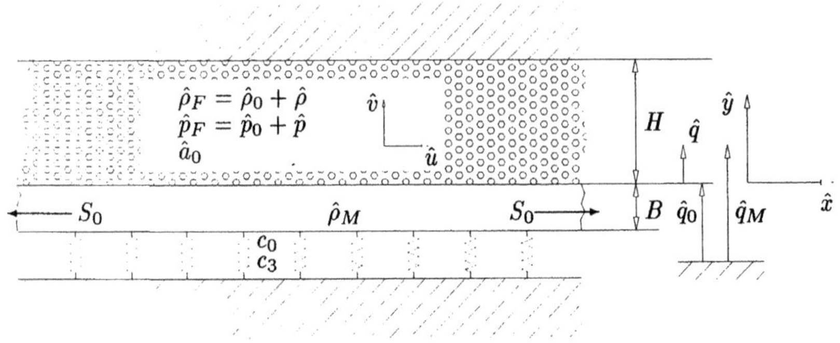

Figure 1: Channel with flexibly supported membrane wall

tial equation for the vibrating membrane (obtained from the dynamic transition condition), and the kinematical transition and boundary conditions at the moving contact interface between fluid and body and the outside fluid border, respectively. To formulate the governing boundary problem in a more compact form, a scalar potential is introduced so that the number of variables can be reduced.

To solve this boundary value problem, a perturbation analysis is performed which allows step by step evaluation. The first–order perturbation problem yields the results characterizing axially propagating fluid–structural waves in connection with small transverse coupled vibrations; from the second– and the third–order perturbation problem, the corrections due to the inherent nonlinearities can be found.

2. Formulation

For convenience, a planar problem is considered appropriately described in a Cartesian $\hat{x}, \hat{y}$ coordinate system. The infinitely long fluid–filled channel is bounded by two parallel walls. One is rigid and the other is deformable as shown in Fig. 1. In detail, it is an elastic membrane structure of density $\hat{\rho}_M$ and constant thickness B (stretched by a constant tension S) embedded on an elastic foundation with a hardening stiffness characteristic (stiffness constants per unit area $\hat{c}_1$ and $\hat{c}_3$).

In the equilibrium state, there is a static deformation $\hat{q}_0$ = const. of the membrane in contact with the fluid layer of thickness H, pressure $\hat{p}_0$ = const. and density $\hat{\rho}_0$ = const. The coupled vibrations of the interacting system are characterized by the space– and time–dependent transverse oscillation $\hat{q}(\hat{x}, \hat{t})$ of the membrane about its static displacement $\hat{q}_0$ along the $\hat{y}$–axis and the time and space–dependent unsteady fluid velocities $\hat{u}(\hat{x}, \hat{y}, \hat{t})$, $\hat{v}(\hat{x}, \hat{y}, \hat{t})$ and corresponding pressure and density fields $\hat{p}(\hat{x}, \hat{y}, \hat{t})$ and $\hat{\rho}(\hat{x}, \hat{y}, \hat{t})$, respectively. The fluid is assumed to be an ideal gas with isothermal state change where $\hat{a}_0$ denotes the corresponding sound velocity.

The description of the moving interface is elementary:

$$\hat{y} = \hat{q}(\hat{x}, \hat{t}). \tag{1}$$

The boundary condition at the rigid wall appears at $\hat{y} = H =$ const.

The mathematical description for the fluid is based on the continuity equation

$$\frac{\partial \hat{\rho}_F}{\partial \hat{t}} + \hat{u}_F \frac{\partial \hat{\rho}_F}{\partial \hat{x}} + \hat{v}_F \frac{\hat{\rho}_F}{\partial \hat{y}} + \hat{\rho}_F \left(\frac{\partial \hat{u}_F}{\partial \hat{x}} + \frac{\partial \hat{v}_F}{\partial \hat{y}} \right) = 0, \tag{2}$$

the Eulerian equations

$$\begin{aligned} \hat{\rho}_F \left(\frac{\partial \hat{u}_F}{\partial \hat{t}} + \hat{u}_F \frac{\partial \hat{u}_F}{\partial \hat{x}} + \hat{v}_F \frac{\partial \hat{v}_F}{\partial \hat{y}} \right) &= -\frac{\partial \hat{p}_F}{\partial \hat{x}}, \\ \hat{\rho}_F \left(\frac{\partial \hat{v}_F}{\partial \hat{t}} + \hat{u}_F \frac{\partial \hat{v}_F}{\partial \hat{x}} + \hat{v}_F \frac{\partial \hat{v}_F}{\partial \hat{y}} \right) &= -\frac{\partial \hat{p}_F}{\partial \hat{y}} \end{aligned} \tag{3}$$

together with a state equation

$$\hat{p}_F = \hat{a}_0^2 \hat{\rho}_F. \tag{4}$$

In addition, a dynamic transition condition at the surface of the moving structure (force balance for the membrane)

$$\hat{\rho}_M B \frac{\partial^2 \hat{q}_M}{\partial \hat{t}^2} - S \frac{\partial^2 \hat{q}_M}{\partial \hat{x}^2} + c_0 \hat{q}_M + c_3 \hat{q}_M^3 + \hat{p}_F(\hat{x}, \hat{y}, \hat{t})\big|_{\hat{y} = \hat{q}_M - \hat{q}_0} = 0, \tag{5}$$

and kinematical boundary and transition conditions for the transverse velocity component at the rigid wall and the contact interface between fluid and membrane, respectively, have to be formulated:

$$\hat{v}_F(\hat{x}, H, \hat{t}) = 0, \quad \hat{v}_F(\hat{x}, \hat{q}, \hat{t}) = \frac{\partial \hat{q}_M(\hat{x}, \hat{t})}{\partial \hat{t}} + \hat{u}_F(\hat{x}, \hat{q}, \hat{t}) \frac{\partial \hat{q}_M(\hat{x}, \hat{t})}{\partial \hat{x}}. \tag{6}$$

It will be noticed that the variables containing both their steady and unsteady part are characterized by the subscript F (for the fluid) and M (for the structural member). The linear version for a one–degree–of–freedom structural subsystem (elastically supported rigid plate performing a purely transverse motion) was discussed by Seemann and Wauer [12].

Assuming an irrotational fluid, we can rewrite the governing boundary value problem (2)–(6) can be rewritten in a more compact form by introducing the velocity potential:

$$\mathbf{v}_F = \operatorname{grad} \hat{\phi}_F. \tag{7}$$

It can be shown (see [15], for example) that the continuity equation (2) and the Eulerian equations (3) can be replaced by

$$\begin{aligned} &\left[1 - \frac{1}{\hat{a}_0^2}\left(\frac{\partial \hat{\phi}_F}{\partial \hat{x}}\right)^2\right] \frac{\partial^2 \hat{\phi}_F}{\partial \hat{x}^2} + \left[1 - \frac{1}{\hat{a}_0^2}\left(\frac{\partial \hat{\phi}_F}{\partial \hat{y}}\right)^2\right] \frac{\partial^2 \hat{\phi}_F}{\partial \hat{y}^2} - \frac{1}{\hat{a}_0^2} \frac{\partial^2 \hat{\phi}_F}{\partial \hat{t}^2} \\ &\qquad - \frac{2}{\hat{a}_0^2} \left[\frac{\partial \hat{\phi}_F}{\partial \hat{x}} \frac{\partial \hat{\phi}_F}{\partial \hat{y}} \frac{\partial^2 \hat{\phi}_F}{\partial \hat{x} \partial \hat{y}} + \frac{\partial \hat{\phi}_F}{\partial \hat{x}} \frac{\partial^2 \hat{\phi}_F}{\partial \hat{t} \partial \hat{x}} + \frac{\partial \hat{\phi}_F}{\partial \hat{y}} \frac{\partial^2 \hat{\phi}_F}{\partial \hat{t} \partial \hat{y}} \right] = 0 \end{aligned} \tag{8}$$

and Bernoulli's law

$$\hat{\rho}_F \left[\frac{\partial \hat{\phi}_F}{\partial t} + \frac{1}{2}\left(\frac{\partial \hat{\phi}_F}{\partial \hat{x}}\right)^2 + \frac{1}{2}\left(\frac{\partial \hat{\phi}_F}{\partial \hat{y}}\right)^2 \right] + \hat{p}_F = \hat{p}_0. \tag{9}$$

The fluid field variables and the displacement of the membrane may be written as the sum of steady quantities and fluctuations:

$$\begin{aligned} \hat{p}_F &= \hat{p}_0 + \hat{p}, \quad \hat{\rho}_F = \hat{\rho}_0 + \hat{\rho}, \\ \hat{u}_F &= 0 + \hat{u}, \quad \hat{v}_F = 0 + \hat{v}, \quad \hat{\phi}_F = 0 + \hat{\phi}, \quad \hat{q}_M = \hat{q}_0 + \hat{q}. \end{aligned} \tag{10}$$

The unsteady acoustic–structural vibrations are of special interest. For this purpose, substitute the assumed solutions (10) into the governing boundary value problem (8),(9),(4)–(6), take the (trivial) steady–state solutions into consideration and obtain, after rescaling,

$$\begin{aligned} &\left[1 - \left(\frac{\partial \phi}{\partial x}\right)^2\right] \frac{\partial^2 \phi}{\partial x^2} + \left[1 - \left(\frac{\partial \phi}{\partial y}\right)^2\right] \frac{\partial^2 \phi}{\partial y^2} - \frac{\partial^2 \phi}{\partial t^2} \\ &\quad - 2\left(\frac{\partial \phi}{\partial x}\frac{\partial \phi}{\partial y}\frac{\partial^2 \phi}{\partial x \partial y} + \frac{\partial \phi}{\partial x}\frac{\partial^2 \phi}{\partial x \partial t} + \frac{\partial \phi}{\partial y}\frac{\partial^2 \phi}{\partial y \partial t}\right) = 0, \end{aligned} \tag{11}$$

$$(1+\rho)\left[\frac{\partial \phi}{\partial t} + \frac{1}{2}\left(\frac{\partial \phi}{\partial x}\right)^2 + \frac{1}{2}\left(\frac{\partial \phi}{\partial y}\right)^2\right] + p = 0, \tag{12}$$

$$p = \rho, \tag{13}$$

$$\frac{\partial^2 q}{\partial t^2} - \beta \frac{\partial^2 q}{\partial x^2} + \lambda_0^2 q + \lambda_3 q^3 + \alpha p(x,y,t)\Big|_{y=q} = 0 \tag{14}$$

and

$$\frac{\partial \phi(x,y,t)}{\partial y}\bigg|_{y=1} = 0, \quad \frac{\partial \phi(x,y,t)}{\partial y}\bigg|_{y=q} - \frac{\partial q}{\partial t} - \frac{\partial q}{\partial x}\frac{\partial \phi(x,y,t)}{\partial x}\bigg|_{y=q} = 0. \tag{15}$$

We have used the following non–dimensional variables and parameters:

$$\begin{aligned} t &= \frac{\hat{a}_0}{H}\hat{t}, \quad x = \frac{\hat{x}}{H}, \quad y = \frac{\hat{y}}{H}, \quad q = \frac{\hat{q}}{H}, \\ u &= \frac{\hat{u}}{\hat{a}_0}, \quad v = \frac{\hat{v}}{\hat{a}_0}, \quad \phi = \frac{\hat{\phi}}{\hat{a}_0 H}, \quad \rho = \frac{\hat{\rho}}{\hat{\rho}_0}, \quad p = \frac{\hat{p}}{\hat{a}_0^2 \hat{\rho}_0} \end{aligned} \tag{16}$$

and

$$\lambda_0^2 = \frac{c_1 H^2}{B \hat{\rho}_M \hat{a}_0^2}, \quad \lambda_3 = \frac{c_3 H^4}{B \hat{\rho}_M \hat{a}_0^2}, \quad \beta = \frac{S}{B \hat{\rho}_M \hat{a}_0^2}. \tag{17}$$

Recall that the governing space coordinate $\hat{y}$ (i.e., y) is measured from the location $\hat{q}_0$ of the membrane surface in the steady state.

3. Evaluation and Results

Following Kevorkian and Cole [6] or Nayfeh and Kelly [9], we assume the structural oscillation to be of the order of a small parameter ε, i.e., $q(x,t) \equiv \varepsilon\bar{q}(x,t)$. To take into consideration that the dynamic characteristics, in particular the eigenfrequencies, will also be corrected by the nonlinearities, a new time variable

$$\tau = \kappa t \tag{18}$$

is introduced. Then, the vibration variables ϕ, p, ρ and $\bar{q}$, and also the circular frequency κ, are expanded in powers of ε:

$$\begin{aligned}
\varepsilon\bar{q}(x,\tau) &= \varepsilon\left[q_1(x,\tau) + \varepsilon q_2(x,\tau) + \varepsilon^2 q_3(x,\tau) + \ldots\right], \\
\phi(x,y,\tau) &= \varepsilon\phi_1(x,y,\tau) + \varepsilon^2\phi_2(x,y,\tau) + \varepsilon^3\phi_3(x,y,\tau) + \ldots, \\
p(x,y,\tau) &= \varepsilon p_1(x,y,\tau) + \varepsilon^2 p_2(x,y,\tau) + \varepsilon^3 p_3(x,y,\tau) + \ldots, \\
\rho(x,y,\tau) &= \varepsilon\rho_1(x,y,\tau) + \varepsilon^2\rho_2(x,y,\tau) + \varepsilon^3\rho_3(x,y,\tau) + \ldots, \\
\kappa &= \kappa_0 + \varepsilon\kappa_1 + \varepsilon^2\kappa_2 + \ldots
\end{aligned} \tag{19}$$

where

$$f[x, y = \varepsilon\bar{q}(x,\tau), \tau) = f(x,0,\tau) + \left.\frac{\partial f(x,y,\tau)}{\partial y}\right|_{y=0} \varepsilon\bar{q}(x,\tau) + \ldots \tag{20}$$

Substituting into the governing boundary value problem (11)–(15) and equating terms of equal powers of ε, we obtain a set of perturbation problems of successive orders. There is a first-order problem

$$\begin{aligned}
\frac{\partial^2\phi_1}{\partial x^2} + \frac{\partial^2\phi_1}{\partial y^2} - \kappa_0^2\frac{\partial\phi_1}{\partial\tau^2} &= 0, \\
\kappa_0^2\frac{\partial^2 q_1}{\partial t^2} - \beta\frac{\partial^2 q_1}{\partial x^2} + \lambda_0^2 q_1 - \alpha\kappa_0\left.\frac{\partial\phi_1}{\partial\tau}\right|_{y=0} &= 0, \\
\left.\frac{\partial\phi_1}{\partial y}\right|_{y=1} = 0, \quad \left.\frac{\partial\phi_1}{\partial y}\right|_{y=0} - \kappa_0\frac{\partial q_1}{\partial\tau} &= 0,
\end{aligned} \tag{21}$$

a second-order problem

$$\begin{aligned}
\frac{\partial^2\phi_2}{\partial x^2} + \frac{\partial^2\phi_2}{\partial y^2} - \kappa_0^2\frac{\partial^2\phi_2}{\partial\tau^2} &= 2\kappa_0\left[\frac{\partial\phi_1}{\partial x}\frac{\partial^2\phi_1}{\partial x\partial\tau} + \frac{\partial\phi_1}{\partial y}\frac{\partial^2\phi_1}{\partial y\partial\tau} + \kappa_1\frac{\partial^2\phi_1}{\partial\tau^2}\right] \equiv F_2, \\
\kappa_0^2\frac{\partial^2 q_2}{\partial\tau^2} - \beta\frac{\partial^2 q_2}{\partial x^2} + \lambda_0^2 q_2 & \\
- \alpha\kappa_0\left.\frac{\partial\phi_2}{\partial\tau}\right|_{y=0} &= -2\kappa_0\kappa_1\frac{\partial^2 q_1}{\partial\tau^2} + \alpha\left[\kappa_1\frac{\partial\phi_1}{\partial\tau} + \frac{1}{2}\left(\frac{\partial\phi_1}{\partial x}\right)^2\right.
\end{aligned}$$

$$+\frac{1}{2}\left(\frac{\partial\phi_1}{\partial y}\right)^2-\left(\kappa_0\frac{\partial\phi_1}{\partial\tau}\right)^2+\kappa_0 q_1\frac{\partial^2\phi_1}{\partial y\partial\tau}\Bigg]_{y=0},$$

$$\left.\frac{\partial\phi_2}{\partial y}\right|_{y=1}=0,\quad \left.\frac{\partial\phi_2}{\partial y}\right|_{y=0}-\kappa_0\frac{\partial q_2}{\partial\tau}=\kappa_1\frac{\partial q_1}{\partial\tau}-q_1\left.\frac{\partial^2\phi_1}{\partial y^2}\right|_{x=0}-\left.\frac{\partial q_1}{\partial x}\frac{\partial\phi_1}{\partial x}\right|_{y=0},\tag{22}$$

a third–order problem

$$\begin{aligned}
&\frac{\partial^2\phi_3}{\partial x^2}+\frac{\partial^2\phi_3}{\partial y^2}-\kappa_0^2\frac{\partial^2\phi_3}{\partial\tau^2}\\
&\quad=\left(\frac{\partial\phi_1}{\partial x}\right)^2\frac{\partial\phi_1}{\partial x}+\left(\frac{\partial\phi_1}{\partial y}\right)^2\frac{\partial^2\phi_1}{\partial y^2}+2\kappa_0\left[\frac{\partial\phi_1}{\partial x}\frac{\partial^2\phi_2}{\partial x\partial\tau}+\kappa_1\frac{\partial^2\phi_2}{\partial\tau^2}\right.\\
&\qquad\left.+\frac{\partial\phi_2}{\partial x}\frac{\partial^2\phi_1}{\partial x\partial\tau}+\frac{\partial\phi_1}{\partial y}\frac{\partial^2\phi_2}{\partial y\partial\tau}+\frac{\partial\phi_2}{\partial y}\frac{\partial^2\phi_1}{\partial y\partial\tau}+\kappa_2\frac{\partial^2\phi_1}{\partial\tau^2}\right]+\kappa_1^2\frac{\partial^2\phi_1}{\partial\tau^2}\\
&\qquad+2\kappa_1\left[\frac{\partial\phi_1}{\partial x}\frac{\partial^2\phi_1}{\partial x\partial\tau}+\frac{\partial\phi_1}{\partial y}\frac{\partial^2\phi_1}{\partial y\partial\tau}\right]+2\frac{\partial\phi_1}{\partial x}\frac{\partial\phi_1}{\partial y}\frac{\partial^2\phi_1}{\partial x\partial y}\equiv F_3,\\
&\kappa_0^2\frac{\partial^2 q_3}{\partial\tau^2}-\beta\frac{\partial^2 q_3}{\partial x^2}+\lambda_0^2 q_3-\alpha\kappa_0\left.\frac{\partial\phi_3}{\partial\tau}\right|_{y=0}\\
&\quad=-2\kappa_0\kappa_1\frac{\partial^2 q_2}{\partial\tau^2}-2\kappa_0\kappa_2\frac{\partial^2 q_1}{\partial\tau^2}-\kappa_1^2\frac{\partial^2 q_1}{\partial\tau^2}-\lambda_3 q_0^3-\alpha\left[\frac{\partial p_1}{\partial y}q_2\right.\\
&\qquad\left.+\frac{\partial p_2}{\partial y}q_1+\frac{1}{2}\frac{\partial^2 p_1}{\partial y^2}q_1^2\right]_{y=0}+\alpha\left[\kappa_1\frac{\partial\phi_2}{\partial\tau}+\frac{p_1}{2}\left(\frac{\partial\phi_1}{\partial x}\right)^2+\frac{p_1}{2}\left(\frac{\partial\phi_1}{\partial y}\right)^2\right.\\
&\qquad\left.+\kappa_0 p_1\frac{\partial\phi_2}{\partial\tau}+\kappa_1 p_1\frac{\partial\phi_1}{\partial\tau}+\kappa_0 p_2\frac{\partial\phi_1}{\partial\tau}+2\frac{\partial\phi_1}{\partial y}\frac{\partial\phi_2}{\partial y}+2\frac{\partial\phi_1}{\partial x}\frac{\partial\phi_2}{\partial x}\right]_{y=0},\\
&\left.\frac{\partial\phi_3}{\partial y}\right|_{y=1}=0,\\
&\left.\frac{\partial\phi_3}{\partial y}\right|_{y=0}-\kappa_0\frac{\partial q_3}{\partial\tau}\\
&\quad=\kappa_1\frac{\partial q_2}{\partial\tau}+\kappa_2\frac{\partial q_1}{\partial\tau}-\left[q_2\frac{\partial^2 q_1}{\partial y^2}+q_1\frac{\partial^2\phi_2}{\partial y^2}\right.\\
&\qquad\left.+\frac{1}{2}q_1^2\frac{\partial^3\phi_2}{\partial y^3}-\frac{\partial q_1}{\partial x}\frac{\partial\phi_2}{\partial x}-q_1\frac{\partial q_1}{\partial x}\frac{\partial^2\phi_1}{\partial x\partial y}-\frac{\partial q_2}{\partial x}\frac{\partial\phi_1}{\partial x}\right]_{y=0}
\end{aligned}\tag{23}$$

and others of higher order not written down explicitly.

To get a systematic solution procedure, it is appropriate to eliminate in a first step the q–variables. Differentiating the equations of motion of the membrane $(21)_2$,$(22)_2$ and $(23)_2$ with respect to time and using the kinematical transition conditions $(21)_4$,$(22)_4$ and $(23)_4$ at $y=0$, we can derive modified transition conditions in which only the ϕ–variables appear:

$$\left[\kappa_0 \frac{\partial^3 \phi_i}{\partial y \partial t^2} - \frac{\beta}{\kappa_o} \frac{\partial^3 \phi_i}{\partial y \partial x^2} + \frac{\lambda_0^2}{\kappa_0} \frac{\partial \phi_i}{\partial y} - \alpha \kappa_0 \frac{\partial^2 \phi_i}{\partial \tau^2}\right]_{y=0} = f_i\big|_{y=0}, \quad i = 1, 2, 3. \tag{24}$$

While for $i = 1$ this modified transition condition is homogeneous, i. e., $f_1 = 0$, the other two are lengthy expressions in ϕ_1 and ϕ_1, ϕ_2, respectively, not written down explicitly.

Now it is straightforward to solve the condensed first–order problem $(21)_1$,$(21)_3$ and (24) (for i=1). For that purpose, assume a solution

$$\phi_1(x, y, \tau) = \Phi_1(y) e^{ikx} (a \sin \tau) \tag{25}$$

and substitute it into the mentioned boundary value problem. This yields

$$\begin{aligned} \Phi_1'' + (\kappa_0^2 - k^2)\Phi_1 &= 0, \\ \Phi_1'(1) = 0, \quad (-\kappa_0^2 + \lambda_0^2 + \beta k^2)\Phi_1'(0) + \alpha\kappa_0^2 \Phi_1(0) &= 0 \end{aligned} \tag{26}$$

where $(.)' = \frac{d(.)}{dy}$. Assuming a solution

$$\Phi_1(y) = A \sin \gamma y + B \cos \gamma y \tag{27}$$

where $\gamma^2 = \kappa_0^2 - k^2$ and applying the boundary conditions $(26)_{2,3}$ leads to the corresponding eigenvalue equation

$$(\lambda_0^2 + \beta k^2 - \kappa_0^2)\gamma \sin \gamma + \alpha \kappa_0^2 \cos \gamma = 0 \tag{28}$$

determining the first–order eigenvalues κ_{0n} $(n = 0, 1, 2, \ldots, \infty)$ to be solved numerically.

Fig. 2 shows these "eigenvalues" κ_{0n} of the vibroacoustic system versus modified non–dimensional stiffness ratio λ_0 for four selected wave numbers $k = 0, 0.5, 2.5$ and 4.0 (here for $\beta = 0$). The coupling parameter α may theoretically range over a wide interval from zero to infinity but is limited here to $0 < \alpha < 1$ which characterizes the region of most practical interest.

It is appropriate first to recall the case of a vanishing wave number $k = 0$ which represents the case of an undeformable structure analyzed in detail by Seeemann and Wauer [12]. Since the frequency curves κ_{00}^* and κ_{0n}^* $(n = 1, 2, \ldots, \infty)$ resulting from $\alpha \to 0$ cross, there is a significant coupling of the acoustic and the structural modes in the regions of frequency coincidence, depending on the magnitude of the ratio α. Instead of curve crossing there is a curve "veering" between acoustic and structural branches of solutions of the characteristic equation (26). In the regions of curve veering one cannot speak about the structural and the acoustic modes separately. If the parameter α is small, then outside these regions the solutions κ_{0n} of the characteristic equation (26) are very close to the solutions κ_{0n}^* for a system

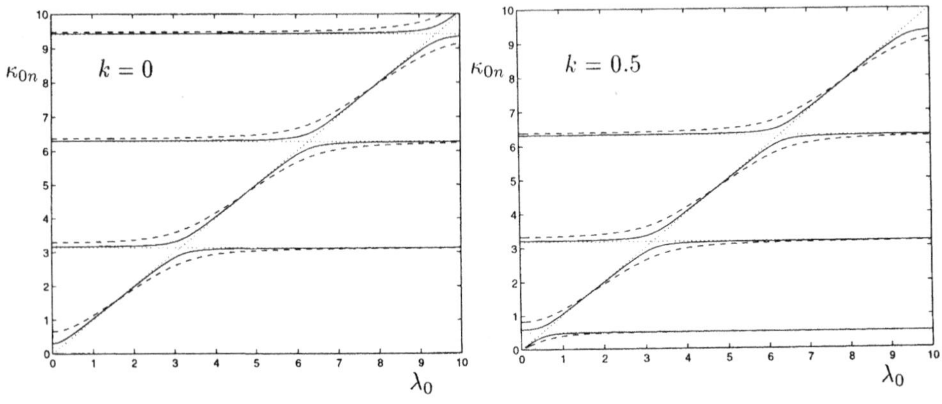

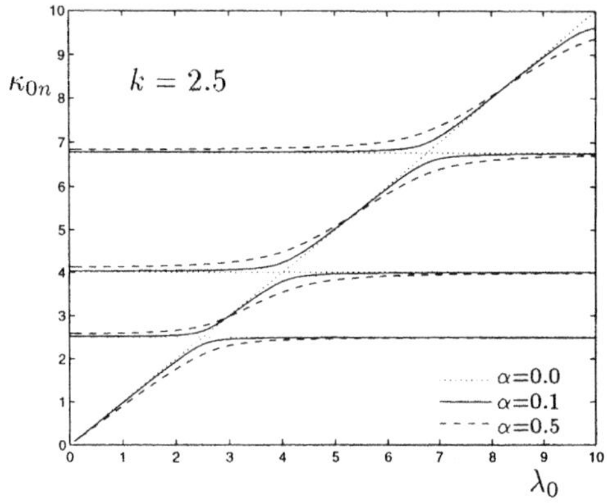

Figure 2: Eigenvalues of the fluid flow–membrane system versus stiffness ratio

without coupling ($\alpha = 0$) and one can speak about the eigenvalues of the acoustic modes and of the structural mode. It will be remarked that for an incompressible fluid in the case $k = 0$ no interaction problem exists because then, no motion of the coupled substructures is possible.

Originating from this, also the cases of finite wave numbers $k > 0$ can be understood (see Simon [14]). In ranges of the κ_{0n}, λ_0–plane where $\kappa_{0n}^2 > k^2$, qualitatively the graph remains unchanged. There are coupled transverse vibrations, and due to the finite wave number $k > 0$, they are combined with travelling waves in the axial x–direction of the channel. But in ranges where $k^2 > \kappa_{0n}^2$, there remains only one transverse vibration which, essentially, is a structural mode travelling along the wave guide in the x–direction. All the "acoustical" modes vanish, which means that the corresponding transverse vibrations (standing waves) change into exponentially decreasing waves in the transverse direction whithout reflexions at the rigid wall. It will be noticed here that the case of an incompressible fluid beneath a massless stiff plate was studied in [5].

Returning the boundary condition $(26)_2$, for instance, we can finally dermine the amplitude ratio A_n/B_n. The eigenfunctions $\Phi_{1n}(y)$ may be normalized

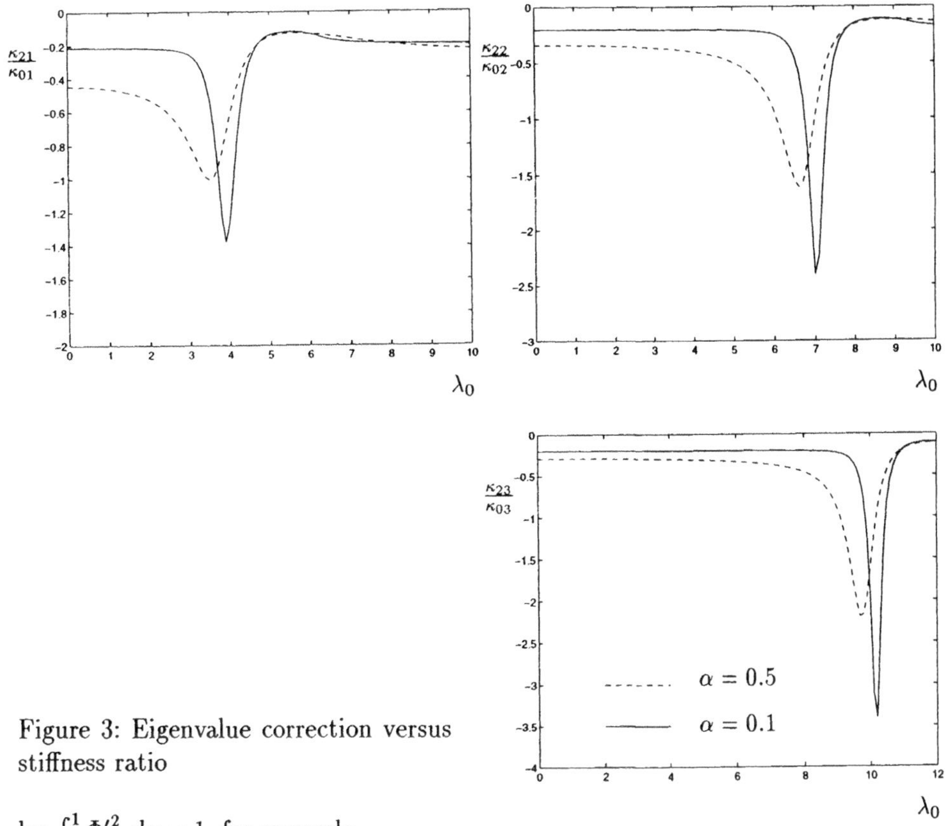

Figure 3: Eigenvalue correction versus stiffness ratio

by $\int_0^1 \Phi'^2_{1n} dy = 1$, for example.

In the next step, the inhomogeneous second- and third-order perturbation problems $(22)_1$,$(22)_3$,(24) (for $i = 2$) and $(23)_1$,$(23)_3$,(24) (for $i = 3$), respectively, have to be solved. The homogenous parts of the field equation and the boundary conditions are identical to those of the corresponding first-order boundary value problem. It follows that the eigenvalue problems and also the eigenfunctions are the same. Therefore, a modal expansion technique can be applied using the eigenfunctions $\Phi_{1n}(y)$. To establish this approach in its classical form, it is appropriate to transform the boundary value problems to ones in which the inhomogenous transition condition turns into a homogeneous one. Introducing Dirac's delta function $\delta(y)$ we can do this, and obtain after same calculation, a representation

$$\begin{aligned} \frac{\partial^2 \phi_i}{\partial x^2} + \frac{\partial^2 \phi_i}{\partial y^2} - \kappa_0^2 \frac{\partial^2 \phi_i}{\partial \tau^2} &= F_i + \delta(0) f_i, \\ \left.\frac{\partial \phi_i}{\partial y}\right|_{y=1} = 0, \quad \left[\kappa_0 \frac{\partial^3 \phi_i}{\partial y \partial t^2} - \frac{\beta}{\kappa_o} \frac{\partial^3 \phi_i}{\partial y \partial x^2} + \frac{\lambda_0^2}{\kappa_0} \frac{\partial \phi_i}{\partial y} - \alpha \kappa_0 \frac{\partial^2 \phi_i}{\partial \tau^2}\right]_{y=0} &= 0, \\ i &= 2,3. \end{aligned} \tag{29}$$

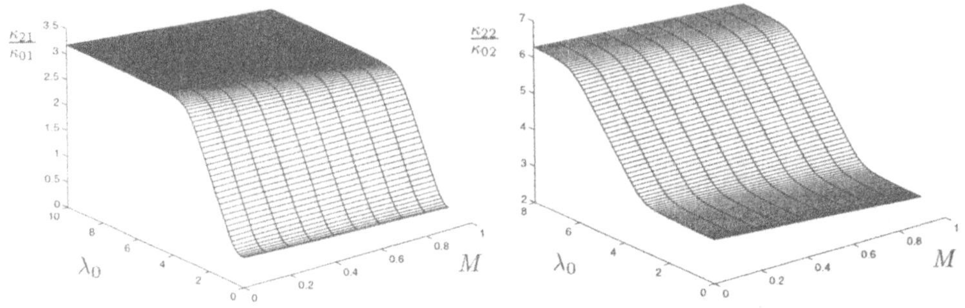

Figure 4: Eigenvalues versus stiffness ratio and Mach number

posessing the desired properties. For convenience, the computation is limited to the case $k \equiv 0$: The x–dependency vanishes but, nevertheless, the full variety of coupled transverse vibrations appears.

Evaluating the second–order problem (29) with $i = 2$, the remaining solutions $\phi_{1n}(y, \tau)$ lead to an excitation $C_{1n} \sin \tau + D_{1n} \sin 2\tau$ where $C_{1n} = \kappa_{1n} C^*_{1n}$ with C^*_{1n} free of κ_{1n}. Thus the eigenvalue correction κ_{1n} must vanish for every $n = 0, 1, 2, \ldots, \infty$ to avoid secular terms. The evaluation of the third–order–problem (29) with $i = 2$ is similar but significantly more time–consuming. In this case, there is a non–vanishing correction κ_{2n} $(n = 0, 1, 2, \ldots, \infty)$ (see Simon [14]) which is shown in Fig. 3 as $\frac{\kappa_{2n}}{\kappa_{0n}}$ versus λ_0, here for $\lambda_3 = 0$ and $n = 1, 2, 3$. For a linear foundation (so that only the non–linearities of the fluid become effective) the corrections are negative, i. e., the eigenvalues decrease with increasing vibrational magnitudes of the flexible wall. For $\lambda_3 > 0$, i. e., a hardening foundation, this effect can be compensated and for larger λ_3–values, the correction may be positive.

For an incompressible fluid, some simplifications appear because the density of the fluid is constant (so that the continuity equation reduces drastically) and there is no additional state equation. An incompressible water layer beneath a massless stiff plate was studied in [5].

4. Extending Remarks

It is straightforward to include an axial stationary flow velocity. $\hat{u}_0 =$ const. Within the dimensionless formulation, there is the so–called Mach number $M = \frac{\hat{u}_0}{a_0}$ as an additional parameter and in particular, the first–order eigenvalue problem can be evaluated without difficulties. In Fig. 4, the first two eigenvalues κ_{01} and κ_{02} are shown versus λ_0 and M for a axial wave number $k = 0.5$. It can be seen that they decrease with increasing Mach number and for larger Mach numbers, they can vanish, i.e., there may be an instability. An interesting aspect is the effect of an external excitation so that forced vibrations of the considered multifield problem occur. For a cylinder–fluid problem (where an incompressible inviscid liquid is assumed) the essentials are addressed by Seemann and Wauer [13] but for a compressible fluid (if two subsystems both able to vibrate are coupled), a corresponding calculation has not been performed yet. The related problem for an unbounded fluid region has been exhaustively discussed by Ginsberg [2] and Nayfeh and Kelly [9]. The most interesting features for that case are distortion effects in the fluid arising from the non–linear wave propagation there. At a sufficiently far distance from the excited structural member, a formation of shock waves will appear and for large values of H (for the channel) and $R_o - R_i$ (for the annulus) such phenomena might occur also for a bounded fluid region.

4. Conclusions

We investigate the finite–magnitude oscillatory motion of a compressible fluid layer bounded by two parallel walls, one of which is a flexibly supported membrane structure. The fluid is assumed to be inviscid and irrotational. A pertubation analysis has been used to calculate not only the dynamic characteristics for small coupled oscillations but also the corrections due to the inherent nonlinearities of the vibroacoustic problem.

The results for the eigenvalues demonstrate that there is only a correction of higher order but for a linear foundation there is a softening behavior and this correction is negative, i.e., the eigenvalues decrease with increasing vibrational amplitudes. The absolute values of the corrections are the largest for $\kappa_{0n} \approx \kappa^*_{0n}$ and are also influenced by the coupling parameter α.

References

1. Chen, S.S. Wambsganss, M.W., Jendrcejczyk, J.A. 1976. Added Mass and Damping of a Vibrating Rod in Confined Viscous Fluid, *J. Appl. Mech.* **43**, 325–329.

2. Ginsberg, J.H. 1975. Multi–Dimensional Non–Linear Acoustic Wave Propagation, Part I and Part II, *J. Sound Vibr.* **40**, 350–358 and 359–379.

3. Horáček, J., Zolotarev, I. 1991. Acoustic–Structural Coupling of Vibrating Cylindrical Shells with Flowing Fluid, *J. Fluids Struct.* **5**, 487–501.

4. Huber, H.–D. 1997. Calculation of the Acoustic Radiation of a Clamped Plate Covering a Cavity, Doctoral Thesis, University of Marseille, in press.

5. Il'ichev, A., Kirchgässner K. Nonlinear Nonlinear Water Waves Beneath an Elastic Ice–Sheet (to be published).

6. Kevorkian, J, Cole, J.D. 1981. *Perturbation Methods in Applied Mathematics*, Springer, New York, 482–488.

7. Matsuzaki, Y., Fung, Y. C. 1977. Stability Analysis of Straight and Buckled Two–Dimensional Channels Conveying an Incompressible Flow, *J. Appl. Mech.* **44**, 548–552.

8. Matsuzaki, Y., Fung, Y. C. 1979. Nonlinear Stability Analysis of a Two–Dimensional Model of an Elastic Tube Conveying a Compressible Flow, *J. Appl. Mech.* **46**, 31–36.

9. Nayfeh, A. H., Kelly, S. G. 1978. Non–Linear Interactions of Acoustic Fields with Plates under Harmonic Excitation, *J. Sound Vibr.* **60**, 371–377.

10. Païdoussis, M. P., Ostaja–Starzewski, J. A. 1981. Dynamics of a Flexible Cylinder in Subsonic Axial Flow, *AIAA–J.* **19**, 1467–1475.

11. Païdoussis, M. P., Nguyen, V. B., Misra, A. K. 1991. A Theoretical Study of the Stability of Cantilevered Coaxial Cylindrical Shells Conveying Fluid, *J. Fluids Struct.* **5**, 127–164.

12. Seemann, W., Wauer, J. 1996. Fluid–Structural Coupling of Vibrating Bodies in a Surrounding Confined Liquid, *Z. Angew. Math. Mech.* **76**, 67–79.

13. Seemann, W., Wauer, J. 1994. Finite Oscillatory Motion of a Body Immersed in an Inviscid Fluid at Rest, in: *Nonlinear and Stochastic Dynamics, AMD–Vol.* **192/***DE–Vol.* **78**, A. K. Bajaj et al. (Eds.), ASME, New York, 135–142.

14. Simon, M. 1999. Wellenausbreitung in einer kompressiblen Fluidschicht mit einseitig elastischer Berandung, Diplom Thesis, University of Karlsruhe.

15. Timman, R. 1960. Linearized Theory of Unsteady Flow of Compressible Fluids, in: *Encyclopedia of Physics* (S. Flügge, Ed.), Volume IX (Fluid Dynamics III), Springer, Berlin, 283–310.

16. Weidenhammer, F. 1975. Eigenfrequenzen eines Stabes in zylindrisch berandetem Luftraum, *Z. Angew. Math. Mech.* **55**, T187–190.

ADDRESSES of Participants

Awrejcewicz J
Technical University of Lodz
Division of Automatics and Biomechanics
1/15 Stefanowskiego St, 90-924 Lodz
Poland
Tel: 4842 312225 Fax: 4842 361383

Benedettini F.
Dipartimento di Ingegneria delle Strutture,
Acque e Terreno, Universita'
dell'Aquila, Monteluco Roio, 67040 L'Aquila
Italy
Tel: 39-0862-434513 Fax: 39-0862-434548
E-mail: ben@sgol.it

Bevilacqua L
National Laboratory for Scientific Computation
Av. Getulio Vargas 333, 25651-070 RJ, Brazil
Tel: 5521 5905480 Fax: 55 21 290 6626
Email: lbevilacqua@hotmail.com

Bernardini D.
Dipartimento di Ingegneria Strutturale e Geotecnica
Universita di Roma "La Sapienza"
Via Eudossiana 18, 00184 Roma, Italy
Tel: 3906 44585276 Fax: 396 48 84852
Email: davbern@tin.it

Bishop S. R.
University College London
Center for Nonlinear Dynamics and its Applications
Gower Street, London WC 1E 6BT, UK
Tel: 44-171-380-7729 Fax: 44-171-380-0986
E-mail: s.bishop@ucl.ac.uk

Chernousko F. L.
Institute for Problems in Mechanics
Russian Academy of Sciences
pr. Vernadskogo 101-1, Moscow 117526
Russia
Tel: 007-095-4340207 Fax: 007-095-9382048
E-mail: chern@ipmnet.ru

Dabrowski Z.
Institute of Machine Design Fundamentals
Warsaw University of Technology
02524 Warsaw, Narbutta 84, Poland

Tel: 4822 6608276 Fax: 4822 6608622
Email: zdabrow@simr.pw.edu.pl

Dao Nhu Mai
Institute of Mechanics
Department of Technical Diagnosis
264 Doi Can, Ba Dinh, Hanoi, Vietnam
Tel: 84.4.8326140 Fax: 84.4.8333039
Email: dnmai@im01.ac.vn

Dinh Van Phong
Hanoi University of Technology
Department of Applied Mechanics
1 Dai Co Viet, Hanoi, Vietnam
Tel: 84.4.9780799 Fax: 84.4.9780799
Email: phong@mail.hut.edu.vn

Do Sanh
Hanoi University of Technology
Department of Applied Mechanics
1 Dai Co Viet, Hanoi, Vietnam
Tel: 84.4.8680469 Fax: 84.4.8692006
Email: dosanh@vol.vnn.vn

Grebogi C.
Institute for Plasma Research
University of Maryland
College Park, MD 20742, USA
Tel: 301 405 5021 Fax: 301-405-5201
Email: grebogi@chaos.umd.edu

Guo-Kang Er
University of Macao
Faculty of Science and Technology
P.O. Box 3001, Macao
Tel: (853)3974361 Fax: (853)838314
Email: fstgke@umac.moy

Hu H.
Institute of Vibration Engineering Research
Nanjing University of Aeronautics and Astronautics
210016, Nanjing, P.R. China
Tel: 86254893278 Fax:86254498069/ 86254891512
Email: hhyae@dns.nuaa.edu.cn

Kapitaniak T.
Technical University of Lodz
Division of Dynamics
Stefanowskiego 1/15, 90-924 Lodz, Poland
Tel: 4842-6312231 Fax: 4842-6365046
Email: tomaszka@ck- sgp.lodz.pl

Kreuzer E.
Technical University of Hamburg-Harburg
Ocean Engineering Section II-Mechanics
Eissendorfer Strasse 42
D-21073 Hamburg, Germany
Tel: 49 040 42878-3020 Fax: 49 040 4278-2028
Email: kreuzer@tu-harburg.de

Le Luong Tai
Thai Nguyen University
Thai Nguyen, Vietnam
Tel: 84 0280 857391 Fax: 84.0280 852665

Maisser P.
Institute of Mechatronics
Chemnitz University of Technology
Reichenhainer Str.88, 09126 Chemnitz, Germany
Tel: 49-371-5314670 Fax: 49-371-5314669
Email: P.Maisser@IfM.TU-Chemnitz.de

Mitropolsky Y. A
Institute of Mathematics
Ukrainian National Academy of Sciences
3 Tereshenkovskaya St.
252601 Kiev, Ukraine
Tel: 38-044-224-53 16 Fax: 38-04422520-10

Nguyen Cao Menh
Institute of Mechanics
264 Doi Can, Ba Dinh, Hanoi, Vietnam
Tel:84.4.8326132 Fax: 84.4.8333039
Email: ncmenh@im01.ac.vn

Nguyen Dong Anh
Institute of Mechanics
264 Doi Can, Ba Dinh, Hanoi, Vietnam
Tel: 84.4.8326196 Fax: 84.4.8333039
Email: ndanh@im01.ac.vn

Nguyen Hai
College of Engineering
Vietnam National University, Ho Chi Minh City
Vietnam
Tel: 088651211 Fax: 088651211

Nguyen Nhat Le
Hanoi University of Technology
Department of Applied Mechanics
1 Dai Co Viet, Hanoi, Vietnam
Tel: 84.4.8694296 Fax: 84.4.9780799

Nguyen Thi Trung
Institute of Mechanics
264 Doi Can, Ba Dinh, Hanoi, Vietnam
Tel: 84.4.8326234 Fax: 84.4.8333039
Email: nttrung@im01.ac.vn

Nguyen Tien Khiem
Institute of Mechanics
Department of Technical Diagnosis
264 Doi Can, Ba Dinh, Hanoi, Vietnam
Tel: 84.4.8329705 Fax: 84.4.8333039
Email: ntkhiem@im01.ac.vn

Nguyen Van Dao
Vietnam National University, Hanoi
144 Xuan Thuy, Cau giay, Hanoi, Vietnam
Tel.(84 4) 8332015/8340570 Fax. (84 4) 8340724
E-mail: vandao @vnu.ac.vn

Nguyen Van Dac
Institute of Mechanics
264 Doi Can, Ba Dinh, Hanoi, Vietnam
Tel: 84.4.8340119 Fax: 84.4.8333039
Email: nvdac@im01.ac.vn

Nguyen Van Dinh
Institute of Mechanics
264 Doi Can, Ba Dinh, Hanoi, Vietnam
Tel: 84.4.8328013 Fax: 84.4.8333039
Email: nvdinh@im01.ac.vn

Nguyen Van Khang
Hanoi University of Technology
1 Dai Co Viet, Hanoi, Vietnam
Tel: 84.4.8680469
Email: nvankhang@hotmail.com

Nguyen Xuan Hung
Institute of Applied Mechanics, Ho Chi Minh City
291 Dien Bien Phu St., 3 Dist., Ho Chi Minh City
Vietnam
Tel: 84-4-088230577 Fax: 84-4-088232300

Ninh Quang Hai
Institute of Mechanics
264 Doi Can, Ba Dinh, Hanoi, Vietnam
Tel: 84.4.8326196 Fax: 84.4.8333039
Email:nqhai@im01.ac.vn

Osinski Z.
Institute of Machine Design Fundamentals
Warsaw University of Technology
02524 Warsaw, Narbutta 84, Poland
Tel: 4822 491880/490195 Fax: 4822 6608622
Email: zosinski@simr.pw.edu.pl

Pham Anh Tuan
Institute of Mechatronics at the Technical University of Chemnitz
Reichenhainer Str.88, D-09126 Chemnitz, Germany
Tel: 49-371-5314690 Fax: 49-371-5314669
Email: Tuan@IpM.TU-Chemnitz.de

Phan Nguyen Di
Le Quy Don Technical University
100 Hoang Quoc Viet, Hanoi, Vietnam
Tel: 84.4.8346949/8533001

Popp K.
Institute of Mechanics
University of Hannover
Appelstrasse 11, 30167 Hannover, Germany
Tel: 49 511 7624161 Fax: 49 511 7624164
Email: popp@ifm.uui-hannver.de.

Rega G.
University of Rome La Sapienza
Dipartimento di Ingegneria Strutturale e Geotecnica
via A. Gramsci, 53 Rome 00197, Ytaly
Tel: 39 06 44589195 Fax: 39 06 3221449
Email: rega@dsg.uniroma1.it

Samoilenko A.M.
Institute of Mathematics
Ukrainian National Academy of Sciences
3 Tereshenkovskaya St.
252601 Kiev 4, Ukraine
Tel: 38-044-224-53 16; Fax: 38-04422520-10
Email: imath2@mail.kar.net

Schiehlen W.
Institute B of Mechanics
Pfaffenwaldring 9
70550 Stuttgart, Germany
Tel: 0711- 685-6388 Fax: 0711-685-6400
Email: wos@mechb.uni-stuttgart.de

Schulz M.
University Engineering Department
Corporate Research and Development FV/SLT2
P.O. Box 30 02 40, 70442 Stuttgart, Germany
Tel: 49 6151 163385 Fax: 49 6151 164125
Email: schulz@mechanik.tu-darmstadt.de

Spyrou K. J.
Ship Design Laboratory
National Technical University of Athens
9 Iroon Polytechniou, Athens 15773, Greece
Tel: 44 141 5484833 Fax: 44 141 548 4784
Email: k.spyrou@ucl.ac.uk

Tran Kim Chi
Institute of Mechanics
264 Doi Can, Ba Dinh, Hanoi, Vietnam
Tel:84.4.8340119 Fax: 84.4.8333039
Email: tkchi@im01.ac.vn

Tran Van Nhung
Ministry of Education and Training
49 Dai Co Viet, Hanoi, Vietnam
Tel: 84.4.8692479 Fax: 84.4.8693243
Email: tvnhung@moet.edu.vn

Troger H.
Vienna University of Technology
A - 1040 Vienna, Austria
Tel: 0222-588-01 Fax: 43-1-5875863
Email: Hans.Troger@tuwien.ac.at

Ueda Y.
Department of Electrical Engineering
Kyoto University
Kyoto 606-8501
Tel: 81 75 757535336 Fax: 81 757511576
Email: ueda@kuee.kyoto-u.ac.jp

van Campen D. H.
Eindhoven University of Technology
Department of Mechanical Engineering
Den Dolech 2, P.O.Box 513
5600 MB Eindhoven The Netherlands
Tel: (040)479111 Fax: 31-40-2461418
Email: secr@wfw.wtb.tue.nl

Wauer J.
Institute fur Technische Mechanik

Universitat Karlsruhe
Kaiserstraße 12
D-76128 Karlsruhe, Germany
Tel: 49 (721) 608 2660 Fax: 49 (721) 608 6070
Email: wauer@itm.uni-karlsruhe.de

Wiercigroch M.
Department of Engineering
University of Aberdeen
Kings College
Aberdeen AB9 2UE, UK
Tel: 44 1224 272509 Fax: 01 224 272497

Wrobel J.
Institute of Machine Design Fundamentals
Warsaw University of Technology
02524 Warsaw, Narbutta 84, Poland
Tel: 4822 660 8286 Fax: 4822 6608622
Email: jery.wrobel@inmr.pw.edu.pl

List of Sponsors

Generous financial support contributed to the success of the Symposium. The support of the following sponsors is gratefully acknowledged

- International Union of Theoretical and Applied Mechanics
- Vietnam Council for Natural Science, Vietnam
- Institute of Mechanics, National Center for Natural Science and Technology, Vietnam
- Vietnam National University, Hanoi, Vietnam.

GPSR Compliance

The European Union's (EU) General Product Safety Regulation (GPSR) is a set of rules that requires consumer products to be safe and our obligations to ensure this.

If you have any concerns about our products, you can contact us on ProductSafety@springernature.com

In case Publisher is established outside the EU, the EU authorized representative is:

Springer Nature Customer Service Center GmbH
Europaplatz 3
69115 Heidelberg, Germany

Batch number: 10755284

Printed by Printforce, the Netherlands